Powermatics

MARIKE FINLAY

Powermatics

A discursive critique of new communications technology

Routledge & Kegan Paul
London and New York

First published in 1987 by
Routledge & Kegan Paul Ltd
11 New Fetter Lane, London EC4P 4EE

Published in the USA by
Routledge & Kegan Paul Inc.
in association with Methuen Inc.
29 West 35th Street, New York, NY 10001

Set in Times New Roman
by Pentacor Ltd
and printed in Great Britain
by T.J. Press (Padstow) Ltd

Library of Congress Cataloging in Publication Data

Finlay, Marike.
Powermatics: a discursive critique of new communications technology.

(International library of phenomenology and moral science)
Bibliography; p.
Includes index.
1. Communication—Technological innovations.
2. Communication—Philosophy. 3. Communication—Social aspects. 4. Discourse analysis.
5. Power (Social sciences) 6. Knowledge, Theory of. I. Title. II. Series.
P96.T42F56 1987 001.51 86–31524

British Library CIP Data also available
ISBN 0–7102–0761–1

For Ramón

CONTENTS

Acknowledgments

Introduction 1
- 0.0 Prophesy 1
- 0.1 Discourse analysis 2
- 0.2 Progress or jargon? 5
- 0.3 Demystification 8
- 0.4 Discursive criticism – articulation of an approach 11
  - 0.4.1 The object of study – communication as discourse 11
- 0.5 Discursive procedures 14
- 0.6 Content does not equal procedure 15
- 0.7 Procedures and technologies 16
- 0.8 Discourse and knowledge 19
  - 0.8.1 Episteme 21
  - 0.8.2 Episteme and techne 23

Chapter 1 Procedures of discourses on new communications technology: the episteme 25
- 1.0 Bridging the gap 25
- 1.1 Establishing the corpus of discourses on new communications technology 26
- 1.2 Referentiality 29
  - 1.2.1 Referentiality: exchange and reification of communication 34
- 1.3 From procedure to issue 36
- 1.4 Order and analysis 42
- 1.5 Fictive narrative scenario 43

1.6 Futurology: progressive, evolutionary, inevitable view of history 45
1.7 A-contextual euphorization or disphorization of technology 50
1.8 A-contextuality 51
1.9 Reductionist conception of knowledge 56
1.10 Principle of excluded middle: double bind semantics 61
1.10.1 Double bind 61
1.11 Eternal mediation 68
1.12 Anthropomorphization of technology 70
1.13 Competition between man and machine for hierarchical position 72
1.14 Individualization and atomization at the expense of social structure 74
1.15 Ubiquity of surveillance 79
1.16 Valorization of speed, growth, efficiency, progress, productivity and profit 81
1.17 Unilaterality of control by knowing subject over object to be known 84
1.18 Confusion of agents: passification of the public's roles 89
1.19 Usurpation, discrediting, stereotyping of the public(s)' voice(s) 96
1.20 Claim to authority by subjects of enunciation: institutionalized will to truth 102
1.20.1 The controlling voice: reporting vs reported discourse 105
1.21 Trivialization of new communications technology and its functions – depoliticization 107

Chapter 2 New communications technology as discourse: techne 110
2.0 Surveillance: the 'hybrid' net closes in – increased memory and cross-referencing capacity/order 113
2.1 Compu-nications 114
2.2 Discursive cartels: totalizing integration of spheres of discourse 115
2.3 Analysis/atomism 117
2.4 Cordoning off pathways: constraints of accessing and association patterns – order 118

2.5 Closed/static language vs open-ended/dynamic language: how much can a computer move? 121
2.6 On-line-real-time mediation: ersatz reality 126
2.7 Control of the source is control of access and association 128
2.8 'Closed user groups': the electronic incarnation of the procedure of exclusivity 131
2.9 Reduction of knowledge: data base as knowledge base 134
2.10 Unilateral control of voice and perspective 136
2.11 Context-free vs context-bound discourses of knowledge 140
2.12 The elusiveness of data processing to monitoring: unaccountability 143
2.13 Valorization of control for speed, growth, efficiency and productivity 145
2.14 Quantitative over qualitative evaluation: the 'pig principle' revisited 146
2.15 Télématique or privatique: centralization or decentralization 147
2.16 Network privatization: the individualization of subjects of enunciation 157
2.17 Unilateral participation: hardware and software for more than feedback 159
2.18 Systems theory: the ideology of the information society 161

Chapter 3 Powermatics: focusing in on power and social control 168

3.0 168
3.1 Lip service 169
3.2 Occultation 170
3.3 Historical examples of an awareness of control 171
3.4 Two axes of a socio-political critique of technology 172
3.5 A re-examination of the characteristics of power 173
3.6 How does discourse affect knowledge: or how does changing the way we communicate change the way we perceive and conceive? 174

3.7 Panopticism: social control via the internalization of discursive constraints on knowledge 177
3.8 A de-centred non-causalist theory of control 180
3.9 Power as the hub of reason 182
3.10 Power is not only repressive but also conducive 183
3.10.1 Compulsive issue-raising 184
3.11 External procedures of discourse 185
3.12 Control of whom?/power to whom? 186
3.13 Revising the man/machine debate 188
3.14 Beyond the conspiracy theory of power grounded in an economic base structure 189
3.15 Beyond economic determinism: bureaucratic organization and the formal rationality of economic action 193
3.16 Changing more than the machines or the economics 196
3.17 Power: centralization vs decentralization 198
3.18 Decentralization: a contextually realizable possibility 201
3.19 Anonymity, paranoia or the sites of power 203
3.20 Scientific managerialism: systems human engineering 207
3.21 Formalist rationality: its present-day concretizations 211
3.22 Reason's exclusivity: the technostructure: 'discursive hegemony' 214
3.23 The military industrial complex 216
3.24 Extra-discursive relations 218
3.25 The dinosauric strain: power to the status quo 219
3.26 Grounds for a critique of power and domination 221
3.26.1 The role of contradiction in critical theory 221
3.26.2 It does not work anymore 223
3.27 Weber's and Husserl's critique of rationality – the non-pertinence of technology to the social life-world 224
3.28 The irrationality of rationality 226
3.29 Rationality's self-legitimation 229
3.30 A good society for science 230
3.31 Discourses of rationalist classical science are undemocratic 231
3.32 Remedies defunct, tried and failed 234

3.33 Neo-liberalist cynicism 235
3.34 Conclusions 237

Chapter 4 Alternative procedures of discourse: epistemologies of technology 239
4.0 From critique to policy 241
4.1 From legislation of the uses and abuses of technology to the legislation of technology itself 244
4.2 Institutions 246
4.3 To regulate or not to regulate? Presence or absence of discursive constraints in an emancipatory society 248
4.4 Discursive transformations 249
4.4.1 Ontological shift of the definition of communication: from object to practice 250
4.4.2 Beyond control by the individualist subject 252
4.4.3 The privacy debate: an epistemic impossibility 253
4.5 Must we abandon the democratic emancipatory principle once and for all? 254
4.5.1 Autonomy and responsibility 258
4.5.2 Universality of the right to communicate 261
4.5.3 Towards a contextual consideration of differing communication procedures for differing regional or cultural groups 262
4.6 From instrumentalism to interactionism 263
4.6.1 Theoretical background of the shift to interactionism 264
4.6.2 Dismissal of the man vs machine debate: interaction replaces control 265
4.6.3 Why interaction? 265
4.6.4 Beware of false interactive participation 267
4.6.5 Beyond a mere simulacrum of dialogical participation 271
4.7 Degree of interactivity: towards a full-dimensional dialogue 272
4.7.1 The procedures of dialogism 275
4.7.2 Communicational competence: in command of certain discursive procedures 277

4.7.3 'The ideal speech situation' and 'the right to communicate' 279

4.8 Some specific preliminary and very inconclusive policy suggestions for participatory, dialogical communication in the realm of new communications technology 283

4.9 Strong anti-trust legislation: necessary but insufficient 284

4.9.1 284

4.9.2 Refocus away from seeding the hardware industry towards encouragement of pertinent public interest software development 285

4.9.3 Monoperspectivist vs multiperspectivist 286

4.9.4 Centralization to give way to contextualization 286

4.9.5 Respect for regional and cultural diversity 287

4.9.6 Avoid funnel effect in networking 290

4.9.7 Diversified software 290

4.9.8 Identification of the agents involved in communicational interaction: re-evaluation of sites of domination 291

4.9.9 The right and socio-technical feasibility of unplugging 291

4.10 An alternative politics of communication 294

4.11 Public decision-making concerning questions of new communications technology 297

4.11.1 Public inquiry as an exercise in participatory democracy 297

4.11.2 Participants: what constitutes public involvement? 298

4.12 Pilot project or dare: the electronic public inquiry 299

4.13 A bit of irony 304

4.14 Communications/technological ombudsman 305

Conclusion Discursive critique: socio-epistemological foundations for an alternative approach to communication and technology 308

5.1 Alternatives in discovering, isolating, and encouraging irregularities 310

5.2 The definition of the object of study 311

	5.2.1 Disintegration	311
	5.2.2 Integrative framework	315
5.3	The need to surpass positivist empiricism	317
	5.3.1 The empiricist's paradox: confidence in referentiality at all costs	317
5.4	Epistemology based on a recognition of the 'discursive formation of the object'	318
5.5	The industrial revolution – a discursive constitution	318
5.6	The discursive formation of the object	319
	5.6.1 The specific pertinence of a discursive approach to the study of new communications technology	322
5.7	The dynamic historical contextualization of the 'object' of study	323
	5.7.1 A-historicity – technological amnesia	323
	5.7.2 Historical approach	323
5.8	The idealization of technology as an immanent essence abstracted from context	325
5.9	Contextualization	326
	5.9.1 A priori context formation	327
	5.9.2 The changing of context(s)	327
5.10	Power/knowledge – the focus for studying the social impact of new communications technology	328
	5.10.1 Out of focus	328
	5.10.2 In focus	329
5.11	How to study power and social control?	329
5.12	The relationship of knowledge to value, of theory to practice	331
	5.12.1 Positivist separation of knowledge and value, of theory and practice	331
5.13	Empiricist epistemology as a reaffirmation of the 'factual' status quo	332
5.14	An epistemology of the possibility of change	332
5.15	The acceptance or rejection of the pre-given logical, discursive space from within which to talk about new communications technology	334
	5.15.1 Myth as overcoding	334
	5.15.2 Epistemic relativization	335

Notes 341

Bibliography

I Mass media coverage of the electronic revolution (1919–1983) 358

II Research works dealing specifically with new communications technology and with policy 365

III Theories of communication, discourse and technology 379

Index 389

ACKNOWLEDGMENTS

I wish to acknowledge gratefully two generous research grants from the McGill University Humanities Research Grants Sub-Committee which made possible the word-processing of the final version of this book.

I would also like to thank the following people for their constructive readings and criticisms throughout the conceptualization, writing, and revision of the work: Kevin Wilson, Morag Shiach, Karin Holland Biggs, Professor Timothy Reiss, Professor George Szanto, and the members of the CIMED – Centre Interuniversitaire pour l'Etude du Discours Social – led by Professor Marc Angenot.

While researching and writing this book I conducted three graduate seminars on the topic at McGill University. Much credit is due to the students who participated actively in these seminars. They have proven to be a constant source of inspiration, archival information, and devil's advocacy.

Special gratitude is owing to Paul Attallah who edited and formatted this work on the word processor with generosity, intelligence, talent, and conscientiousness. Donna Gill helped greatly with her diligent editorial work on the bibliography and notes.

Also, were it not for John Roston's good-humored help, the conceit of word-processing a critique of communications technology would not have been possible or would, at the very least, have been far more painful.

Finally, Ramón Pelinski has graciously suffered to live with the manic flow of elations and miseries that characterize the moods of scholars in the throes of writing a book such as this. This one's for you Ramón.

INTRODUCTION

0.0 Prophesy

We are on the threshold of a 'communications revolution,' an 'information society,' a 'technetronic age,' a 'post-industrialist society,' a 'knowledge-based economy,' (. . .) All of these phrases refer endlessly to the conviction, now rampant amongst academics, politicians, and journalists alike, that some fundamentally revolutionary change is occurring in our society and that information and/or communication (the two not always being distinguished) is at the root of it. The deployment and development of what is called 'new communications technology,' more specifically of electronic equipment for linking computers and telecommunications systems into networks, is seen to be at the foundation of this new society. These declarations are also always accompanied by numerous futurist declarations concerning the *inevitable* social impacts of this change: the 'end of ideology,' the 'third wave,' 'disemployment,' 'bio-genetic hybrid populations,' an 'age of leisure,' a 'return to feudal-style cottage industry,' and so on.

This book explores those propositions, shows the interests and forces that subtend and motivate them, the forms of reasoning which underpin them and which they propagate, and formulates a critique of this way of talking about new communications technology. In short, this book undertakes a discourse analysis of new communications technology.

0.1 Discourse analysis

What is discourse analysis and why is it appropriate to the study of new communications technology? Stated in its briefest, most simple form, discourse analysis is the study of the way in which an object or idea, any object or idea, is taken up by various institutions and epistemological positions, and of the way in which those institutions and positions treat it. Discourse analysis studies *the way in which* objects or ideas are spoken about. New communications technology is one such object/idea.

Let us give one brief example. The computer is no doubt the single most important element in the field of new communications technology. It is the machine upon which everything else, increased productivity and efficiency, telecommunications hook-ups, networking patterns, data storage and transfer, etc., depends. Given its centrality, it might also be granted that there can exist a number of ways of talking about the computer. One could adopt a luddite position and condemn the machine for what it will do to existing social structures. One could say that the machine is fundamentally irrelevant because it has nothing to say about the important existential questions of the day. One could say that, on the contrary, the computer brings with it certain capabilities that will allow us to reorder our lives, to delegate certain uninteresting tasks to it in order to free up our lives for more leisure time. One could talk about the changes it will bring about or allow in any number of fields, such as health care, education, banking, and so on.

It should be obvious, however, from these and other potential ways of talking about the computer, that the empirical object itself, the machine, the computer, is being taken up and integrated into any number of other interests. Next to the empirical object is being constructed a 'discursive' object. It is being spoken about in certain ways. Qualities, faults, and capabilities are being attributed to it. Whether it does or does not do any of the things said about it is irrelevant. The important thing is that people believe that it can and that they take the time to say so.

What people say, however, is not totally arbitrary. It is governed by certain rules and procedures. Whatever our position on the computer, we would probably all agree that if someone said computers will be great because trolls can live better in them or because they taste best on salad, that person would not have said something worth remembering. That is because such a statement would not correspond to our commonly recognized and accepted rules and procedures of pertinence, seriousness, and rational thought.

The interesting thing about rules and procedures, however, is that they change historically. They change across time and place. It is obvious to even the most cursory student of history, that what people thought important at different times and in different places has varied wildly. The question which must therefore be addressed is: 'how come we now hold to be true those things which we hold to be true?' And it is to this question which discourse analysis addresses itself.

Discourse analysis attempts to uncover the rules and procedures which subtend and legitimate the things we say and believe. It attempts to demonstrate not what a statement means, though that is not unimportant, but rather why that statement was produced, when it was, and as it was. It holds that if we can understand why and how statements are produced, we will have gained a new insight into their meaning.

The procedures of the modern world, those rules of thought and presentation which legitimate statements and beliefs for us, are the ones which will interest us most in this book. We shall try to show the procedures that legitimate what is said about new communications technology. And what is said can be called *the discourse of new communications technology*.

We will therefore have to look at a large body of writing on new communications technology and see what it actually says about its object. This writing can come from a variety of sources: learned articles, the popular press, technical documents, policy documents, critical articles, and so on. Across all the documents, the procedures of rationality which legitimate their statements and their positions, will emerge. Some of these procedures will have to do with instrumentalist rationality, with means/end logic, with the rule of excluded middle, with rules of logical and ordered presentation, and so on. Any statements conforming to these procedures will be held to be true. Statements deviating from them will be suspect. Any group of statements using the same procedures will be said to constitute a discourse.

It should be obvious that the procedures are of many types.

They are basically, however, nothing more than the rules which legitimate statements and beliefs. Obviously, some procedures will be very powerful and others will be secondary. Procedures which can draw upon rational scientific proof will, for us, carry a high charge of validity. Procedures which merely draw upon the good will of the speaker, though not insignificant, will in many cases be less powerful.

If statements exhibiting the same procedures can be called a discourse, the set of all procedures taken together constitutes an episteme. Discourses are variations upon the episteme. The episteme also cuts across disciplinary boundaries. Quantum mechanics and the popular newspaper science column, though of clearly different genres and based in markedly separate institutions, both draw upon the same sort of rationality for their proof and in order to legitimate their claims to truth.

These, then, are the procedures of the discourse on new communications technology. They are of many types, they cut across disciplines, they are rooted in different institutions, and they constitute the basis of modern belief. New communications technology, however, is not only talked *about*. It also instantiates a discourse of its own.

The extent to which a machine instantiates a discourse can be grasped easily if we recognize that machines do not happen by accident. Machines are designed, invented, perfected, brought about, discovered, etc., for a reason, at specific historical moments, and in order to serve specific needs. This is not to say that they cannot be used for purposes which were not originally intended, it is only to say that some original intention, however diffusely manifest and hard to pin down, presides at the birth of any machine.

Let us take the computer again as our example. It is clear that the computer was not invented by accident. It was perfected in a highly determinate institutional site for highly determinate reasons. As such it was designed for procedures of order, analysis, speed, efficiency, productivity, and so on. Those who perfected it were situated in a history of similar research and intentions. The computer has designed into it procedures of instrumentalist rationality. And for its purposes, it is difficult to think of more appropriate procedures. But it is not an accident, consequently, that computers lend themselves very well to management goals, with their rationalist demands for speed, productivity, efficiency, analysis, and order, and not very well at all to composing poetry. Clearly, poetry does not require those procedures.

It is, therefore, a comment upon our society to note the sums of money which are invested by governments and corporations into the propagation of certain procedures and not of others. We do not wish to fall into a romantic demand for greater appreciation of the procedures of poetry. We only wish to note that some procedures enjoy greater legitimacy than others and have, consequently, a greater power to speak the truth.

Any machine is the embodiment of social interests and forces. It contains and reproduces certain procedures. It lends credence and legitimacy to certain discourses. It is not just an empirical object but also a discursive object.

Discourse analysis attempts to show also the procedures propagated by and contained within the objects which it studies. Just as there are discourses on new communications technology, so are there discourses of new communications technology.

0.2 Progress or jargon?

With each new work on new communications technology that falls into our hands, it is as though the ante were being increased on a dare; try to make some sense out of the constantly proliferating host of discourses in the field of knowledge known as new communications technology or the 'communications society.' The articles and books dealing with the social implications of new communications technology run the gamut from the absurd to the threatening:

'Well Done Gabrielle Computer Tells Student' (*Toronto Star*, 13 March 1979, 11)

'Computer Ready to Run Your Life' (*Winnipeg Free Press*, 14 August 1979, 1)

Giant Brains of Machines That Think (Berkeley 1949)

Let Erma Do It: The Full Story of Automation (Woodbury 1956)

Monster or Messiah? (Matthews 1980)

'You'll Talk to Your Briefcase' (*Toronto Star*, March 1979, 1–11)

'A Revolution is Brewing in Your TV Set' (Ostry, in *Toronto Star*, 24 February 1979, III, 4)

'Sex, Money, Politics – Computers Know All' (*Toronto Star*, 17 March 1979, 4)

'Never Stumped' (*New Yorker*, 4 March 1956, 20–21)

'Labor of a Year is Done in 400 Hours' (*Life*, 18 February 1957, 92)

'Electronic Intricacy in Compact Design' (*Life*, 21 May 1956, 108–9)

What are we to do with this ever-growing body of literature on the communications revolution, which is to be found in newspapers, government policy documents, and academic publications in the pure, applied, and social sciences? Do these messages refer us to some actual change in society, pointing to the essential issues that need to be dealt with, or do they serve merely as a form of *noise* which interferes with other important social messages about crucial and pressing matters?

The debate has penetrated into just about every sphere of the social sciences ranging from business, through education, development, human rights, international law, and the culinary arts. On a more general plane, we are also inundated with endless clichés apparently not based on anything more thoughtful than dogma: 'technology is inherently good (or bad),' 'it all depends on how you use it,' 'man can (or cannot) survive without machines,' 'man and machines are fundamentally alike (or different).'

In the spheres of sociology, economics, and policy-making we find such 'gurus' of the information age as Porat, Machlup, Bell, and Valaskakis stating that we are in an information society because over fifty per cent of the gross national product is devoted to the information sector. They encourage us to believe that 'information is our new-found wealth.' But how are we to react when, looking more closely at their classifications of 'information workers,' we find that they include preachers, teachers, and perhaps even the oldest profession in the world?

In development studies, the ideological lines of the debate have already been drawn. The Servan-Schreibers declare that developing countries can simply skip over the industrial age by leaping headlong into the information age where peasant farmers can solve their essential problems with a home computer. And the cost would be only slightly greater than the gross annual income of most South American farmers. The Mattelarts counter that information technology is inherently designed to serve the interests of international capital situated in developed countries. Both sides now refer to developed and developing countries as the 'information poor' and the 'information rich'.

Parents are beckoned to rush out and buy a home computer so that their children can compete for computer literacy with their peers and their teachers. Teachers spend their summers in school learning computer techniques lest they lose their jobs to the information age.

The very poignant debate for developing countries and for developed but dependent ones such as Canada over national autonomy and cultural identity has now come to include the issue of new communications technology. To ensure the survival of threatened cultural identities, should we adopt, reject, adopt and regulate, design and develop, or simply ignore the onslaught of new technologies?

On the international front, even the issues of human rights is touched by the prophesies of the information age. New phrases are coined for new rights: 'the right to information' and 'the right to communicate.' Yet they are so riven with euphemism and bureaucratic convenience in order to suit all ideological camps that one wonders whether or not they retain any meaning whatsoever.

'Access' to new communications technology has suddenly become a major element in the improvement of the 'quality of life,' while civil rights groups warn that it threatens 'privacy'.

In management studies, technology is said to facilitate the management of business by increasing control over 'variables,' such as the market, distance, time, and the workforce. Governments, unions, and employers alike all panic about whether or not we will be able to jump on the new communications technology bandwagon quickly enough to avoid what is now euphemistically called 'disemployment.' Will we continue to be hewers of wood and drawers of water for Japanese tourists grown rich from the sales of new communications technology on our markets which we failed to capture? Promises are held out, like carrots on a stick, of new information-related work in a cleaner more stimulating, less alienating environment.

In all of these debates, one finds a constant turning towards the future, a sense of urgency associated with the technological imperative, a hopeful fear of the unknown, an undeniable belief that change is progress and that progress is good. And one also finds a concomitant turning away from the past and the history of another technological revolution, the industrial revolution.

0.3 Demystification

> The first task is to *demythologize* the rhetoric of the electronic sublime. Electronics is neither the arrival of apocalypse nor the dispensation of grace (. . .) As we demythologize, we might also begin to dismantle the fetishes of communication for the sake of communication, and decentralization and participation *without reference to content or context*. (Carey and Quirk 1970, 423)

If we could talk to the information society, we would ask it: 'Does the communications society/revolution really exist?' It, of course, would answer: 'Yes.' But this would be an answer given from within an already technologically-biased society. One would have to wonder whether this revolution really existed or whether the society were trying to bring it about. One simply cannot take the answers of technological society at face value when technology itself is concerned. Alternative ways must be found of getting answers to our questions about the 'communications revolution.'

For the purposes of this investigation, we will assume that in order to know something about the interface of new communications technology and society, we can only inquire into the ways in which both society and new communications technology itself 'speak' (about) this interface. The way in which society receives new communications technology, the social impact, must be perceived in and through the ways in which society talks about the social impact of new communications technology; *this talk is the trace of social impact*. If we wish to know something about new communications technology as it exists in society we must also look for *traces* of it. Such traces of new communications technology are to be found both in the documents about this technology as well as in the actual communications practices of it, i.e., extant networking systems, materials and organization of data banks, etc.

Not only must we observe new communications technology as discourse but its social conduct as well. In answer to the question:

what is society made of? or how is it constituted? we would suggest that without communicational interaction there would be no society. Society is at least partially constituted in and through communication. Therefore, if we wish to discern social configurations or social change, we must look less to percentages of GNP and more to types of communicational relations and rules. Of course, one might argue for the primacy of other technological and social activities which are not purely communicational, such as working on an assembly line. To this, it could be replied that communication is at the heart of assembly line production, that we can learn a good deal about non-communicational activities simply by observing the communication within them. Or again, one could maintain that telematics is at any rate nothing more than a bunch of silicon chips and telecommunications cables, so why study it as discourse. But is not informatics based on or made possible by, first and foremost, a discourse of knowledge commonly referred to as information theory? The point is: discourses of knowledge are at the heart of technology.

Nevertheless, discourse is not merely an accurate reference to or reflection of reality. 'How' we speak becomes just as important as, if not more than, the content of 'what' we say. Social communication is not merely a content, a set of messages, but rather first and foremost a set of ways and rules of interacting, the 'how' of communication. Of course, the content of a discourse is important. We cannot simply ignore the themes in all this talk about new communications technology. However, given that these statements are more an indication of our patterns of social interaction than they are adequate references to objects in the exterior world, a study of technology and society must concentrate primarily upon the 'how' of communication. For example, if we wish to know something about the power relations that will characterize society with the advent of new communications technology, it is not sufficient to say, as does Mattelart, that in capitalistic, monopolistic society, capitalistic and monopolistic power relations will prevail. Instead, we must look at who has which rights to speak and to dictate to others how they may speak at various levels of society. Furthermore, we must ask: what powers accompany knowledge and which types of rules may be followed in order to generate statements that are accepted as true in any given society? Rather than assuming that the power that accompanies technology lies in the hands of a bureaucratic communist State or in those of a capitalist oligopoly, we must first investigate actual practices of power as practised in and mediated by communicational interaction.

Not only must the relationship between technology and society be seen as discursive, but to it must also be added an historical perspective. For example, if we were to speak about a communications revolution, or any other revolution for that matter, we would first have to know the state of affairs which preceded the situation which we are calling revolutionary in order to know whether or not there had been any diachronic change. Consequently, it does not suffice to look at current ways of communicating in order to evaluate the importance of the development of communications technology. We must relate these patterns of interaction, these cultural symbols, these rhetorical tropes and mythoi to past ways of speaking and communicating. Should the statements, the ways of communicating and the prescriptive rules for making 'correct' statements about technology and for 'correct' practices of technology, be seen to alter radically, then there would be both epistemological and sociological grounds for stating that social reality has changed, that a communications revolution has actually occurred.

To summarize, then, an alternative method for assessing the social impact of new communications technology would have to do four things:

(a) go beyond apparent content since communication is not a mirror reflection of brute unmediated reality;
(b) study social order and social impact in terms of communicational relations which both mediate and constitute society;
(c) understand communication as more than content, more than 'what' is said, and rather as the rules for 'how' to communicate which actually underpin the types of statements that can be made and how society receives them.
(d) relate the findings of the investigation of how today's 'information society' relates to a *history* of how various societies communicated in the past, and more specifically to industrial society since it is the implicit background against which the 'communications revolution' is supposed to have occurred.

0.4 Discursive criticism – articulation of an approach

0.4.1 The object of study – communication as discourse

In the preceding paraphrase of the many things that have been and continue to be said about new communications technology, one finds a dilemma. On the one hand, the 'reformers' are saying quite different things. Servan-Schreiber says that it will solve the problem of underdevelopment. Mattelart, however, says that new communications technology is imperialist. Some say that communications technology will cause education to advance. Others claim that students will cease to be able to think for themselves. All of these statements, then, possess a different content. And yet, and this is a major point which we will pursue throughout the book, the way in which these 'reformers' look at or study new communications technology is fundamentally similar. Both the Servan-Schreibers and the Mattelarts base their statements on certain similar assumptions and presuppositions about the nature of society and about how we know the world. For example, both cite economic statistics as the major indicator of the nature of the information society, though the former uses them to disprove exploitation and the latter to prove it. Both those who advocate and those who oppose computer-communications in education share the presupposition that more information is better. They merely disagree about whether the information should be stored in the head of the student or in a data bank. Recent letters to the editor on the occasion of *Time* magazine's declaration of the computer as man of the year (1982), whether they opposed or welcomed the computer, argued their position in terms of the metaphysical debate around the similarities/dissimilarities between man and machine and their hierarchical relations to each other. Even though the contents of these statements may be diametrically opposed, they nonetheless operate on the basis of the same presupposition and consequently move within a very restricted domain of what statements may be made. Our study of new communications technology and of communication about this technology wishes to study these presuppositions and assumptions as they are manifest in the ways in which people communicate about and via new communications technology. It is these sets of presuppositions and assumptions as they are manifest in communication about and of communications technology that constitute the object of our study: discourse.

A discursive approach studies communication. Communication, though, is not reducible to the information model:

Sender	Message	Receiver
	Code	
	Channel	

Nor is communication merely a set of signs with both syntactical organization and a semantic correlatum, as the semiotic model of communication would hold. Communication is something more than these two models would allow. The term ‘discourse’ accounts for this something more.

In the most simple terms *discourse is merely language practised*. A discursive analysis of the ways of speaking about new communications technology would not merely analyze the formal linguistic features or the semantic content of statements. It would analyze the actual practice of making statements in society. A discursive analysis of new communications technology *per se* would not simply describe the machines or talk about the content in data banks. Rather, it would examine new communications technology as a practice of communicating, of using verbal and non-verbal languages.

Discursive analysis, then, does not limit its object of study to the customary elements of linguistics (semantics and grammar or syntax) or to basic units (the sentence, the proposition, the speech act). Discourse has its own rules of operation above and beyond the structures of linguistics. But this is nothing new. The term discourse has always been used to refer to something broader than language or the sentence. Discourse has always been associated with the enactment of a practice.

What is original in Michel Foucault’s understanding of discourse is the way in which he relates the practice of language to other practices in social life: politics, culture, economics, social institutions such as schools and prisons, etc. *Discourse itself is a specific social practice*. Furthermore, discourse is understood by Foucault as perhaps the central social practice lying at the heart of all other non-discursive social practices. And it will be seen that Habermas' understanding of communication is very close to this. Foucault cannot specify exactly where discourse begins and social life ends, or vice versa, but this lack of specificity is precisely due to the fact that both discursive and non-discursive social practices are inextricably bound together.

A discursive analysis of the social impact of new communications technology, then, would have as its object of study the social practices of language that form the environment of this technology as well as the social practice of communication carried out by new communications technology itself. Also, it would have as its task the establishment of the relationships between these practices of language and other social practices. Also, we will argue that new communications technology itself can be considered to be nothing more nor less than sets of discursive practices.

In short, then, discourse may be understood simply as the act of communicating, as the set of practices of speaking. Discourse is communication, not an object, but rather an interactive activity.

But discourse is not simply an amorphous mass of practices. These practices have certain empirical forms which are neither more nor less than the rules or procedures that are constituted as the regularities of discursive practices that society condones as correct or suitable. We will refer to these rules or regularities as 'discursive procedures.'

Before specifying in more detail what discursive procedures are, we would wish to insist that *discourse is a thing of this world*. It would be a grave error indeed to believe that the discursive approach is somehow an abstraction from things in the world. Discursive practices are just as evident as, if not more so than economic statistics, although neither can be known with the unproblematic and unmediated ease which positivist empiricists would wish. To believe that discourse is not a thing of this world is to hold what Foucault would call a classical view of language whereby discourse is thought to be nothing more than a mere reflection of the real world without being a part of itself. We cannot emphasize sufficiently the fact that discourse is itself a *social practice* which is a part of the world, indeed perhaps the central part of the social world. The social world, especially, has been shown by Foucault and Habermas to be constituted mostly by activities of communicational interaction, or by what Foucault calls 'interdiscursive relationships.' Discourse, then, is as much a social, political and economic activity as manufacturing.

To summarize, we may consider the discourses that surround new communications technology to be social practices in the real world rather than mere semantic reflections of the world.

0.5 Discursive procedures

As stated earlier, discourse is not some nondescript, uncircumscribed practice. Discourse is practised with a certain degree of regularity. These regularities can be called discursive constraints or discursive procedures. All societies practise, have practised, and will undoubtedly continue to practise discourse according to specific constraints or procedures, which nevertheless vary from society to society. Societies either accept or reject various specific discursive practices. The ones they accept are reinforced and become rules of what Habermas would call 'communicational competence.' There is no such thing as an anarchic discourse which follows absolutely no constraints:

> I am supposing that in every society the production of discourse is at once controlled, selected, organized and distributed according to a certain number of procedures, whose role is to avert its powers and danger, to cope with chance events, to evade its ponderous, awesome materiality. (Foucault 1972, 216)

These discursive procedures are the rules which must be followed in order to generate statements which would be accepted as 'true' and 'correct' within a specific society. In order for specific statements to be made and to be accepted, certain discursive rules must be functioning. For example, in order for troubadours in the middle ages to sing a repertoire of songs which were never written down or individually composed but merely transmitted orally with minor variations from person to person, there had to exist a discursive procedure of the collective and anonymous subject of discourse (Zumthor 1975). And today, as we will argue, for so many books on new communications technology to find significant the listing and naming of machines and their 'inventors,' the procedures of referentiality and individualization must be operating to regulate discourse. Thus, we may define discourse as a body or set of regularized practices of language (verbal and non-verbal) condoned by a given society. These practices underpin and render possible certain statements and communicational practices while disallowing others:

> We shall call discourse a group of statements insofar as they belong to the same discursive formation; it does not form rhetorical or formal unity, endlessly repeatable, whose appearance or use in history might be indicated (and, if necessary explained); it is made up of a limited number of statements for which a group of conditions of existence can be defined (. . .) Lastly, what we have called 'discursive practice' can now be defined more precisely. It must not be confused with the expressive operation by which an individual formulates an idea, a desire, an image; nor with the 'competence' of a speaking subject when he constructs grammatical sentences; it is a body of anonymous rules, always determined in time and space that have defined a given period, and for a given social, economic, geographical or linguistic area, the conditions of operation of the enunciative function. (Foucault 1972, 117)

0.6 Content does not equal procedure

At this point we may now more fully appreciate the distinction between content and discursive procedure, between the 'what' and 'how' of discourse. For example, many discourses on new communications technology will be seen to state as their overt semantic content that new communications technology will further democratic communicational participation in society. At the same time, however, this same discursive practice follows the discursive procedure of *hierarchically excluding* certain speakers, of refusing them the right to speak or at least, though they may be allowed to speak, of claiming any 'scientific truth' for the form and content of their statements. Often in current debate on new communications technology engineers and bureaucrats enjoy such hierarchical exclusivity at the expense of those they often call 'naive and unpragmatic humanists.' Furthermore, these same engineers who claim that new communications technology will lead to the democratization of society are also busy devising networking procedures of new communications technology to allow for what is commonly called 'closed user groups,' a system of networked time-sharing of computers which excludes certain people from access to certain information, reserving it exclusively for others who can profit from it either politically or economically.

What we are faced with in these examples is a subtle and usually hidden contradiction between, on the one hand, the overt semantic content of statements about new communications technology, and, on the other hand, the discursive procedures that are practised in both talk about technology and in technology itself. In the above case, *the contradiction lies between a content of democratization and a procedure of hierarchical exclusivity*. Throughout the book we will concentrate on exposing many more of these contradictions between procedures and content. First, however, we shall have to describe many of these procedures in detail.

0.7 Procedures and technologies

While it is easy to understand why we would consider talk about new communications technology as discourse, it is perhaps more perplexing to think of technology itself as discourse.

What exactly do discursive procedures have to do with technologies and more specifically with new communications technologies?

In a recent lecture at the University of Vermont in Burlington (Fall 1982), Michel Foucault used the terms 'procedures' and 'technologies' interchangeably, though with a degree of insistence. Elaborating upon the title of his lecture, 'Technologies of Self and of State,' Foucault defined 'technologies' as a set of procedures put into operation by the sovereign or the State which the 'polis' (citizenry) was supposed to follow. These procedures ensured a certain philanthropic treatment of the citizenry while simultaneously recording their whereabouts, organizing them and obliging them to serve (in the army) and to protect the State or the sovereign. Foucault's well-documented historical case studies all serve to illustrate how the discursive procedures of hygiene, recreation, jurisprudence, education, psychiatry, etc., were all adopted as social practices from the seventeenth century onwards, and how the citizenry had in turn to pay for them by being registered in the service of the State, i.e., it must surrender a degree of control over itself to the sovereign or the State.

'Technologies,' then, does not merely designate concrete machines. It means rather 'machinery' in the sense of means to an end, or procedures of institutionalized knowledge which serve the ends of social order and of sovereign or State power over the 'selves' that make up the public. In other words, technologies are the rules of how to do or how to make. In the case of Foucault's examples, technologies are the rules of how to organize 'selves' into State – or sovereign-serving body politic. These rules govern not only the selves' use of State-run machinery, such as courts of law or prisons, but also their every minute gestures, such as personal hygiene, these latter being designated by the term micropractices.

Technologies are sets of rules for how to do or make things: how to speak, how to organize a citizenry, how to manufacture a car, how to eat, how to communicate, etc. As such, technologies may be related to what Max Weber termed 'instrumental knowledge,' meaning knowledge of the means that best meet the desired ends. Technologies are procedures of knowledge which are means to an end. They are procedures of instrumental knowledge. For the time being we will use 'technologies' and 'procedures' synonymously.

There may still be some confusion as to why we wish to treat new communications technology not merely as procedures but specifically as discursive procedures. For one thing, the etymology of technology, 'techne,' meaning knowledge of how to do or to make, is closely related to 'poiesis,' the Greek word for making, which is also the etymology of the art of making with language – poetry (Castoriadis 1978, 223–24). Thus it would seem fitting to link technologies to procedures of language practice given their close etymological link. Indeed, as we will argue in this study, discursive procedures of knowledge are indispensable to any and all knowledge of how to make.

The subtitle of this work, 'A discursive critique of new communications technology,' should now indicate clearly the central theme of this study. In order to study the social impact of new communications technology we will look not only at the traces which the impact of this technology has left in the discourses of the society that talks about it. We shall also look at how new communications technology itself functions in society as practices of communicational interaction that follow certain rules, i.e., as discursive procedures. Both talk about new communications technology and new communications technology itself may be considered to be social discursive practices. The only difference is that new communications technology as discourse is a communicational practice that occurs in media other than, though not exclusive of, the verbal medium. The rules of discourse of new communications technology for example might include certain regularities of networking such as 'exclusivity,' or 'hierarchization.' The procedures of 'order,' 'classification,' and 'analysis' will occur not as written texts but in the data banking operations of storage and cross-referencing. The procedure of participant a-symmetry may occur simply by virtue of inequality of the capacity to program.

We treat new communications technology as a discourse whereby the machines themselves, the combination of hardware and software such as Telidon (Canada), Antiope (France), Prestel (Britain), Hi-Ovis (Japan), are merely practised concretizations of sets of procedures for how to communicate. Any and every machine of new communications technology, we will argue, has designed and deployed within it certain communicational procedures which allow for certain ways of communicating, certain uses, and not others. It is the procedure, not the machine, that counts. The procedure, however, is designed directly into the machine. For example, there is much talk about participatory media in new communications technology. Participation refers to the discursive procedure that we will call 'dialogue.' The kinds and degrees of dialogical interaction permitted by bi-directional communications technology will be seen to depend on the levels of dialogue permitted by the procedures which are actually designed and built into the technology. A two-way channel merely permits a two-way flow of messages. There is nothing built into this technology to ensure that each partner shares the same degree of stylistic or rhetorical competency. Nor is it likely that each partner would enjoy the same ability to program the computer-communications system to best suit his or her needs. Thus, we will show how the discursive procedure of dialogism entails many more procedures (or sub-procedures) of communicational competency than have so far been built into two-way systems. The bi-directional systems alone do not contain the inherent procedures of participatory dialogue. We will show that new communications technology has certain procedures of communication, and not others, designed and built into it. The result of this is that it permits certain practices of communication and not others. For other practices to exist we will see that other discursive procedures would have to be built into the technologies. This, of course, would require other technologies.

0.8 Discourse and knowledge

Thus far we have only alluded to the connection between discursive procedures and technologies, on the one hand, and knowledge, on the other. This relationship needs to be made more explicit.

We have defined technology as knowledge procedures, more specifically, those of instrumental knowledge, which means knowledge of how to do or make. But how do we define knowledge? Earlier we stated that our epistemological assumption was that we could have no perfect, unmediated knowledge of the exterior world. This makes a definition of knowledge rather delicate.

In classical times, knowledge was defined as the content of our cognition that corresponded perfectly and adequately to the object we wished to know, to the real world whether it be 'natural' or 'mental.' However, once the accuracy of our cognitive apparatus as well as of the discourses that express our knowledge is thrown into doubt, knowledge comes to mean something else. Knowledge is no longer defined in terms of its content. It becomes rather the set of statements about the world that are accepted by any given society at any time and place as 'true.' No absolute proof exists of the truth of these statements since we have no unmediated knowledge of pure reality by which to judge the accuracy of the statements. Rather, the statements are judged to be 'true' on the basis of whether or not the community that receives them believes them or not.

Foucault has shown that it is not so much the content of the statement that determines whether it will be considered to be valid or not as it is the way in which the statement is made. If a statement has been made using the set of procedures which a given society condones as 'reasonable' or 'scientific,' then chances are that the content of the statement will be found to be 'true.' Knowledge, then, is the set of statements that is underpinned by the set of discursive practices that particular societies condone as acceptable. A perfect example of how this works is to be found in the anecdote about how a group of doctors was tricked at a medical convention. The tricksters took an actor and initiated him to the procedures of speaking at a medical convention. The actor then put any content he wanted into his speech. He stood in front of the convention claiming to have made some medical discovery which he presented using the appropriate procedures of Latin nomenclature, classification, presentation of slides and specimens, citing other experts, etc. Most of the doctors, when questioned, had complete confidence in the scientific validity of the medical 'knowledge' they were being presented. Intellectual history provides ample evidence of how the rules of scientific activity are fundamentally rules for ways of practising the language of scientific investigation and that as these rules change so do the set of statements known to be true, i.e., so does knowledge (Feyerabend 1975). Knowledge depends on the procedures of discourse that society accepts as 'reasonable.'

Whoever controls or adheres to the discursive procedures condoned by social consensus as 'scientific' or 'reasonable' is already 'within the field of truth' and hence enjoys the acknowledged capacity to generate statements that are accepted as 'true' by that community. Those who do not follow the 'right' procedures are often called 'unreasonable,' and sometimes even 'mad' (Foucault 1973).

The implications of this definition of madness for a critique of new communications technology are enormous. First of all, when one claims to make 'true' or scientific statements about new communications technology or its social impact, one must follow the acceptable procedures of discourse. For example, it lends scientific credibility to one's statements if one follows the procedure of 'referentiality' according to which one lists, before saying anything at all about new communications technology, many machines along with illustrations, in order to show that one is indeed referring to an object in the real world. One must follow the 'instrumental' procedure whereby growth and increased productivity are valued above all else. Those who do not and who suggest that other values are often more important, often come under attack. They are said to be 'unreasonable,' 'naive,' or even 'paranoid.' The parallel with the seventeenth century when those who were not 'right' were said to be 'mad' is unmistakable (Foucault 1973). Futhermore, the rules that are sanctioned as knowledge-generating must also be followed by new communications technology if it is to be considered to be a correct knowledge – knowledge of how to do or make. For example, correct technologies of communicating are supposed to classify. Thus, new communications technology concentrates much effort on designing and deploying data storage and retrieval systems with elaborate, cross-referenced classificational capacity. This insistence occurs, of course, at the expense of other procedures that one might conceivably suggest as correct for knowing how to communicate.

Not only Foucault but also James Carey stresses the strict relationship between rules of communicating and how knowledge is defined and accepted by society.

> Modern computer enthusiasts like Pool may be willing to share their data with anyone. What they are not willing to give up so readily is the entire technocratic world view that determines what it is that qualifies as valuable fact. What they wish to monopolize is not the data but the approved, certified, authorized mode of thought, indeed the very definition of what it means to be reasonable (Carey 1975, 45).

As well as attempting to show which procedures of discourse restrict and compel both the formation of certain statements about new communications technology, and certain practices of this same technology, as 'reasonable' or 'unreasonable,' it will also be essential to show how these accepted and restricted procedures in the domain of new communications technology proper relate to the larger context of the knowledge procedures that our society condones. Foucault and others refer to this larger context of 'scientifically acceptable' procedures thought to generate true statements in any given time and place as the 'episteme.'

0.8.1 Episteme

The concept of episteme explains how specific practices of discourse are situated within a larger socio-historical context and how this context dictates to a large degree the acceptability of the procedures by a society as 'reasonable' or scientific. By studying disciplines such as psychiatry throughout history, Foucault discovered that the procedures of 'scientific' investigation changed from society to society and from epoch to epoch. He also found that in the same society and at given times, different disciplines tended to practise the same procedures of knowledge. For example, in *The Order of Things* Foucault showed how, in the seventeenth century, the sciences of grammar, economics, biology, and pictorial art forms all practised their discourses according to the procedures of exchange, referentiality, order, analysis, hierarchy, exclusivity, etc. This epoch, and the group of procedures that characterized discursive practices of knowledge during it have been called the *'classical episteme.'* Foucault names other epistemes in which the procedures of discourse sanctioned in that time and place were distinct: the Renaissance, the modern age, and the post-modern age.

> By *episteme* we mean, in fact, the total set of relations that unite, at a given period, the discursive practices that give rise to epistemological figures, sciences, and possibly formalized systems; the way in which, in each of these discursive formations, the transitions to epistemologization, scientificity, and formalizations are situated and operate; the distribution of these thresholds, which may coincide, be subordinated to one another, or be separated by shifts in time; the lateral relations that may exist between epistemological figures or sciences in so far as they belong to neighbouring, but distinct, discursive practices. The episteme is not a form of knowledge (connaissance) or type of rationality which, crossing the boundaries of the most varied sciences, manifests the sovereign unity of the subject, a spirit, or a period; it is the totality of relations that can be discovered, for a given period, between the sciences when one analyses them at the level of discursive regularities. (Foucault 1972, 191)

It is possible to explain the emergence of certain practices of new communications technology as well as of certain statements about this technology by situating their procedures within the larger context of the dominant episteme out of which they arise, i.e., the larger set of procedures that our present day society accepts as scientific procedures generative of true statements. By so doing, we can begin to see if there is anything new emerging in the 'communications revolution,' any transformation in relation to the historical, epistemic background. We can look at the greater social context of practices of knowledge in an attempt to discern the mutual interdetermination of new communications practices and other social knowledge practices of the time.

We might add one note of warning here, though. Just as we began this investigation with the intention of not idealizing new communications technology, but rather of judging it by virtue of the regularities found in concrete, empirical practices of it, so too must we not idealize the episteme. The epistemic context must be investigated, not assumed. In order to do that one must search for the regularities of our most dominant knowledge practices in society throughout history. This we will try to do within the limitation of our study. Also, one should not equate the derivation of an epistemic context with a history of ideas. This approach is historical but it does not articulate ideas. It articulates concrete discursive practices, procedures not content. Foucault insisted very much on this distinction when he wrote a rebuttal to George Steiner's interpretation of his work as a history of ideas (in *Diacritics* 1971, 60). This is an enormous distinction, especially when it comes to defending the discursive approach in the face of attacks from positivists that it ignores empirical reality. It does no such thing. It studies empirically evident traces, often in the form of documents, of actual, concretized discursive social practices. It studies non-discursive social practices in the traces that they leave in discourse. How else can one study social practices, especially past practices, except in and through the traces they leave?

0.8.2 Episteme and techne

If the temporally and spatially delimited sets of procedures of knowledge are referred to as episteme, we might take this categorization a step further to introduce another distinction. We suggested earlier that technologies were procedures of knowledge of how to do or make, of instrumental knowledge. Could we not then call the set of procedures that a given society in a given epoch sanctions as correct the *dominant 'techne'*? This distinction, of course, will only add to the argument if there exist other procedures of knowledge that are not instrumental. These would simply be epistemic procedures.

One of the main tasks of this book will be to study the relationship of 'techne,' the procedures of instrumental knowledge examined in chapter two, to the procedures of non-instrumental knowledge. We will investigate whether the ways of knowing about new communications technology are themselves a case of instrumental knowledge – techne – or whether other forms of knowledge exist in today's environment for the new communications technology debate. In short, we will investigate the relationship between techne and episteme. Are they kept distinct? Does one subsume the other? Is one collapsed into the other? And what are the implications of their relationship for the type of society in which we live? These are some of the fundamental questions to which any examination of technology must reply but which have been neglected in recent discussions of new communications technology. We insist so thoroughly on episteme and techne as historical circumscription of knowledge contexts because any approach, discursive or other, cannot understand what is happening in today's society without looking at the social practices which used to occur. To shed light on past social constraints and compulsions which may still be operative is to understand better how the less desirable ones might possibly undergo transformation, i.e., how our society might change for the better.

CHAPTER 1

Procedures of discourses on new communications technology: the episteme[1]

1.0 Bridging the gap

This chapter provides a continuous paraphrase and quotation of the familiar ways of talking about new communications technology and its social implications.

It is the starting point of any discursive analysis to allow the discourse under study to speak for itself. As a result, it may seem at first that we are simply adopting, entirely and uncritically, the pat jargon and clichés of these discourses. This is only an impression. Gradually, a critical voice will insert itself. It will expose the procedures of the discourses as well as the contradictions between their overt claims and their covert rules of operation. Gradually, it will begin to point to alternative procedures for understanding and practising new communications technology. At the end of this examination one thing at least will become apparent: the internal contradictions of the traditional approach have brought it to a point of crisis. The beginnings of an interdiscursive 'challenge' to traditional discourses coming from the irregularities of alternative discourses will become apparent.

Eventually, the innocence of paraphrase and the 'principle of commentary' (Foucault 1971) will give way to 'critique.' First, however, the issues and techniques will be allowed to plead their own case. This will make evident the procedures that make them possible as statements and, as a result, their claims to 'naturalness' and 'universality' will have to be abandoned. It will become possible to distance oneself from the usual episteme that positions contemporary debate simply by demonstrating that episteme's mode of operation. This exercise could be called a demystification of statements concerning new communications technology, a hermeneutics that operates partly by juxtaposing their claims to universality with the limited conditions of their possibility.

Why bother, however, to describe or reconstruct these conditions of possibility? Simply because there is more than a descriptive telos in operation here. If a set of old though still current procedures have made it possible for new communications technology to function in a certain way as a discourse of knowledge then *alternative procedures would make possible an alternative practice of new communications technology*. Taken at face value, or for their content, the host of messages concerning new communications technology and society will tell us absolutely nothing. Taken as traces of a dominant episteme or as a sort of 'worldview,' however, these discursive fragments may hold the key to solving the issue of whether new communications technology represents a revolution or conservatism, of whether it represents control or emancipation.

1.1 Establishing the corpus of discourses on new communications technology

The choice and organization of the discourses to be studied must obviously be justified before the discursive analysis can begin. It is customary for communications studies to segment their corpus according to traditional divisions of genre, type or medium of discourse.[2] Consequently, a study of news would differ from a study of popular literature and these generic boundaries would further be distinguished by whether they were written, televised or transmitted by radio, and so on. These boundaries have often formed unquestioned, water-tight compartments. Such divisions, however, do not really aid the search for common discursive procedures which cut across boundaries and types.

The discourses on new communications technology also contain certain traditional delineating categories. But as these categories are themselves the result of society's discursive procedures, they also prove to be unfruitful for what this investigation seeks.

Many theorists make two separate bibliographies in the field of new communications technology. On the one hand are the pro-technology texts, advocating it for the improvements in life-style and security that it will bring. On the other hand are those texts which, for a variety of reasons, condemn new communications technology. This study will show that both the pros and the cons raise much the same issues and use or recognize the same discursive and epistemological procedures. For example, some texts will condemn the video revolution as economically unlikely while others will encourage it as an economic boom. Their common ground is the procedure of instrumentalism which judges communication and knowledge according to a means-ends logic aimed at profitablility/productivity.[3]

As Carey and Quirk have argued, both the advocacy and denunciation of new technology alike often operate according to the same rhetorical procedures. What is needed is to transcend or transform the procedures of discourse and knowledge that make the myth of technology possible in the first place. 'Paranoia about mass media and a sense of powerlessness are the simplistic obverse of the *mythos* of electronics.' (Carey and Quirk 1970, 422)

Objects taken as referents (referent-objects) have also often provided the grounds for typologizing texts on new communications technology. Most 'state of the art' articles break their subject down into sections which correspond to various techniques ranging from satellites and coaxial cable to fibre optics and videotex. This taxonomy relies strictly on an acceptance of the discursive procedure of referentiality. Once this procedure is questioned, however, a certain strategic 'découpage' of the corpus ceases to be justifiable.

Various issues also serve as lines of demarcation for the corpus. One text will choose to work on the issue of international communication, another on pay tv, and yet another on national content, and all will divide their corpus accordingly.

We prefer, however, to group all of these 'issues' together and to show how specific discursive procedures cause general introductory issues, issues of national sovereignty, and issues revolving around the metaphysics of man-machine relations to be raised in a certain way. Other discursive procedures would permit different sets of issues to be raised in different ways.

In other words, a discursive approach considers the issues but does not organize itself or its corpus according to them. A discursive approach is more concerned with demonstrating how and why various issues are raised and how they could be dispelled or raised in an alternative manner.

While this study's choice of corpus recognizes the existence of traditional types of discourse, and while it strives to represent each, it does not organize itself along their lines. Theory, journalism, advertising, popularized scientific reporting as well as policy and mandate documents will be grouped together. Any ensuing typology will not necessarily follow traditional guidelines. Instead, it will be shown that certain procedures cross all generic, media, and readership boundaries. For the time being though, it will be sufficient to view all of these particular discourses *simply as social discourse*.

It would also be incorrect to assume that because it ignores traditional typologies this study presumes to identify only the similarities and the continuities of discourses on new communications technology. While many regularities will be uncovered, so too will many *irregularities, discontinuities* and *differences*. Indeed, it is the emergence of a set of irregularities defined on a background of regularities that will be seen. While engineers and bureaucrats conform to the dominant episteme of classical scientificism and instrumentalism, Carey, Foucault, Habermas, Innis, Castoriadis, Cohn-Bendit, and Weizenbaum all begin to elucidate and to follow discursive procedures which are radically discontinuous to the traditional approach to new communications technology. A schism will thus emerge between an episteme articulated along the lines of power and control and one functioning on principles of interaction and democracy. Indeed, even some national and international policy documents and projects concerning new communications technology will be seen to partake of these irregularities.

Finally, a note of warning. It will not always be possible to uncover explicitly all of the procedures in all of the discourses under study. Often the procedures are obscured by the explicit semantics of the text even though they continue to underpin the discourse which the text conveys. Occultation, however, should not deter us from the search for common procedures. Let us now turn to the discursive procedures themselves and to their detailed analysis.

1.2 Referentiality

'Something new to talk about.'

Why start speaking in the first place? Usually, it is because there is a new object to talk about. This object is technology. Technology, however, can mean two things: it can be either a tool or a machine or it can mean a form of knowledge or of social organization, an 'idea-object.' For example, a technology of control could mean the machines and devices that will actually do the controlling, but it could also mean the sets of social relations necessary for the machines to operate and to come into existence in the first place.

Referentiality, however, as a discursive procedure is hardly restricted to texts on new communications technology. It is part of an academic convention within the classical tradition that cuts across every field of discourse. We are simply singling out discourses on technology here. We are showing that they fit into the referential paradigm and secondly that there is a contradiction between them and their futurological procedures.

The epitome of referentiality is achieved by those texts known as 'state of the art' texts. Here, referentiality as a procedure demands and makes possible endless lists of machines and techniques. The purpose of these texts is to list, describe or comment upon, in a 'value-free manner,' the objects that are supposed to exist out there in the world and that are just waiting to be named and investigated by journalists, technicians, and scholars alike. Almost every anthology on new communications technology includes at least one chapter of this kind. Walter Baer's 'Technology in the 80s' serves this referentialist function in G.O. Robinson's anthology (1978, 3–55), while Madden's 'Simple Notes on a Complex Future' does the same in Godfrey and Parkhill (1980). Books such as Mumford's *Technics and Civilization* (1963) rely heavily on this very mechanical use of referentiality. The book is packed with lists and descriptions of equipment which are consolidated by hundreds of pictures and drawings of machines. Power ratings, photos, names of machines and inventors abound.

Journalists frequently begin their articles on the information age with the statement that there exists some new gadget that needs to be talked about. And they often include a picture of the gadget which turns out to be quite unidentifiable. The 'thing' has to be represented and it sits there on the page as a 'concrete' legitimation of the beginning and continuation of discourse. A corollary of this obsession with the referent might be called 'gadget-philia.'

'Gadget-philia' is a term which we coin as a derivative of the term *'scopophilia.'* A film semiotician, Christian Metz, used the term 'scopophilia' to denote the compulsive pleasure derived from watching images on the screen.[4] Similarly, 'gadget-philia' refers to the compulsive and undeniable fascination and love that many documents on new communications technology and on technology in general exhibit towards machines, gears, micro-chips, and circuits.

The adjectival index, 'marvellous,' in Ellul betrays such a fascination:

> one has to understand that the computer is simultaneously manichean, repetitive and non-comprehensive. Still, how can one, after a prolonged frequenting of this marvellous apparatus, not come to absorb this mode of thought? (1977, 118, author's translation)

The lists of machines in 'state of the art' articles and books on new communications technology are a further indicator not only of the procedure of referentiality but of fascination with the referent. Mumford's insertion of sketches and pictures of machines that have little to do with the accompanying text can perhaps only be explained by his utter fascination with the gears, circuits and complexity of it all.

'Gadget-philia,' however, is not as innocent a fascination as it might sound. *When something fascinates us it holds a certain compulsive rather than merely repressive power over us.* The captivated expressions of actors in Atari tv ads for home video games and the expressions of rapt attention reflected in video game arcades might also bear witness to the gadget's compulsive power of fascination.

Technology is generally associated with two types of referent-objects:

1 Technology refers to techniques or machines, inventions, and tools
2 'Technology' dating back to Aristotle and Marx refers to 'techne' or 'know how.' Rather than remain a material object, technology becomes an idea-object, a set of abstract concepts of how to make

Mumford distinguishes three or four levels which can be seen as constituting referents for technology, namely technology, technics, and machines and tools:

> The essential distinction between a machine and a tool lies in the degree of independence in the operation from the skill and native power of the operator. (1963, 10)

The machine is 'a shorthand reference to the entire technological complex.' (Mumford 1963, 12) And *technics:*

> is a translation into appropriate practical forms of the theoretical truths implicit or formulated, anticipated or discovered, of science. Science and technics form two independent yet related worlds: sometimes converging, sometimes drawing apart. (Mumford 1963, 55)

Ellul also begins his study with a definition of the referent or rather with a distinction between object-machines and object-ideas as referents of the term 'technology.' Firstly, there is technology as a technical operation which covers all works accomplished with a certain method in order to obtain a certain result, that is to say as a form of knowledge which 'measures the results and will consider the precise end of technique which is efficiency.' (Ellul 1964, 18, author's translation) On the other hand, technology is defined as a sort of grid of knowledge or cognitive filter which will 'make it clear in the eyes of all men the advantages of techniques' and which 'thus produces the rapid and almost universal spread of techniques.' (Ellul 1954, 18, author's translation)

More recent messengers of technology continue to insist on the referent although in a slightly more convoluted way. In his book *The Third Wave* (1980), Toffler adopts as his main referent, even while talking about specific techniques such as the computer, the *metaphor* of the *wave,* which is borrowed from Mumford's text.

Mumford spoke of 'the wave' to describe the 'neotechnic age,' which in Toffler becomes the 'third age.' Here, the referent has become a metaphor which stands for an epoch which, in turn, illustrates a form of technology in operation. It could be suspected that the metaphor is only an emotive gloss for an undefined and undefinable referent. Quite despite itself, Toffler's insistence on the metaphor as his referent might indicate a crisis of referentiality at least as regards the firm grounding of correlata associated with the term 'technology.' Nonetheless, Toffler also uses the procedure of referentiality to legitimate his discourse.

Other dilemmas surface as a result of a reliance on the procedure of referentiality. Many texts, even when they set out to treat a referent which is not a machine, such as the educational aspects of technology, also begin with a 'state of the art' account of new communications technology. These accounts do not follow the pragmatic postulate of contributing to the receiver's stock of information since they all tend to repeat over and over again exactly the same information. It therefore becomes a legitimate question to ask why these obviously borrowed lists are repeated yet again. One might hazard the reply that these texts repeat the lists in order to legitimate their own discourses. The lists provide an opening onto the field of technology about which they want to talk. But presumably they also repeat them in order to bring some new knowledge into the world, as this is one of the pragmatic postulates of scientific speaking. If a referent is constantly repeated, however, this would suggest one of two things: either the referent has not been grasped or properly represented in earlier discourse and/or the nonsense of total redundancy does not bother the authors. Whatever the case, the adequate representation of the referent is certainly called into question here by virtue of the constant narration that tries *ad infinitum* to fix it. Were it not called into question there would be no need to introduce the information society with repetitive lists of machines. The French psychoanalyst, Jacques Lacan, suggests that we keep on speaking because we are never sure that we have correctly referred to the object ('signified') we intend (Lacan 1966).

A second twist to the procedure of referentiality, linked to the procedures of fictive narrative scenarios (3.0) and of futurology (4.0), occurs when authors begin by saying that it is difficult to talk about the referent (new communications technology) since it is in the future, a part of the unknown, and then proceed to do precisely that which was so difficult: they talk about the referent, the wired society, the home computer, etc., with a confidence that belies their earlier misgivings: 'There is no "maybe" in the world of digital communication.' (*New York Times,* 30 June 1968; Proulx 1982).

It is easy to see how these discourses assume such contradictory positions. Apart from the QUBE system and a few isolated field trials carried out under optimal conditions, there is yet no 'telematization of society,' no 'wired city,' no 'global village,' no universalized two-way incasting. And yet, adherence to the procedure of referentiality demands that they be spoken about. A classical 'scientific' discourse that legitimates itself by virtue of the procedure of referentiality falls into difficulty precisely because of the procedures yet persists in clinging to it. The contradiction usually solicits a concluding platitude about how difficult or even impossible it has been to talk about what they really could not talk about but have done us the favour of talking about anyway. For example, Toffler confesses that 'systematic research can teach much, but in the end we must embrace – not dismiss – paradox, contradiction, hunch, imagination and daring.' (1980, 402) Dertouzos in 'Individualized Automation' concludes a long referential description with a remark that delegitimates his own undertaking, although he appears not to realize it: 'the leap in imagination is greater than the leap in practice.' (1980, 54–55) So then, what is the sense in talking about industrial practices in computer-communications if these practices have no ontological status?

For other texts the legitimating referent is neither a machine nor a form of social organization or knowledge but rather the notion of 'newness.' The discourse is legitimated because it speaks of the 'new.' The long lists of inventions, including the telephone, the light bulb, the Turing machine, etc., that open many texts, would seem to substantiate the notion that, in order to be talked about, the referent must be an invention, i.e., it must be new. This fetishism of the new also explains the continuous harping back to the term 'revolution' in most writings on new communications technology.

However, as Weizenbaum has shown, the words 'new' and 'revolution' also have a somewhat dubious ontological status as concerns computer-communications. In this article, 'Once More – a Computer Revolution' (1982), he shows how this 'newness' has been foisted upon society for years. Despite the slight variations in the specific techniques that are new, the referent 'newness' perseveres with remarkable tenacity:

> As you all know, the computer revolution has been announced a great many times. We are now being bombarded by a still new announcement – the micro-computer revolution, the home computer revolution. (Weizenbaum 1978, 12–13)

Indeed, as an article entitled 'The Communication Revolution' (Albion 1931–32, 718–20) illustrates, 'newness' in the realm of communications is a misnomer. Furthermore, while past articles show the link of electronics to the 'newness' of electricity, more recent articles have dropped the electrical connection altogether, since, for us, electricity has lost its aura of novelty. Early forerunners of computers, called 'electric calculating machines' were touted in *Science Newsletter* as available for the 'first time in history' (14 September 1940). This electrical connection is present in articles from the *Scientific American* (1935) and from the *Literary Digest* (1930) as well.

James Carey and John Quirk suggest that the rationales behind an insistence upon newness are the attempt to redeem in electricity or in the electronic mythos the unfulfilled dreams of industrialism, as well as to disarm the fears and disappointments that the earlier new technological invention may itself have caused:

> But reality was unable to reverse rhetoric and in the last third of the nineteenth century, as the dreams of a mechanical utopia gave way to the realities of industrialization, there arose a new school of thought dedicated to the notion that there was a qualitative difference between mechanics and electronics, between machines and electricity, between mechanization and electrification. In electricity was suddenly seen the power to redeem all the dreams betrayed by the machine. (Carey and Quirk 1970, 226)

1.2.1 Referentiality: exchange and reification of communication

Referentiality, as a procedure, does not stand alone and isolated in the world of discourse. It is bound and sustained by numerous other procedures. The two most important are exchange and reification. The ability to refer to things, independently of their ontological status, is based upon the 'exchangeability' of those things. Their exchangeability in turn constitutes them as isolated objects thereby reifying them and isolating them from their social context.

'Information is wealth.' Such is the constant refrain about new communications technology. In it can be heard the reference to exchange. French scholars of the group 'Tel Quel' have shown that referentiality, a discursive procedure which substitutes a signifier for a signified, is strongly linked to the economic function of *exchange-value,* whereby exchange-value is substituted for work – or authentic-value. The procedure of exchange to which referentiality is so closely linked finds an easy and obvious ally in the discourse of economics. It operates in other fields as well, however. In grammar, as equal sign is placed between two unequal terms, signifier = signified, and in economics x number of hours is said to equal $y (x hours = $y). Referentiality as exchange operates a substitution which is no more than 'false identity' posing as true identity (Adorno 1966).

Michel Foucault in *The Order of Things* (1970) has compared the seventeenth century discourses of Port-Royal *grammar* and of *economics* and finds them to function according to at least one common procedure: *exchange*. In many of the remarks we read about new communications technology, it would appear that this classical discursive procedure is still in full force today. First of all, it is very important to note that they treat new communications technology as well as communication and knowledge as *objects* which can, according to commodities theory, be replaced by 'wealth.' Consequently, communication is *reified* in much the same way that a worker's production is said to be reified by exchange-value. An almost brutal analogy between communication and exchange can be found in the issue of billing. If any single issue takes up the largest amount of space in texts on new communications technology, one could hazard to guess, even without undertaking a quantitative content analysis, that it is the issue of billing. In many Canadian and American regulatory and policy documents it is suggested that communication be defined as a 'commodity' and that it be regulated according to conventional trade agreements. 'Information is wealth.' 'What is its rate of exchange?' 'Who should pay for it?' 'Who should own new communications?'

1.3 From procedure to issue

At this stage of the investigation it becomes possible to demonstrate how discursive procedures such as referentiality and exchange actually operate as the conditions of possibility for the raising of certain issues as well as for the way they are treated with regard to the debate on the social impact of new communications technology. Certain discursive procedures render possible and necessary the raising of both traditional and current issues in very specific ways.

Referentiality and its corollary, exchange, are two of the discursive procedures that act as the conditions of possibility for the raising of the following issues: free flow, participation, vulnerability, content restriction, artificial intelligence vs human intelligence. Both procedures are also highly determinate of *how* these issues are raised.

The *access* debate hinges on a consideration of communication as a set of objects, instruments or bits of information that must be equitably distributed to all citizens. Godfrey and Parkhill (1980) raise this question in terms of the right to possess the communicational object 'information': 'Access is the ability to obtain information from and place information into storage.'[5]

The notion that information and communication are object-referents to be equitably distributed at a fair exchange rate is linked to another which, tongue in cheek, might be dubbed the 'pig principle.' The 'pig principle' refers to the belief that has haunted much liberalist thinking about communication and which holds that the more communication people have the better off they will be. The economic principle, maximization of goods and profit (J.S. Mill), finds a counterpart in this principle of maximization of communication. Quality of life is measured in terms of the quantity of possessions, and communication would seem nowadays to be a desirable possession. This economic factor in the reification of communication as a referent-object to be possessed is placed into relief by the current obsession with issues of free flow and billing. Gordon Thompson, a spokesman for Bell-Northern

Research, speaking on the CBC *Ideas* series (Fall 1980), suggested that one of the principle challenges of the communication revolution would be to decide how much information is worth and how much people are willing to pay for it. R.A. Russel in a study for the Institute for Research in Public Policy echoes this exchange-value orientation when he suggests that:

> Information is wealth and can be created, improved or squandered (. . .) governments should encourage our information companies. (1978, 42–43)

The coupling of information and economic vocabulary is perhaps no coincidence but indicates rather the predominance of *exchange* as a procedure that gives direction to issue treatment. The very terms *'information rich'* and *'information poor'* so frequently used to discuss the implications of new communications technology for the Third World are a further indication of this phenomenon. Even development issues do not escape the discursive constraints of referentiality and exchange.

A corollary of the 'information is wealth' formula is the constant demand for 'free flow' of information, made increasingly by developed Western countries in the face of growing Third World pressure for a redistribution of the world's wealth, be it informational or other (MacBride 1980). Free flow, as has so often been demonstrated, is nothing more than the communicational counterpart of the 'laissez-faire' economics formulated by Adam Smith. 'Laissez-faire' is based on and legitimated by the liberalist belief that when left to its own devices the market will find an equilibrium which will ensure a free and just exchange-value for each person's work and products. Many scholars and politicians in the field of communication do seem to tout blanketly the free flow principle.

The belief in 'laissez-faire' and free flow also explains a good deal of the OECD literature on new communications technology (Pool and Solomon/OECD 1980, 79–140). Information is wealth and international and national legislation on new communications technology suggested by the OECD is primarily concerned with ensuring the 'natural human right' to private property. Thus Pool and Solomom, relying implicitly upon an epistemology of factuality that reinforces the status quo, suggest in their study for the OECD that nothing can be done to ensure the privacy and property rights of smaller countries or of citizens since the technology for monitoring or preventing 'illegal' data transfers is simply unfeasible. The corporations must be left alone except in cases of 'negative exteriority' when they hurt the occasional unassuming private citizen. For example, Pool and Solomon do not ask what technology or legislation would be socially desirable or possible. They ask only what technology exists factually in the corporate world. As a result, the corporate status quo is affirmed by the dominant discursive procedure of referentiality which constitutes, in part, the scientific episteme.

One traditional broadcasting issue, the regulation of communication as content, has also been carried through to new communications technology. And, of course, it is also dependent upon a 'thingification' – *Verdinglichung* – of communication. Furthermore, by increasing distribution capacity without augmenting the product to be distributed, new communications technology further heightens the acuteness of this product-oriented issue. In Canada, for example, communications issues are still often reduced to the separation of content and carrier:

> Since, however, there is a serious gap between the rapidly expanding distribution capacity of our delivery systems and the availability of programming from domestic producers, the department judged it essential to encourage the development of additional Canadian television productions of a quality that will make viewers both at home and abroad watch them regularly and eagerly. (Fox/DOC 1979–80, 10)

The question of 'Canadian content,' a household word in Canada, constantly resurfaces, now with reference to cable tv, to pay tv, to transborder data flow, to the electronics industry, to satellite earth stations, and so on. Once again, the issue of content is a case of referentiality whereby communication is reified, treated as a signified or a content, instead of as a practice or an interaction.

In the field of new communications technology the way in which the issues are raised seems to indicate that we are on the threshold of a similar referential approach to national identity. Of most concern is the wish that the referent-objects, the hardware and content in data banks, originate in the country that uses them.[6] On the other hand, the software, the aspect of new communications technology that governs the communicational capacity for patterns of interaction, association and networking is likely to be developed elsewhere, with little concern for specifically national communication needs beyond the need to access information objects.[7] This concern with hardware and content is only another off-shoot of the referentialist approach to communication. It is interesting to note that Sweden has made an alternative choice in this area, deciding that it cannot compete in hardware development with foreign powers and that it should concentrate its energies on developing software for new communications technology systems and uses pertinent to Swedish needs (Gotlieb and Zeeman 1980). In the referential procedures of discourse which equates communication with an object, the whole *participation debate* is also obscured. Articles such as 'Interactive Cable TV and Social Services' (Brownstein 1978) confidently assume that social interaction and democratic participation will flow automatically from the availability of new communications technology. 'Install a bi-directional cable and you'll have a democratic society.' Licklider and Toffler echo this naiveté, though with perhaps less than total sincerity, when they declare that new communications technology will give 'Computer power to the people.' (Licklider 1980, 124) Once again, the actional aspect of communication is reduced to the possession of or access to hardware. 'Put cameras in courtrooms and you'll have a just and participatory legal system. Give each child a terminal and they'll all communicate better and on an equal footing. Give people skills to operate the new communications machines and they'll all be creators.' Once again the insistence upon access to hardware – the objects – is accompanied by an acceptance of the discursive procedure of exchange at both commercial and social levels.

The issue of *artificial intelligence (AI) research* is an off-shoot of referentiality whose social content is less obvious but whose social implications may be just as momentous. Artificial intelligence research is, for the most part, an attempt to build the perfect representation of human intelligence. It seeks to reconstruct the operations of human intelligence by using electronic circuitry. The circuitry is therefore the medium for the representation of human intelligence and the machine is a reference to and a metaphor for the human mind. In some cases the representation is justified almost metaphysically. References are made to a supreme being which created these two minds, natural and artificial, of the same order. Herbert Simon's justification of AI research epitomizes referential epistemology since his project is based on a confidence that the order of his discursive models in electrical circuitry reveals or mirrors the order of man's thought which in turn reveals the order of the social and mental universe:

> What the computer and the progress in artificial intelligence challenge is an ethic that rests on man's apartness from the rest of nature. An alternative ethic, of course, views man as a part of nature governed by natural law, subject to the forces of gravity and demands on his body. The debate about artificial intelligence and the simulation of man's thinking, is in considerable part, a confrontation of these two views of man's place in the universe. It is a new chapter in the vitalism/mechanical controversy. (Simon 1981, 431)

Interestingly enough, this supposedly empirical referentialist position is underpinned by a metaphysical stance which assumes that the link between man's mind and machine is pregiven. Without this assumption it would be impossible to assume that the scientifically discovered order of AI reflects or refers us to the order of the human mind. The authority of science's project in the seventeenth and eighteenth centuries to discover and dominate the order of the universe was based on the assumption (first principle) that the universe, the mind, and scientific laws constituted homologous orders which were all grounded in a Divine Order. Descartes' *Discours de la méthode* and the Port-Royal grammarians are as good an example of this as one may hope to find. Elsewhere, in historical anthropology, this metaphysical principle underlies the argument that man cannot divorce himself from technological progress because his own bodily and mental functions have evolved to depend upon it (Gehlen, in Mumford 1963, 26).

Not only are AI models, according to AI researchers themselves, transparent referential models of human intelligence, but they also claim to be a species akin to the human species. The order of language/electronic circuitry equals the order at least of the mental universe. Such is the adage that might summarize their experiments, an adage that mirrors the claims Descartes made for his own discourse of scientific method in the seventeenth century:[8]

programmed computer and human problem-solvers are both species belonging to the genus. When we seek to explain the behavior of human problem-solvers (or computers for that matter), we discover that their flexibility – their programmability – is the key to understanding them. Their vitality depends upon their being able to behave adaptively in a wide range of environments (. . .) For, as we have seen, we need postulate only a very simple information processing system in order to account for human problem-solving in such tasks as chess, logic and cryptarithmetic. (Simon and Newel 1972, 870–71)

Denicoff, speaking of Winograd's AI experiments with computers and context, draws the referential equation even further. He states that AI model reveals the nature of the whole social universe to us:

In terms of a complexity measure, the upper end of the natural-language understanding scale is concerned with machine representations of knowledge about the total human social framework – the knowledge base essential to computer encoding and treatment of narrative related to people's purposes, beliefs and intentions. One researcher (Winograd) put forth the notion of computer programs that could demonstrate common-sense reasoning abilities in responding to questions about programs to deal with the human employment of language to convey both conscious and unconscious intentions of flattery, exaggeration, deceit, emphasis and spite. (1980, 378)

The term 'like' in the following remark by Winograd is yet another though somewhat occulted assumption of referentiality: 'If a program is to be useful, it must operate in a mode much more like a human consultant' (1980, 68–69).

However, 'likeness' to human behavior is not an adequate reference to or representation of human behavior as indicated by Weizenbaum's demystification of his own 'Eliza' computer program, a simulacrum of Rogerian psychoanalytical techniques (Weizenbaum 1976). Weizenbaum suggests that computer programs only imitate the most simple communicational gestures of human beings, imitations which are no more than simulacra devoid of human emotions and pertinent presuppositions.

The whole debate concerning whether or not AI models represent human intelligence would be absolutely meaningless, indeed would not even have been raised, without the fundamental underlying discursive procedure of referentiality. Indeed, the whole practice of making models of human intelligence would be senseless.

1.4 Order and analysis

Analysis is a discursive procedure that treats its object of study as a complex that must be broken down into its smallest parts in order to be understood. Analysis leads to an individualist (cf 11.0) and atomist approach to science. A look at many of the texts on new communications technology, especially the so-called 'state of the art' texts, shows them to be broken down according to individual techniques, specific functions or particular issues which are all separate from each other. To talk about the wired city, the texts begin by breaking it down into all of the particular techniques that constitute it. Analysis is the only procedure that can explain the long lists of particular techniques, machines, issues and users, each discussed in its own, unintegrated, unfocused, water-tight compartment. The alternative to analysis would be to see new communications technology as a whole within social processes rather than as a part sectioned off from them. New communications technology would be integrated within a greater schema such as the dominant episteme rather than being cut off from social processes and divided into its minutest parts. For example, the problems of employment would have to be treated globally rather than on an industry by industry basis (Gotlieb 1978, 43).

Issues, techniques, and uses are often ordered hierarchically as a list of priorities. Yet it is not usually made clear whom this order serves or why it does. For example, as regards mechanical developments, the top priority in Western developed countries is military, as expenditures reveal. However, with few exceptions, the literature on the development of new communications technology ignores this, seeming to argue instead that all developments are hierarchized on the basis of whether or not they increase our quality of life (cf 18.0). Where quality of life is not the supposed ordering principle, chronological linearity once again occults the actual operative ordering principle.[9]

1.5 Fictive Narrative Scenario

'Once upon a time. . .'

In 'Julia's Dilemma' (Madden 1980) the narrative opens with the possibility of holocaust unless two factions of society, those who want new communications technology and those who do not, agree to live on separate continents. They finally do agree and Julia's dilemma is to decide where she will go with her son. We then flash back to an earlier scenario: Julia's attempt to go 'back to nature' and its fatal consequences for her husband and first child. Both died because they turned their backs on technological progress. Returning to the first scenario, the AGEE, a nickname for the home computer terminal, helps Julia talk to her son in order to reach a decision. The story's semantic value structure proposes a middle-of-the-road ethics which argues that technology is good but also that the fanatics need to use it with moderation and to curb their inherent tendencies toward conflict. The overall message would appear to be that Julia's difficult choice is due to society rather than to technology.

The following narrative scenarios are typical of many found in texts on new communications technology:

> Quentin R. Smith wrenches the top off another beer, hunches over the console of his home communicator and spins the television dial through a circle of channels (. . .) Upstairs Nasturtium is finishing her last class of the day on the educational channel (. . .) Across the street, the Wencelas T. Jones family is busy at the communicator too (. . .) Oh, didn't I tell you? The Jones and the Smiths live in Frobisher Bay, NWT. (Stewart, in *Maclean's,* December 1968)

> Scenario I: Each of us loading 500 pound rolls of newsprint into our electronic home-newspaper delivery devices, in order to save (. . .)
> Scenario II: Millions of men in their undershirts, beer cans in one hand, digital response pads in the other (. . .)
> Scenario III: Similar throngs eagerly spending \$25–100 to

purchase a video-cassette or a videodisk (. . .)
Scenario IV: . . .
Scenario V: . . .
Scenario VI: . . . (Berkman 1982)

An interesting twist in the narrative scenario procedure may be found as early as 1927. In a *Scientific American* article on the film *Metropolis* we find an intriguing mixture of 'scientifist referentiality' and fiction. It is argued that fictional portrayals of the technological explosion, such as Frankenstein, have been rendered more vivid and faithful by the actual technology of the movie industry itself. Fiction is legitimated by technology and technology is given a referent in fiction. Fiction and referentiality reinforce each other:

> 'Feats of science help movies give vivid picture of a world ruled by machines'
> A new German Film Based on the Old Robot Story, in which a mechanical person, created by an inventor becomes a Frankenstein Monster has been produced in Germany, utilizing novel electrical effects in an unusual manner (*Scientific American* 1927).

One might go so far as to see in this article the procedure whereby referentiality in technology relies upon fiction to make it real-seeming and fiction relies on referentiality for its claim to be an 'adequate' portrayal. Such an admixture of fiction and referentiality, however, flies in the face of the empirical truth of so-called referential, scientific documents on new communications technology.

These are just a few of the many articles and texts on new communications technology which have failed to resist the temptation of a free imagination and of concocted scenarios of the new communications world, all the while speaking in the present tense and the indicative mode in order to give their foresights the air of certainty and actuality.

The narrative fictive scenario may also be a strategic discursive procedure which allows the portrayal of a very crucial popular message: 'Yes, things will change in your world because of these new referents about which we are not entirely certain but of which we can give a fairly good description.' 'Anyway, relax, enjoy the story, and don't worry, because even in this new technology there will still be many familiar elements to allow you to get and keep your bearings.' This is the underlying message of the fictive scenario's constant reference to familiar items such as beer, television, families, kitchens, living rooms, and happy endings.

There is a pragmatic strategy of reader/receiver seduction in operation when fictive scenario is used. Dr Johnson's dictate was 'To Teach by Pleasing,' and as any teacher knows, one of the best ways to gain attention is to tell stories. Fictive narrative scenario makes for a change from the dryness of scientific, referential discourse to the seductive familiarity of narrative plot.

Nevertheless, despite all of these auxiliary functions of the fictive narrative scenario the degree of its inclusion seems to be inversely proportionate to the capacity of the discourse to find and to cope with a referent for the signifier 'new communications technology.'

1.6 Futurology: progressive, evolutionary, inevitable view of history

'Promises. Promises.' and 'It's coming like it or not.'

'State of the art' articles tend to list in chronological, linear order a progressive generation of technologies. We have 'evolved' from the electric lightbulb to the satellite via the telephone and the television tube: 'One small step for a man, one giant leap for mankind.' Mumford, and later, Toffler, enumerate three of these giant leaps for mankind: Mumford's 'eotechnic, paleotechnic and neotechnic' ages and Toffler's 'first, second and third waves.' One thing is clear, the history of technology is conceived of as *the history of a linear march,* be it toward our salvation or our doom.

The discursive procedure of evolutionism has as its first important consequence *the euphorization of progress for its own sake.* Progress is necessary and ineluctable but it is also desirable because it fulfills the evolutionary march toward the absolute.

Along with the euphorization of progress comes another twist: even if progress were not positively valued there would be no alternative. Many of these texts remind one of the old maternal trick whereby a certain morsel of food is highly praised as a treat but should the child not want it he is forced to eat it anyway. There is no other alternative. Progress is inescapable. 'We cannot undiscover what we know.' (Toffler 1980, 164)

> The system is so great that one cannot backtrack: to try detechnization would be the equivalent of burning the native forest for primitive peoples. (Ellul 1977, 9, author's translation)

And Ellul goes on to state, relying upon more metaphysical assumptions from historical anthropology, that if we can build a car that goes faster we will necessarily want it and it will therefore be built (1977, 46). Fulford's article 'How Telidon May Change Your Life' has a picture of a monkey with the shadow of a man behind it using a computer. The Darwinian evolutionist overtones are hard to miss here (Fulford, in *Saturday Night Review,* September 1980).

These procedures of continuity and linearity have as their principle strategy the creation of an impression that computer-communications is upon us and is as irrevocable as biological evolution: 'The public should know that the information revolution is already underway.' Dagenais reinforces this feeling when she writes variations on the phrase: 'The era is on our doorstep' (Dagenais, in *Le Devoir,* 19 November 1980, author's translation).

The myth of progress has been pushed so far as to change the etymology of the word 'revolution.' Before the French revolution the term meant a turning back. Since then it has taken on the connotations of progress, newness, change for the future, and movement toward a perfect world (Kumar 1978, 19). This same futurological bent has been evident in speculation about the new communications technology known as television, surfacing in the following article written in 1951:

> The prediction, to be opened 100 years from this week, generally foresaw a rosy future for the medium. But the World Telegram and Sun Critic Harriet Van Horne took bitter exception. ('Dark Screen Future,' in *Time,* July 1951)

Interestingly enough, in some writings, the discursive procedure of progress is quite closely linked to the discursive procedure of exchange. The link euphorizes any increase in the quantity of possessions which can be substituted for the sensuous activity of interaction. *Progress is as inevitable as man's desire to maximize profit,* the (metaphysical) first principle of this rather vulgarized form of historical anthropology:

> To deny that we shall strike out in such new directions is to deny the fundamental adaptability of human beings and our inherent desire to possess more than we own and achieve more than we have already accomplished. (Dertouzos 1980, 55)

The discursive procedures of evolutionism, continuity, and inevitability of progress provoke and maintain an issue treatment of new communications technology based on a confidence in and a reaffirmation of that which Licklider, a researcher in new communications technology and defence for the Pentagon, laments all people will not accept: 'the technological imperative.' (1980, 125) Mankind must accept that its capacities be technologically amplified: there is nothing more natural than this: 'just as the engine of a bulldozer amplifies the muscle power of the man who controls it,' ('Intelligence Amplifier,' in *Time,* May 1956)

Future Shock (1970) by Alvin Toffler sums up in its very title the attitude towards history adopted by most discourses on new communications technology. Everything is happening in *the future* which is *both upon us now and still to come.* One thing is certain, new technology can be accounted for neither in the past nor in the present. This merely pushes problems and problem-solving into the promising future. Gee Whiz. The future will be great. 'There is, likewise, a great realm of potential before us and it is going to be very exciting to explore and develop it.' (Whittaker/OECD 1980, 39)

Tv ads also conform fully to these futurological procedures. In a Bendix ad, the earth on our screen whisks off into the distance while another planet fast approaches in the absence of any familiar reference coordinates of proportion or of recognizable landmarks. And in an Atari ad, a golden triangle floats through space across our screens. The intertextual reference is to the tablet-covenant in *2001.* You can play infinite variations of space wars on a host of available video game systems. CNCP telecommunications ads open onto 'office of the future' with streamlined consoles and desks, open space, miniskirted attendants, and space odyssey-type sound effects.

Futurology is a dominant undercurrent in much advertising for new communications technology. Those who do not indulge in futurological symbolization, such as Bell telephone ads, usually resort to its nostalgic counterpart, the other side of the coin, whose aim is to avoid accountability for the present. Although new communications technology is touted as a thing of the future, any of the public's potential futurist anxieties are abated by the frequent use of traditional images such as the archetypal family of the Bell, Bendix, and Atari ads.

If we look also at a few expressions or syntagms from Dagenais' article 'L'information à domicile: l'ère du petit ordinateur relié à la télévision est à nos portes,' we find that the futurological bent is imprinted in language: 'the era is on our doorsteps,' – 'will turn out to be' – 'old fashioned naiveté' – 'since last September a three-year program' – 'ever more rapidly' – 'even now the next decade is already prepared.' Once again most of these expressions point to the line of continuity into the future, a future which contaminates the present, which derides the past as old-fashioned but which does not seem to let either the past or the present contaminate its own utopianism.

When the euphoric possibilities of new communications technology are spoken of, the indicative mode is used. The future *will be* good. 'There is (. . .) a great realm of potential before us and it is going to be very exciting' (Whittaker/OECD 1980, 30). On the other hand, when possible nefarious effects are spoken of it is always in the conditional or subjunctive modes. There *could be* possible negative effects but. . .

> Failure of an international dialogue *could* reverse the positive benefits which information technologies offer to the world. (Piera/OECD 1979, 237)

> That the future of automation *will* take this direction, with its attendant humanizing consequences is more *probable* on technical, economic and social grounds than the more spectacular and more-touted dehumanizing alternative. (Dertouzos 1980, 55)

> Dangers, in the sense that the Orwellian prophesy for 1984 *might* still come true; however, the fact that privacy protection – the most advanced form of human rights – is taken seriously in the advanced Western democracies, *is a certain guarantee* against such dangers. (Gassman/OECD 1980, 61)

Thus it is quite common in discourses on new technology to label technology 'the realm of the possible' – 'das Reich des Moelglichen' and 'the fate of the future' – 'das Schicksal der Zukunft' (Dessauer 1927, 135).

A recent ad for Fujitsu echoes this futurological fetish, adding a spatial dimension:

> We at Fujitsu spend most of our time in tomorrow. It's an exciting place, an untapped country of heretofore undreamed potential. And with technology we are perfecting today we can take you there. (*Japan Echo* 1982)

Although there are a few exceptions to the rule of futurology, such as a 1977 article in *Le Monde* (20–21 March 1977, 128–30) on CB radios, which asks why we always speak of two-way communications in the future when CB has been around for quite some time, most texts are indeed guilty of a very severe neglect of the present.

Carey suggests that futurology has been a dominant aspect of the rhetoric from the industrial revolution through the invention of electricity to electronics. The futurological bent is, in effect, a way of making promises which never have to be kept since past performance will always be suspended by yet another futurological discourse.

> Electricity promised, so it seemed, the same freedom, decentralization, ecological harmony and democratic community that had hitherto been guaranteed but left undelivered by mechanization. (Carey and Quirk 1970, 228)

Much earlier Mumford had also criticized the prevalent futurism for being unable to remember the lessons of the past or to make the present responsible to both the past and to the historical trends that indicate the probabilities of the future (Mumford, in Carey and Quirk 1970, 237). Futurology is a ticket to ignore the past and carte blanche for actions in the present which need never be accounted for: if everything is just about to happen rather than having already happened or being in the process of happening, then the need for policy and decisions can constantly be put off until tomorrow. Fortunately for the procrastinators, the future never does come since then it would be the present, the day of reckoning.

While some may believe that futurology is a type of a-historicism, such is not the case. Futurologists neglect any close study of the past but do not exorcise it from their discourses. Where the past is included it is in the form of a blanket and priori theory of simple *causality*. Technological determinism still thrives today in many discourses on new communications technology. Ellul, at one point, describes the French revolution as *the result* of a long technical experience (1954, 49). Indeed, it is often difficult for those who would avoid causalism to evict all traces of this attitude toward history from their own discourse. Toffler, for one, certainly does not avoid technological causalism in saying that computers will change society and even speed up history. He delivers the dynamic of history unto the purview of technology. But even Mumford, in his long treatment of coal mining, leaves some doubt as to whether he is saying that the form of coal mining is socially organized or that the act of coal mining itself induced certain forms of social organization: 'During the last thousand years the material basis and cultural forms of Western Civilization have been profoundly *modified* by the development of the machine.' (1963, 3, emphasis added) However, by stressing the term 'modified,' Mumford avoids falling entirely into the causalist trap. He does not state that past technology caused present technology nor that technology causes history or molds society.

Very often, these causalist discourses turn the deterministic procedure off and on at will as it suits their argument. For example, A. Smith in *The Geopolitics of Information* (1980) suggests that technology will cause great international inequality and dependence for the Third World. He then turns around to say the technology *adapted to* the Third World can and must be shared. Here, technology is simultaneously determinate and neutral. Dessauer (1927), some sixty years earlier, uses much the same shift: on the one hand, one should not interpret victims of social organization as victims of causalist technics, and yet, on the other hand, technics has an inherent order that preconditions man (for the better in Dessauer's view).

1.7 A-contextual euphorization or disphorization of technology

'The technological sublime' (Leo Marx in Carey and Quirk, 1970); 'Gee Whiz.' and 'Woe betide.'

Discourses on new communications technology rarely make explicit dependence of values upon the content of enunciation. Technology is assumed to have certain values or certain use-values regardless of the context that speaks about technology. Technology is either euphoric or disphoric, either good or bad. 'There is no "maybe" in the world of digital communication.' (*New York Times,* 30 June 1968)

Dessauer is one author who euphorizes new technology out of context precisely because of his self-avowed idealist epistemological stance.

> The ideal subject of technics. In each of technic's manifestations we are not only presented with the whole of nature but also of supernature, the realm of the possible and the destiny of the future. (1927, 135, author's translation)

Dessauer's a-contextual axiologization of technology is quite exemplary because he admits that such an evaluation is idealistic. He evaluates technology as an *essence* independent of any material or historical context.

1.8 A-contextuality

'The *essence* of technology is beneficial/malevolent/neutral/.'

The way in which Mumford tries to solve the dilemma of causality is to be praised. He seems to have realized that most discourses on technology are forced to be deterministic or even a-historical because they cannot cope with the contextual relationship of technology to society in history.

> No matter how completely technics relies upon the objective procedures of the sciences, it does not form an independent system, like the universe it exists as an element in human culture and it promises well or ill as the social groups that exploit it promise well or ill. (Mumford 1963, 6)

Mumford is genuinely concerned with trying to show the rootedness of technics, not only its uses but also its structures and forms, within the forms of social organization. Although he does not always succeed, he is to be commended for having tried to recontextualize technology in principle if not in practice. Indeed, technics is just that, the realized and socially conditioned forms that technology, as a pure possibility of knowledge may take (Mumford 1963, 6).

Later on, the possibilities of contextualizing technology will be considered, but first, let us turn to some of the most flagrant discursive practices of a-contextuality. Apart from the priori assumption of a singular historical procedure, such as causality or futurist evolutionism, classical texts on technology are also guilty of pure metaphysical, idealist abstraction from historical context.

Friedrich Juenger in his book *The Failure of Technology* (1949) exposes the metaphysical abstraction and universalization of technology in more recent discourses. Stereo equipment and electrical guitars as much as airplanes are held in awe. He refers to that attitude as the 'Realmetaphysik' of the twentieth century, whereby technology is ideally and universally deified at the expense of man.

One index of decontextualization found in many documents but especially in diplomatic, policy jargon is the profusion of universal, generalized subject categories such as 'mankind' and 'society.' This generalization will resurface later in this study with regard to the confusion of actantial roles (cf. 15.0). For the time being, however, we may note that representatives of both East and West resort to the vagaries of such euphemisms:

The Prague Symposium of September 1976 was organized in order to elucidate various modes of meeting challenges posed for *mankind* by science and technology. (Cohen/USA, UNESCO 1981)

> Modern science and technology have given *mankind* powerful instruments. (Fedoseyev/USSR, UNESCO 1981)

Ellul also decontextualizes technique, describing it as a self-contained system which is autonomous and therefore quite independent from contexts other than that of its own functioning as a mechanism:

> technique has become a reality in itself which is self-sufficient, which has its own laws and determinations. (1954, 121, author's translation)

> Each element of the technical ensemble follows laws which are determined by the relation with other elements of the same

> ensemble, internal laws within the system which consequently are not influenceable by external factors. (1954, 251, author's translation)

Such an a-contextualization of technology produces two very familiar statements: 'Technology is inherently good (or evil),' and 'Technology is neutral, it all depends on how you use it.' Variations on these two themes occur so often that it is really unnecessary to cite specific examples here. It suffices to open any text, article or television channel to run across it. Schumacher's description of technology as something inherently powerful but able to be molded by man combines both of these clichés under the label of 'powerful house pet':

> At the moment the most important thing for us to do is to prove that technics opens possibilities which do not imply the bondage of our everyday life. On the contrary, it must be proven that with the help of technics new liberties can be created, with the delimitations which are necessary to control chaos. This means that the human being is no longer a slave of technics but rather its master whose carefully planned goals it obeys like a powerful domestic animal. (1932, 26, translation Angela Haberman)

Such a-contextual evaluation is extremely prevalent in the treatment of 'quality of life' issues. It is often declared that technology will either humanize or de-humanize the workforce. This declaration does not attempt to consider the specific workforces within specific contexts.

It would be more pertinent to consider 'quality of life' issues within specific contexts of the workplace. For example, why did the air traffic controllers go on strike in the USA in 1981? What were their grievances about stress in the workplace as related to its computerization? And again, how did the Bell employees react to the informatization of their workplace? (cf 'Quarterly Report,' CBC, 6 June 1982) Why the strikes and labour unrest? Marx advocates a contextualized view of the role of technology in the workplace. For him technology is not only a force of production but also one of the conditions of production related to other conditions of production and therefore highly contextual, even in the materialist sense of this term.[10]

A slight contradiction does however seem to arise when the same voices first state that 'technology is neutral' and then, in the next breath, suggest that 'technology is after all inherently good' or will cause something good to happen. Simon, more recently, falls into this trap, stating, on the one hand, that 'depending on how we program computers they will lead to centralization or decentralization,' and on the other hand, that 'computers encourage centralization' (1981, 426).

Such de-contextualization affects the 'referent.' Lack of context results in a failure to distinguish between 'technology' in the general sense of the term and specific technologies, e.g. fibre optics. In each case, the correlata of the term 'technology,' and their associated values depend on context. Hence it is necessary to see the framework within which communications technology takes place as an event in order to associate meaning and value with it. Thus, when Brooks defines technology as 'a reproducible and publicly communicable way of doing things' (*Daedalus* 1980, 10–11), we must further nuance this definition by seeing the context of reproduction and of communication.

Carey and Quirk have also criticized a-contextuality in debates concerning the role of new communications technology in education and centralization/decentralization, suggesting that each debate can only be carried out meaningfully if situated in its pertinent context (1970, 424).

Contextualization is not only a question of temporarily situating a debate. Spatial or regional concerns are also all-important as Harold Innis' work on communications has shown (1951–77). A contextual definition and axiologization of new communications technology would have to recognize spatial or regional concerns and differences.

In some cases, the major concern is one of extending technology blanketly to all areas, something that in Canada has raised strong disputes between the federal and the provincial governments.

Where regional differences in profile and needs are neglected by the *centre with regard to the periphery*, a framework for the issue of new communications technology and development must break away from such a-contextual discursive procedures if it is sincere in avoiding the phenomenon of 'monopoly of knowledge' which Innis has shown to have existed throughout history on both spatial and temporal planes (Innis 1951–77 and 1950–72).

If we look closely at the details of the discursive practices that declare new communications technology to be inherently good we can discover their implicit or occulted contextualizations of this value judgment. *Discourse sometimes betrays its own contextualizations.* New communications technology will be a wonderful thing to have in the home or school. It will be designed to detect fire and theft, to bill for the use of utilities, to test intelligence quotients and compare with the average. It will also be a fine tool for government, giving it: 'the important data that it needs to understand its own social processes and to analyze its problems.' (Simon 1981, 430) Perhaps Simon has just given the secret away here. Technology is not neutral or inherently good or evil. The social context conditions what technology becomes including its structures and values. The social context is oriented around the protection of private property, the division of labour, competitiveness, the need for surveillance and testing which are the social conditions of possibility of the way that new communications technology will be conceived, designed, structured, built, and used. We might refer to these conditions of possibility as the *extra-discursive relations* of discourses on new communications technology to society, although they are discursively quite manifest. Weizenbaum indicates, if somewhat functionally, the contextual structuring of new technology when he asks: 'Can a rope be used as a hammer?'

Inherently neutral or inherently valorized clichés have their impact on policy treatment of new communications technology. Regarding issue-treatment, Parkhill (1980, 70–71), Pool and Solomon (OECD 1980, 79–139), and the Canadian Telecommission Study *Instant World*(1971), have all advocated that government leave off regulating the design, conception and manufacture of new communications technology and concentrate strictly on its uses and applications. What better *practical,* non-discursive statement could one hope to find of the cliché that: 'Technology is neutral, it all depends on how you use it'?

To recognize that the values associated with the various procedures of new communications technology are relative to the episteme is to contextualize the axiology of technology. This whole investigation has so far been an attempt to do just that, to explain the evaluation of technology in relation to the discursive procedures that assume it within a certain time and space. Carey and Quirk (1970) have also tried to contextualize the values of technology by illustrating how the euphorization of electronics is relevant to the rhetoric of the mechanical revolution and of electricity, a rhetoric that creates a mythos particular to the USA but not necessarily to post-industrialist Europe.

1.9 Reductionist conception of knowledge

'Oh come let us calculate.'
'Solve the problem.'

Computer-communications facilitates problem-solving and decisionistics. Calculation replaces understanding and judgment, as indicated by the subtitle of Weizenbaum's book: 'From Judgement to Calculation.'

However, this reduced conception of knowledge, based primarily on the unilateral action of a totalizing subject upon its object, is still more nuanced and complex than knowledge as portrayed in discourses on computer-communications in education and the business place.

When computer-communications first became popular there was a sense of euphoria about all that they could do and know. Huge, expensive projects of computer translation were undertaken until it became apparent that computers could not translate 'alligator shoes' without committing the error of translating 'the shoes belonging to the alligator.' Nevertheless, the debate raged on about how and how much computers could know.

The standard data banking procedures reduced knowledge to memory and limited patterns of association or cross-referencing. It is this reduction of knowledge to memory that begins to indicate how the episteme, the procedures of knowledge, has been reduced to techne, the procedures of knowing how to do. Knowledge procedures are reduced to technologies. An article entitled 'Sacred Electronics' reporting Vatican attempts to use a computer to constitute a lexicon for the Dead Sea Scrolls by situating every word within the six other words preceding and following it evidences this quantitive and denotative version of knowledge. (*Time,* December 1956). The text 'Sacred Electronics' is a perfect example of traditional forms of knowledge as hermeneutic, contextualized, historically conscious interpretation having been replaced by a-contextual, rigid, denotative forms of interpretation.

More recently in Godfrey's 'No More Teacher's Dirty Looks' in

Godfrey and Parkhill, 1980 (151–160), knowledge and education are equated with the *learning of skills,* especially *memory skills* such as grammar and spelling, which the computer is capable of teaching.

These implicit or explicit definitions of knowledge are closely tied to the discursive procedure of *referentiality* in that knowledge is first and foremost defined as the acquisition of a homogeneous body of givens – things – such as grammar rules or spelling. Secondly, it is associated with *analysis* and *ordering* in that it is the task of the computer to take large bodies of information and to order and break them down in such a way as to allow for re-ordering, fast retrieval or cross-referencing of quantities of atomized items. Such is how computer-knowledge is often defined.

At this point it is necessary to say that knowledge, for discourses on new communications technology, is actually implicitly constituted by the sum of all the discursive procedures (1.0–1.21). What we are now dealing with however is the thematic or explicit conceptualization of knowledge that these discourses portray. Implicit and explicit conceptions of knowledge do not necessarily contradict each other. It would be more correct to suggest that the implicit concepts of knowledge subsume the explicit ones, elaborating and redefining it as one of the discursive procedures that make up the whole episteme. It must be recalled though that the episteme is a sort of complete definition of the rules of discourse that constitute knowledge in any given epoch.

In this case, the *explicit definition of knowledge* by discourses on new communications technology is often one of reduction to statistics, calculations, and denotation at the expense of qualitative contextual definitions.

However, as though motivated by a guilty conscience about such reductionism, certain discourses on computer-communications have insisted on the 'more human aspect' of electronic intelligence. AI researchers try to insist that computer knowledge is as sophisticated as human knowledge. A further addition to the computer capacity for knowledge is the capacity to reflect and to self-program, according to spokesmen for AI as quoted in 'Teaching Computers to Learn: progress in creating artificial intelligence' (*World Press Review,* May 1981, 21–23). Toffler is yet another who tries to make computer and human knowledge members of the same species. Indeed the former will even improve the latter: 'computers will, by handling complex data, allow for decision-making of a higher type; they will radically change the social memory and make possible new ideologies, theories and artistic insights.' (1980, 310) Here we have it, straight from the horse's mouth, so to speak. Computer-communications knowledge is far more refined than in its early days, but it is still a form of problem-solving and decision-making. And, as such, knowledge, even the human knowledge to which computer knowledge is here likened, is reduced to decision-making and problem-solving. While previously computer-communications knowledge was reduced to memory functions, it has now progressed to problem-solving functions. But, as a glance at any intellectual history will demonstrate, such a view of knowledge obliterates centuries of historical development in the conceptualization of knowledge. For example, knowledge, understood as mere problem-solving can in no way account for the episteme in the sense in which it has been used throughout this analysis.

While attempts are underway to see if computers are capable of what Bateson (1979, 287–301) calls 'knowledge of knowledge,' the fact still remains that the first level of object knowledge has been reduced to calculable problem-solving. Computerized knowledge is simply designed to make first level knowledge, i.e., problem-solving, more efficient. Computer programming might be capable of reflective knowledge but it is still primarily concerned with calculation, problem-solving and decision-making at the object level of knowledge.[11]

Furthermore, attempts to teach computers to deal with context and reflection have really been very limited. Even Winograd himself, the father of contextualized programming, admits as much (1980). The main accomplishment of the computer remains to classify, handle, order and break down huge quantities of contents – information.

While some business reviews corroborate the definition of computer knowledge as problem-solving and decision-making, others speak of the primary function of knowledge as complexity *reduction*. The most cited example is that of the use of computer-communications for the NASA space shots, which, it is argued, could not have 'got off the ground,' without a computer to reduce the sheer volume and complexity of calculations required. Knowledge, then, becomes explicitly reduced to the reduction of complexity. But even this definition of knowledge does not really suffice to make computer-communications obligatory and omniscient since, as Weizenbaum demonstrates, the atom bomb was as complex as the space shots and was constructed without the computer. Indeed, Weizenbaum asks whether this knowledge as complexity reduction did not simply preserve complex institutions, such as banks, which would otherwise have been too cumbersome to survive without it. Technology is thus more explicitly preserving than reducing, in that complexity is perhaps not to be found in the life-world environment but rather in an artificial system of which technology is one.

And where one does concede that computer technology is complexity reducing, one still must ask whether what is reduced is not actually the complexity of what it means to know. It is indeed very complex to study the millions of lexical terms known in the Bible within their six word context but to hold this to be knowledge of the Bible is merely to simplify what knowledge and biblical exegesis really are. There is a similar example in the reduction of knowledge to problem-solving and decision-making in the business climate. Treating knowledge in this reductionist manner ignores contextual and ethical aspects of knowledge, not to mention history, and suggests easy fixes which do not work for complex human problems in the business world.

To reduce the complexity of the world would be to solve or simply to understand such complex problems as political struggle, ethical conflict, or paradoxical epistemologies. To reduce the complexity of knowledge is to throw these problems out and to reduce all knowledge operations to building quantifiable models, to calculation and to either/or type solutions which always serve the instrumentalist aim of efficiency of production and maximization of security and profit. In what follows, Weizenbaum points out the double edge of the complexity/knowledge debate:

> At least one component of the fuel for the development of this technology is the complexity of society. This alleged complexity is said to be so enormous that the computer is necessary just to make it possible to maintain the most basic societal functions (. . .) This complexity is really of two kinds. There is structural or organizational complexity, where there are principles of organization by means of which the complexity can be

comprehended: one has a theory: one can understand it. Consider for example, the street map of Manhattan. It does not take too much to understand that, even though it is in a certain sense complex. Then there is the kind of complexity that is real incoherence. This derives in society from a lack of tradition or from an impermanence of values. It seems to me that it is precisely in the latter case that technological fix becomes most attractive and it is at the same time the most tempting invitation to a Faustian bargain. (Weizenbaum, in *Daedalus* 1980, 4)

What we have here, in the exchange between Weizenbaum and the community of AI researchers, is a conflict between two different conceptions of complexity and knowledge. For Simon, complexity is not valorized and knowledge must reduce it. However, he does so by reducing knowledge itself to the status of a model or a system as represented by AI models. For the business counterparts of computer-communications technology, knowledge is simplified to mean reduced complexity and problem-solving, 'either/or, how much systems.' Consequently, when Simon criticizes Weizenbaum for questioning the pertinence of AI models of intelligence, he accuses him of seeking to reduce knowledge.

For Weizenbaum, on the other hand, the world, its problems and knowledge itself are complex and many faceted, requiring many perspectives and models, not a reduction to one:

It is those who seek to understand the world from a number of different perspectives, including the scientific one, who prefer ignorance to knowledge. It is those who, blinded by their faith that science can yield 'full' explanations, prefer to remain ignorant of whatever knowledge's other ways of knowing the world have to offer (. . .) *They are really questions about power and the limitations of a variety of ways of knowing the world.* (Weizenbaum 1976, 437)

The question is whether or not every aspect of human thought is reducible to a logical formalism, or, to put it into modern idiom, whether or not human thought is entirely computable. (Weizenbaum 1976, 12–13)

Finally, then, the question of knowledge is a discursive procedure of closure for classical discourses on technology, whereby one procedure is singled out as the definition of knowledge, such as logical calculation or problem-solving. Traditional discourses on new computer-communications technology do not recognize that an episteme is composed of many procedures, that epistemes are relative, and, consequently, so is knowledge.

1.10 Principle of excluded middle: double bind semantics

'Damned if you do. Damned if you don't.'
'Objectivity is presenting both sides of the story.'

'Either/or,' this is the logical trap set for the reader of discourses on new communications technology. It is the discursive procedure and logical rule that has governed Western reasoning, with few exceptions, since at least Aristotle's time. The principle of excluded middle is most prevalent in the various texts that line up two oppositional categories: A/ adopt new communications technology vs B/ reject new communications technology. These two camps are equally trapped within the principle of excluded middle at a meta-level in that they see no alternative to an either/or position.

Apart from this major opposition between adoption and rejection, there exist many sub-oppositions where the reader becomes embroiled in having to make a choice between two opposing and usually equally displeasing stances, with no logical allowance for either a middle term or a term from outside the system.

Based on the logical procedure of excluded middle the discourses on new communications technology seem to function principally by producing either/or situations where there is no real possibility of making the 'right' choice and yet where the discourse provides a compulsion to do so. These discourses trap the reader in what such Palo Alto psychotherapists of communication as Bateson and Watzlawick call a 'double bind' or 'catch 22' unless one can recognize the logic that traps one within it and escape from it altogether.

1.10.1 Double bind

One of the inspirations for the writing of this study was the initial puzzlement experienced by a perusal of many of the texts on new communications technology. This puzzlement was due to the formation of major redundant themes – 'isotopies' (Greimas 1966) – which were simultaneously affirmed even though they were semantic opposites. We might refer to this situation as a semantic structure of non-synthesized oppositions. Because it places the reader in a position of having to make a choice which he cannot logically make, this semantic structure leads to a double bind situation whose pragmatic effects on the public will be suggested later.

For example, in the Dagenais article, 'L'information á domicile' (*Le Devoir,* 19 November 1980), two pairs of irreducible themes are affirmed. First of all, technology is 'foreign,' 'highly scientific,' and 'inaccessible.' Secondly, technology is 'domesticable' and 'familiar.' The second pair of oppositions includes initial assumption that technology will lead to citizen participation in all walks of life, while the other assumption is that the citizen should just wait, indeed can do no more than wait until the technology comes to him. We might express these oppositions in the form of an irreducible square of simultaneously affirmed semantic opposites:

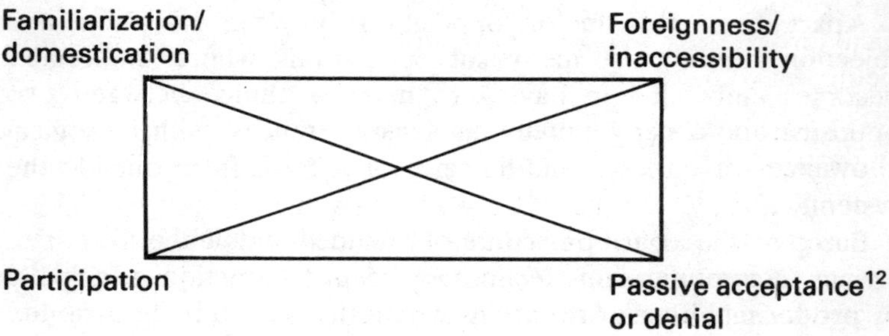

Many double binds plague the debate on new communications technology and its social implications. We will concentrate on three:

1 urgency to adopt vs need to reflect,
2 choice vs inevitability,
3 bad if you do adopt vs worse if you do not.

1 Urgency vs reflection

Gotlieb and Zeeman best exemplify the urgency/reflection double bind. They express it in economic terms. All but the industrial giants are fast losing ground in relation to other economically developed countries in the research and development (R&D) of new communications technology. They will lose the hardware market and national autonomy in data processing, in much the same way that they lost the television industry to Japan, unless they rush headlong into the development of government-subsidized implantation of this technology in society. 'We must capture our share of the market.' (Gotlieb and Zeeman 1980, 'Introduction')

On the other hand, there is an awareness that imposing new communications technology will lead to grave dangers for which society needs time to prepare. For example, the debate around computer-communications in the workplace is riveted with this double bind: 'Hurry up and deploy new communications technology or you'll lose your job.' vs 'But this technology is bound to cause disemployment as it is, retraining and readaptation programs must be set into motion before the cataclysm.' This is a particularly malicious discursive strategy because it situates the problem in an a-contextual future. It never questions the technological imperatives and it allows for no alternatives but to be caught in it. We are trapped by a false technological imperative where technological progress is being discursively created rather than simply being unavoidably present.

> the introduction of microelectronics technology into a wide range of information processing applications will produce two conflicting effects: the creation of new industries and occupations and the destruction of those it renders obsolete (. . .) it appears likely that the service sector growth will not compensate for the labour displaced from the manufacturing sector. In fact, more than one half of Canadian employment is attributable to the production, distribution and use of information, the area in which microprocessors will have their greatest impact. Whether the net effect is positive or negative, a massive redeployment of workers from paper-work based to technically based jobs must occur, with major social consequences. (MacClean/Institute for Research on Public Policy 1979, ix)

In terms of policy-making, this double bind is an 'obvious' argument against government 'interference'/regulation of new communications industries since legislation slows things down. And yet, governments are called upon to plan, project, and prepare for the consequences. Even government is placed in a double bind by this rhetoric. In Canada, for example, while MacClean, in the above, calls for government planning, Gotlieb and Zeeman threaten that the government will place the country out of the running if it tries to regulate the industry (Gotlieb and Zeeman 1980, 'Introduction'). Even the Canadian DOC argues against government intervention saying that it would:

> fragment and hinder the optimal development of data communication, so that the long-term technical quality and economic viability of Canadian Telecommunications networks become jeopardized. (*Green Paper* 1973, 5)

Regulation and any reflective slowing down would cause 'grave difficulties in obtaining necessary management resources and capitals.' (*Green Paper* 1973, 5) The double bind is more thoroughly indicated by the fact that in Britain, Prestel, a government-owned corporation, suggests that it is ahead because no one has interfered with its urgent and speedy development (Reid/OECD 1980, 42).

Some texts have provided alternatives. Friedrichs (UNESCO 1981) describes how a joint commission composed of government, worker and management representatives discussed guidelines for the introduction of technology within the workplace in order to avoid the rush situation. Rather than rushing to develop hardware, it is rushing to determine the communications needs pertinent to the country and to develop specific technologies to meet those needs. Japan as well, with its non-tariff trade barriers, has sought to give its industry and its society some breathing room from the urgency of competition. There are alternatives to the double bind options.

The urgency/reflection double bind betrays its own lack of choice in such remarks as: 'We are slipping rapidly from the post-industrial age into the age of electronics – *like it or not.*' ('High Tech at Home,' in *Ms.*, 12 October 1982) Even if the need to reflect exists the new technology is rapidly and inevitably arriving and one really has no choice.

Weizenbaum demystifies this urgency/reflection double bind by suggesting that urgency is a trumped up imperative used to justify computer technology. He suggests that no matter what invention one speaks of it is always touted as urgent although quite often society could continue to manage without it since it merely serves to conserve dinosauric institutions rather than to aid progress:

> Yes, the computer did arrive 'just in time.' But in time for what? In time to save – and save nearly intact, indeed, to entrench and stabilize – social and political structures that otherwise might have been either radically renovated or allowed to totter under the demands that were sure to be made on them. The computer, then, was used to conserve America's social and political institutions. (Weizenbaum 1976, 31)

2 Dangerous to adopt/not to adopt

This double bind is expressed in the following capsular quotation:

> To impose informatics obligatorily would be a grave mistake. But it would be equally serious to ignore it, leaving it to an elite to develop. ('Vivre avec l'ordinateur,' in *L'Express,* 23 December 1978, 67, author's translation)

Herbert Simon in 'What Computers Mean for Man and Society' (1981, 212–218) and in 'What Computers Mean for Society' (1981, 430ff) provides some of the best examples of this paradox:

> Making the computer the villain in the invasion of privacy or encroachment on civil liberties simply diverts attention from the real dangers. Computer databanks can and must be given the highest degree of protection from abuse. But we must be careful also that we do not employ such crude methods of protection as to deprive our society of important data it needs to understand its own social processes and to analyse its problems. (Simon 1980, 430)

This double bind resurfaces in development issues as well. Anthony Smith's *The Geopolitics of Information* is caught with the argument that Third World countries are menaced by the adoption of new communications technology while it is equally threatening for them to be left on the sidelines as concerns the development of this technology and its implementation. The common title for this paradox is that of the trade-off between *dependence* and *development* (Cardoso and Faletto 1979).

If this double bind sounds familiar, it might be because one of its main variations lies in the arms race issue of 'deterrence.' It is dangerous to have x number of nuclear missiles pointed at you. But it is equally dangerous not to. This double bind procedure seems to serve the military-industrial complex well whenever it is faced with public scepticism. We find a happy combination of the double bind mechanism in Gilpin's 'The Computer and World Affairs' (1980, 241–248) when he argues that information technology will result in a type of struggle and conflict that will replace armed conflict: 'data warfare which will replace warfare.'

3 Choice/inevitability

This double bind reminds one of the biblical phrase 'God made Adam free to stand, destined to fall.' Discourses on new communications technology insist that society is faced with making many crucial choices that will affect its development for years to come. Indeed, one of the common alibis for writing documents in the first place is to contribute to the knowledge and information required for decision-making on the part of society. Already there is a slight problem in equating decision-making with choice since the former presumes that the alternatives are already extant whereas the latter would imply that alternatives may be freely proposed. The sacredness of choice is reflected in the most dominant and sacred of maxims of the communications debate: free flow. A recent CBC news reel of the debate over licensing satellite earth stations features angry citizens marching with placards yelling: 'We'll watch what we want.' and 'Freedom of choice.' In the same vein, the Headingley 'Ida' Telidon field trials set out to discover what people's chosen information needs would be and whether they would choose to use new communications videotex services at all. As it turned out, the top item on their list was a request for information updates concerning 'grocery bargains' (Foidart/DOC 1980). Yet despite all of the choices that face us, the same documents talk of the inevitability of the communications revolution. This talk is partly linked to the earlier discussed notion of temporality whereby the communications revolution is inevitable because it is a step in the progressive march of man's intelligence. The development and deployment of new communications technology must and will happen, like it or not.

Walter Stewart's article 'The Next Big Revolution Will Happen in Your Livingroom,' (*Maclean's,* December 1968), speaks of 'helping Canadians to keep pace' and 'igniting Canadian reaction to a succession of revolutionary developments' which, though it speaks superficially of choice, really means that citizens are not choosing, not 'acting,' but are 'reacting' to and keeping up with events that are simply out of their control.

If we have choice at all we certainly do not have it as regards change which is described as some sort of self-propelling motor force and occurrence entirely out of the hands of the reader let alone of the inventors of new technology: 'Telidon may well do the same: change our lives in ways that no one, not even its creators, can now imagine.' (Fulford/*Saturday Night Review,* September 1980)

The double bind between choice and inevitability, a discursive procedure that circumscribes a certain view of history, fails to define choice at all in communicational terms. Choice is defined by this procedure as a selection amongst limited objects presented by others, as decision-making amongst limited and preconceived options.

But perhaps choice cannot be measured entirely as statistics of decision-making. This double bind could only be resolved by redefining choice in communicational and political terms to mean more than a preference amongst pre-selected items, e.g., between weather reports and grocery bills. Choice is only meaningful as something more than decision-making between 'both sides of the story.' Perhaps choice should also apply to whether or not one wants change at all. Choice must provide the possibilities, educational and other, permitting choosers to suggest alternatives other than those on the marketing questionnaire.

As regards the double bind mechanism in general, there is no question that, throughout history, double binds have been techniques of political and psychological blackmail. Watzlawick and Bateson describe double binds as the rocks upon which many a human relation has foundered. Hitler often used the unemployment double bind as a menace against objectors to his regime. Corporations use it to blackmail governments into tax credits and the like with little assurance that employment will really increase because of them. The alternative to the ridiculous either/or position of saying, for example, that jobs are not important, is to break the double bind by finding a third term between blanket technologization and unemployment.

Finally, though, the only way to get out of the double bind is to go to a higher level than the two terms of the catch 22 in order to recognize the logical trap and to hunt for a third term. Choice must not be identified with simple decision-making (Weizenbaum 1976, 260).

1.11 Eternal mediation

> Technology is usually conceived as a mediator – for all things. The technocratic consciousness, then, fantasizes science and technology as the utopian absolute, being the perfect fusion of both unlimited power and goodness, and to which all willingly submit hence reconciling all social conflict. The technocratic then becomes the secularization of duty. (Gouldner 1976, 261)

Mediation is one of the most prominent discursive features of classical discourse. Foucault, throughout his studies of discourse, reiterates that classical discourse conceives of language as an 'eternal mediator.' The mediation occurs between signs and meanings, between words and reality, between man and universe, between man and man, etc.[13]

New communications technology seems to insert itself within this discursive tradition of mediation. In 'Julia's Dilemma' the AGEE videotex system mediates between Julia and her son, hence between man and man. In the UNESCO symposium *The Social Implications of the Scientific and Technological Revolution,* technology is spoken of as a mediator of ideological differences and of inequalities in standards of living between developed and underdeveloped countries.

Dessauer suggests that technics mediates between man and God: 'Technics is a meeting with God. Through technics appears his creative spirit in our time, the resurrection of mankind.' (Dessauer 1927, 31, author's translation)

Technology as the sacred mediator between God and man is the major theme of a short and unintentionally hilarious article from *Time* (December 1956) entitled 'Sacred Electronics.' It concerns a computer programmed to do lexical studies of the Gospel. New communications technology is portrayed as a great discursive mediator. The Archbishop is quoted as saying: 'These machines become a modern means of contact between God and Man.' The computer is seen as translating or rewriting, with non-human, divine capacities, the word of God. Machines come to man from God so that man may know God. What is more, the machines, as a creation of man's intelligence which is made in the image of God, are thus a mediated image of the minds of man and God. Thus the machine mediates between human and divine intelligence as both an interpreter of the Gospel and as a model of God's mind. 'I'm praying to God,' said Father Busa last week, 'for ever faster, ever more accurate machines.' Thus it is no wonder that in today's jargon computer scientists are often called the 'secular saints' of the age.

Mumford also speaks of technology as a universal mediator. It mediates between man and nature, man and man, reader and book, and even various historical epochs. For Toffler technology mediates humanity's move from one age to the next.

An interesting variation on the mediation procedures is an IBM advertisement where a specific technique, the 'helpful button' of the IBM keyboard, mediates between man and that which was formerly inaccessible, namely complicated technology.

> In praise of the humble button (. . .) IBM makes more than 18 000 different buttons. Push one and touch technology behind it. (*The New Yorker,* 1982)

Carey and Quirk also touch upon the mythos of technology in America as a myth of mediation. One of their findings consisted in picking out a difference between the European and American attitudes towards technology. Whereas Europe had consciously suffered the scars of industrialism on its cramped quarters, America had promulgated the myth that the New World, with its vast natural frontiers, could use technology to mediate between man and nature without alienating the environment from man (Carey and Quirk 1970, 396). They describe this myth of mediation as the middle-of-the-road belief that Nature could absorb technology without being destroyed and that it could be used to mediate between man and nature. Indeed, they say, the mediational myth went so far as to *identify* nature/environment and technology: technology was the environment. The mediation, however, failed. The two sides – organic natural harmony and technological expansion – were quite mutually exclusive. For the moment, what interests us here is yet another testament to a discursive procedure that belongs to classical liberalism, namely mediation: nature can absorb anything and technology as the mediator of man and nature will not unbalance the environment. Technology is touted as a mediator and as such it simultaneously marks off the difference between the two elements it is supposed to mediate. So, when discourse on new communications technology boasts of the immediacy of satellite data transfers – 'on line, real time' – one should not forget that it is a mediator and as such is neither immediate nor identical. The distance is still there between all of those things that technology claims to mediate: the rich and the poor, left and right, man and man, man and God, man and nature. The distance must be there for technology to pose as a mediator in the first place. It is perhaps better to admit the distance than to pretend that technology makes an immediacy where there is none.

1.12 Anthropomorphization of technology

'Talk to your toaster.' (*Ms.,* 12 October 1981)
'Hello, Mr Pump, Get Busy.' 'All right, Chief, I'm working.' (*Literary Digest,* 25 October 1930)
The five machines stood rectangular, silver and green, silent. They obviously were not thinking about anything at all (. . .) ('Sacred Electronics,' in *Time,* December 1956)

Computers think like man. There is no difference between the way the thought processes of man and those of computers operate. Such is the assumption of the Artificial Intelligence Research Community (Newell and Simon 1972, 870–71). This belief is nothing more than an electronic variation on the Newtonian world machine symbol.

Not only is it too late to turn back from technology but technology and man have evolved in complementary ways, according to anthropological historians such as Gehlen, and, to a degree, Mumford. Man has taken on traits akin to technology and vice versa. This might also be the interpretation of the message of

'Julia's Dilemma' since the back-to-the-earth community needed technology to survive.

Technology is not only anthropomorphized, it is also discussed by authors such as Dertouzos, Moses, and Margaret Boden as further *humanizing* man by individualizing him (Boden 1981, 445). This argument is an attempt to reply to those who use the anthropomorphization procedure to argue against new technology, i.e., machines dehumanize and alienate man (Reisman 1950–7).

Jaques Ellul, although he refers to technology as 'impersonal' actually anthropomorphizes technology through the verbs that predicate it and the animate traits that he bestows upon it. He speaks of it as 'this completely impersonal force whose trend we follow.' (1954, 351, author's translation). Technology is described as an active force rather than as a passive object. It is characterized by traits associated with human beings: self generation (1954, 82), unity (1954, 93), autonomy (1954, 121, 257).

For Dessauer (1927, 171), technology is not exactly anthropomorphized, it is *deified*. Work will ascend to religious trancendence once man internalizes the technical process.

Whether the positive or the negative side of the anthropomorphization equation is evoked in technological discourse, it is interesting to note that Robbe-Grillet (1963), Nathalie Sarraute (1956), Roland Barthes (1953), and Stephen Heath (1972) have all argued that the procedure of anthropomorphization in the classical novel is a trait of discourse particular to the bourgeois discursive procedure of exchange within the rise of industrialist society. The house is exchanged for the man and vice versa rather than recognizing the authentic value of each. Anthropomorphization is a case of false referential exchange whereby man is equated with machine and machine becomes a signifier of humanity.

A very interesting irregularity of the anthropomorphization process involves current discourse on new communications and electronic technology in the arms industry. The armaments themselves are anthropomorphized: we find mention of 'killer satellites,' of 'killing off machines,' of missiles called 'Popeye,' etc. On the other hand, and equally indicative of the socio-ideological implications of anthropomorphization, human death is deanthropomorphized. Men are never spoken of as killed, only machines are knocked out (Mattelart 79, 69–72).

In the seventeenth and eighteenth centuries, when Bacon, Descartes, Hobbes, and Locke were developing their model of new science, man was placed at the centre of the universe and given domination over it. Foucault refers to this epoch as the anthropocentric episteme. By anthropomorphizing technology, discourse on new communications technology assures a place for technology at the centre of the universe arising out of the anthropocentric episteme. Which nature will it dominate, though?

1.13 Competition between man and machine for hierarchical position

'Anything you can do tech can do better.' or vice versa (. . .)

In an article entitled 'Matching Wits with the Computer' (Walker, *Psychology Today,* June 1981), we find an illustration of a race between a robot and a human. The human is just barely ahead of the robot on his way to the finishing line but is looking anxiously over his shoulder at the robot which is fast gaining on him. E.H. Walker, a research physicist for the US Army Ballistic Research Laboratory, writes his text as a scenario of increasing competition between man and machine for superiority in quantity and speed of calculation:

> Thus the human brain has the advantage at present, although it will probably be surpassed in speed sometime during this decade by a new generation of computers. (Walter 1981, 108)

Again and again we find evidence of this concern with competitiveness.

> Will thinking machines ever be as intelligent as the human brains that create them? ('Intelligence Amplifier,' in *Time,* May 1956)
>
> 'Scientist Says Man Outpaces Himself'
> 'Super-Ape Needs Intelligence to Best Machines' (*New York Times,* April 1954)

In yet another article entitled 'Tinker Toy TV' (*Newsweek* 1954) we read of how technology compensates for the weakness of man in assembly line technology.

The fact that so many texts express the relationship between computer-communications and man as a struggle for superiority and for the capacity of one to subsume the intelligence of the other, is again a procedure of classical discourse. It is an interdiscursive procedure based on exclusivity whereby exclusivity is a competitive will to be more truthful and more powerful that the discourse of the other (cf chapter four).

This insistence on respective domination and exclusivity is nothing short of a throw-back to the Feuerbachian statement of the relationship between man and God. The greater the image of God that man creates, the lesser and more alienated the image of man. To paraphrase Feuerbach, the greater the image of new communications technologies created by discourse on new communications technology, the lesser the image man will project of himself. This ratio is only possible in a situation in which exclusivity and will to power reign as the dominant discursive procedures.

Margaret Boden expresses both sides of the ratio. She suggests that since man created this technology, it affirms both his greatness and his humanness in opposition to machines:

> But, Dr Boden concludes, 'the prime metaphysical significance of artificial intelligence is that it can counteract the subtly dehumanizing influence of natural science (. . .) Far from showing that human beings are 'nothing but machines,' it confirms our insistence that we are subjective creatures living through our own mental constructions of reality (among which science itself is one). (Boden, in Owen, 'Teaching Computers to Learn,' *World Press Review,* May 1981)

However, concentrating on this exclusive, competitive aspect of the man-made machine relationship, Boden falls into a trap. She has just stated that machines may serve to augment man's image. However, speaking of computer-communications in education she contradicts herself by stating that the 'undermining of one's experience of oneself as responsible may therefore be exacerbated by clever programs.' (Boden 1981, 444)

Later, Boden's expression of this competitive relationship takes the form of a subsumption of intelligence functions. Will man develop, program and subsume machine's processes? Boden seems to forget her earlier remark:

> Computerized psychological and pedagogical programs may be more humanized than behaviorist cognitive models in that the former are adapted to specific subjectivities rather than to a blanket pattern of the human race. (1981, 445)

This competitive and inverse relationship between man and technology will resurface in the discursive procedure known here as the instrumentalist view of man's relationship to nature. In that technology is instrumental in helping man to control nature, man subsumes nature. But in that man is used to meet the ends of technology, a relationship of control is established on a higher level: technology subsumes man.[14]

William Leiss has shown in *The Domination of Nature* (1972) that there is a far greater relationship of competition and domination involved with technology. The man/machine debate occults the darker side of the moon of the myth of Enlightenment. Technology was to have aided man in dominating nature at the service of mankind. What is never discussed however is whether it is often the case that the nature dominated is human nature.

1.14 Individualization and atomization at the expense of social structure

Individualization is both a theme and a discursive procedure. Throughout the documents on new communications technology, when authority is called for, the singular voices of Mr So and So, inventor of such and such or head of such and such company or institution are heard. Most texts on new communications technology are the product of a single individual. Edison invented the light bulb, Bell the telephone, Turing the prototype of the computer, Eckhart and Motely the real computer, and an Ottawa civil servant the second generation videotex hardware known as Telidon. Gordon Thompson is described by Walter Stewart (*Maclean's,* December 1968) as a 'one-man think-tank.' In the same article, Eric Kierans, the postmaster general, is described as the one man who must find solutions to a situation that is out of control. Dertouzos (1980) suggests that society will be better off with computer-communications technology because we will be more individualized. We will be more individualized because we will be able to have custom-made products. Earlier, Boden argued for computer-communications in education so that it too could be individualized. In 'Sacred Electronics,' Father Busa is the young, individual intellectual who is able to develop uses for technology to mediate between God and man. Computer-communications systems are portrayed as individuals racing against other individuals, namely men.

It is not surprising to find talk of new communications technology positioned by discursive procedures of individualism. First of all, individualism correlates very well with a procedure of analysis or atomization whereby the whole is broken down into small parts. In this case, the social totality is broken down into isolated individuals. The single subject is considered to be the source of power and the origin of truth and discourse. Foucault has shown that classical discourse clings to the singular subject as the source of all discourse and knowledge, thus explaining classical discourse's insistence on authors, biographies, and the cult of the genius (Foucault 1977, 113–38).

Developments in technology, especially new communications technology, are portrayed as emerging spontaneously from the mind of a genius or individual creator. What is more, only the individual authority, only the Gordon Thompsons and Daniel Bells, can speak correctly about this new communicational technology. Very rarely do we find an attempt to represent in a comprehensive and nuanced way what various social groups or publics think and say about new communications technology, nor do we find any treatment of how the social and corporate structure of society might be a condition of possibility for new communications technology.

Jennifer Daryl Slack in her excellent study of patent law for new communications technologies convincingly illustrates how histories often insist on the individual inventor whereas the success or failure of an invention really depends on corporate struggles and juridical patent protection (1984, 93–138).

The all-persuasive way in which the individualist perspective reigns is responsible both for the emergence and the treatment of certain issues. The most important of these individualist issues is *privacy*. If privacy is raised as an issue at all it is because, at least since liberation and the 'laissez-faire' protection of private property, discursive procedures have insisted on the individual's right to be alone, and to have his property left alone. The concern for privacy is the major issue raised by the 'humanist' critics of new communications technology. H. Vandenberghe in 'La vie privée et les banques de données' (OECD 1979, 249–56) argues that, apart from Sweden and Germany, most developed countries have little or no legislation to protect the privacy of citizens, and sees this as a failure requiring legislation. Pool and Solomon (OECD 1980) argue that the only public concern for new communications society should be protection against 'negative exteriority,' by which they mean rights of privacy for vulnerable individuals as opposed to corporations. Corporations or even social groups should not be regulated unless they threaten individuals and their privacy rights. No mention is made of accountability to social groups or community structures or of the impact on them. Futhermore, whenever a struggle is mentioned it is once again one of individuals against individuals. We as individuals must mobilize to ensure that an 'elite' group of individuals does not monopolize new communications technology. Legalistic texts in the OECD studies carry individualism so far as to speak of 'corporate individuals.'

Another major debate that stems greatly from the individualist discursive procedure is that of *homogeneity/heterogeneity* or *individualization/massification* (Riesman 1950–7). One of the major concerns of the critics of new communications technology is that it will lead to a massification of society, to a homogeneity of cognitive processes and cultural production. Those who argue against the likelihood of massification simply reverse the coin of the individualist procedure. Dertouzos and Boden argue that computers will customize production, be it educational, industrial or cultural, thereby furthering individualism. Each person, they claim, will have his own console and will be able to be his own publisher or creator. We will wear robot-built, made-to-measure alligator shoes. In both cases the standard of measurement for praise or criticism is the degree of individualism permitted. In the following examples of the individualist bent, individualism is linked almost exclusively to consumerism and to competition in the market and educational spheres.

> mass-production will be replaced by an individual oriented manufacturing industry. Computer-based automation can help reverse this dehumanizing trend without decreasing the cost

benefits of mass-production because the computer is capital – rather than labour-intensive and can be programmed to meet the needs of consumers. (Dertouzos 1980, 53)

Since individuals with special needs can draw on information pools in different ways and for special purposes, the power to the individual is enhanced far more significantly compared to what a sudden expansion of energy resources would do for him. (Lowi 1981, 456)

Another issue that cannot be divorced from individualist discursive procedures is that of free flow. Freedom is defined in terms of the absence of constraints upon the individual (Jenson 1976). Freedom is not defined in terms of any positive capacity of society but in terms of negative theology: one cannot say what freedom is, only what it is not in relation to the individual:

To be free is to have one's power of action i) without restraint or control from outside and ii) with whatever means or equipment the action requires. (Hocking 1947, 54)

Parkhill also defines free flow in terms of the absence of any constraint on the individual's right to send information (not to be confused with the individual's and society's 'right to communicate'):

in a free society, surely, (. . .) no control is either needed or desired other than that which ensures the free exchange of information in a truly open marketplace of ideas. It is therefore of vital importance that we establish now, while systems like Telidon are still in their infancy, those fundamental principles concerning freedom to publish and freedom from censorship that lie behind the time-hallowed slogan, 'Freedom of the Press.' (Parkhill 1980, 80)

Most US policy and regulatory statements on transborder data flow tend to insist unequivocally on the supreme right of the individual to transfer data whenever and however he/she/it sees fit.[15] Free flow is simply defined as the absence of any interference with the transfer of information by any individual. Once again, no mention is made of the rights of certain groups to protect themselves from other groups or individuals.

Whether one is for or against free flow, one thing is certain, no other issue of new communications technology is so evidently a child of the discursive procedures that make up the episteme. One of the first things that an alternative episteme would trigger, then, is a reevaluation of the free flow doctrine.

The surveillance debate, part of the privacy debate as well, also operates to the discursive procedure of individualism. First of all, those who criticize new communications technology's surveillance capacity usually do so on the basis of their belief that the supreme human rights issue is *privacy*. Secondly, those who require surveillance and push for its development in the field of computer-communications do so on the basis of a firm belief that the supreme right is that of protection of private property (CBC, 'Quarterly Report,' 6 June 1982). *Instant World* has criticized the obsession with surveillance in the free flow debate, suggesting that its proponents operate under the false assumption that there is a clause guaranteeing the supreme right to private property in the charter of human rights (*Instant World* 1971, 16ff). And yet when asked, most Canadians still feel that the single most important issue of new communications technology is privacy.

Why choose privacy as the battle horse for social consciousness in relation to the new communications technological revolution? Anyone who opposes the surveillance techniques of new communications technology is immediately criticized on individualist grounds: he has something personal to hide. The result of this obsession with individualism is an absence of participation in the debate by public and social groups. As content and as participants, public groups are conspicuously absent from texts on these issues. There is a tendency to make individuals both responsible for and victims of all problems or benefits stemming from new communications technology. The notion of social responsibility or impact on the social structure is never raised. Perhaps surveillance is not a threat because it threatens privacy but because it will produce a paranoid-type social structure whereby not just individuals but whole groups are looking over their shoulders, and whereby authorities have incredible information about behavior patterns, such as the consumer habits of specific groups. Or again, perhaps the real threat is not to individual privacy but in the form of far enhanced 'public management.'

1.15 Ubiquity of surveillance

'New eyes for computers: chips that see.' (*Popular Science,* January 1982)

First one grows accustomed to the ubiquity of exterior supervision, e.g., as a school child, and gradually one interiorizes the rules of behavior sanctioned by the supervisor, to the extent that an exterior supervisor is no longer necessary. The supervisee becomes his own supervisor by virtue of interiorizing the procedures.

Early uses of electronic technology are noted not only for their surveillance capacity but also for the omnipresence of this capacity. Surveillance by machine is far more efficient and ubiquitous than human surveillance: 'Light chart of 'L' system shows flow of power' 'In the supervisor's office of the Chicago Elevated Lines; on the Wall is a chart of the entire system; tiny lamps indicate conditions on the routes by day and night.' (*Popular Mechanics* 1927, 747)

Historically, one of the functions of electronic business machines, the precursors of business computer-communications, was to centralize surveillance and record-keeping of workers, payroll, and customers ('Parade of Business Machines,' in *Business Week,* October 1935).

The desire for control via surveillance and record-keeping is quite unocculted in several early discourses on data processing. For example, as late as 1955, we find the use of technology described as follows:

> to throw some light on some of the great mysteries of why we behave the way we do (. . .) psychologists may some day be able to come up with scientific means of setting problems for workers (. . .) We may be able to devise methods of presenting data so that it can be most easily transferred from one fellow's mind to another, for example, in advertising or safety. ('Information,' in *Business Week,* 30 July 1955)[16]

Surveillance is, first and foremost, *ubiquitous.* If we take another close look at the Dagenais article we can see that indicators of space are constantly expanding or exploding. She mentions 'télématique' as a system that will be part of 'l'univers familier.' The expansion of technology subsumes the house and the universe: technology subsumes house subsumes universe. Technology, she says, will link two hundred and fifty homes, hence subsuming them within the central data bank and computer. These machines 'integrate' homes and machines, hence technology subsumes the family and society. Finally, geographically, the USA is described by Dagenais as subsuming Quebec which in turn subsumes information which subsequently subsumes private life. Today, we are told, technology is in our kitchen, in our child's school, in our study, in our toaster, in our doctor's office, in our briefcase, in our mail box, and in our government. And nothing seems to have changed: 'New Nationwide TV Net would include nearly 1400 stations using seventy UHF channels and would cover areas enclosed within the clustered circles above. . .' (*Life,* v 31 (2), 1951)

New communications technology is described as expanding, as reaching out its tentacles to embrace or subsume every place of social life. What is more, new communications technology is described as a seeing eye, as a recording, remembering, watch-dog technology. Rather than worrying about privacy one should perhaps keep a close watch on the type of social space where technology or surveillance communication is ubiquitous and still expanding. This does not imply procedures of an individual internalization of surveillance but of a whole society internalizing surveillance. We might coin the terms social panopticism, a new type of surveillance which like other forms of panopticism is the cheapest, most expedient way to maintain social control.

Perhaps the early electronic invention that best exemplifies a valorization of internalized surveillance is the lie detector:

> Electricity is used to solve crime mystery. By quickened heartbeats, and by a change in the electrical resistance of his skin, due to an effort to conceal the truth, the 'culprit' was easily detected (. . .) (*Popular Mechanics,* April 1927, 604)

What is important is not so much the external supervision, the FBI agent hovering over you. What is at stake here is that you know that something can get inside you to survey you, something over which you have no conscious power and to which you must acquiesce by behaving constantly as your own check. This internalized checking mechanism is further provoked by the following caption: 'Spy satellites: somebody could be watching you' (*Electronics and Power,* August 1978, 573). Such captions are also implicit in the whole debate around billing procedures and legislation of the records kept of citizens' information consumption habits. However, while some texts do not occult the ubiquity of surveillance they still insist on the philanthropic outcome of this surveillance. Surveillance makes your life safe by fighting crime and by deterring Soviet aggression, etc. Forget what it might do to the structure of your community.

1.16 Valorization of speed, growth, efficiency, progress, productivity and profit

'If a little's good, a lot's better.'
'More, faster, cheaper, always. . .'

In the article 'Sacred Electronics' even the priests espouse the instrumentalist values of speed, cost-efficiency, and correctness: 'I am praying to God (. . .) for ever faster, ever more accurate machines.' And for the espousal of these values, T.J. Weston, an American computer manufacturer (IBM), bestows his praise and blessings on the Italian fathers: 'You seem to be more go ahead than we are.' Once again, growth and progress are sacred in the world of discourse on technology.

Instrumentalist rationality is akin to the 'pig principle' – more and faster is better. 'Quality of life issues' are often reduced to these values. Take, for example, Madden's article of the 'Gee Whiz' variety when he compares the present state of computer-communications technology to 'a beautiful mansion which used to cost $1M and now only costs $1.' (1979, 40) John Garret and Geoff Wright reiterate this valorization of the faster, the cheaper and the more productive in their article so appropriately entitled 'Micro is beautiful' (81, 488). Even Dave Berkman, who argues against the adoption of new communications technology, does so with these values primarily in mind. His argument is that the venture is not economically productive or profitable:

> A major error which underlies so much of this hyperbolic forecasting is the assumption that just because a technology

exists, people will rush automatically to purchase it. (reference unavailable)

One debate that is highly determined by the instrumentalist discursive procedure is the debate on accessing. In yet another variation of the 'pig principle,' it is simply assumed that if a little communications hardware is good, a lot is better. The more it can produce the better. Access and choice are questions of available quantity not of qualitative difference.

Cost efficiency is the primary consideration in assuring access, a concern in keeping with the economics of 'substantive rationality,' whereby maximization of profit is the main goal:

> Quality and cost, in particular, are vital considerations in achieving the overall objective of making computer-based services widely and equitably available. (*Green Paper*/DOC 1973, 4)

Dertouzos, Moses (1980), and Garret and Wright (1981) also fall squarely into the procedure of instrumentality. They consider new communications technology to have utilitarian value because it will manufacture goods and itself be manufactured in greater quantities, at greater speeds, and more economically.

Although it is neither within the aim nor the scope of this study to discuss the economic issues surrounding new communications technology, any more than it was to discuss artificial intelligence, it is nonetheless possible to indicate how the framing of some of the economic issues most evidently reflects the instrumental focus on technology.

The notions of *growth and progress,* while a partial off-shoot of the linear view of history, also derive very much from the instrumentalist paradigm where technology is a 'storing up' of power and production potential. This storing up is precisely what the corporation dedicated to technologically encouraged growth is doing under the mantle of free flow. 'Progressive thinkers realize that the best market in the world is the world.' By the 1980s estimates indicated that the multinationals will contribute about one quarter of the Gross World Product.' (Gotlieb and Zeeman 1980, 19–20)[17]

The upholding of the principle of 'free flow' also stems in part from a rationalization of communication. 'Laissez-faire' capitalism guarantees growth, control and profit:

> The capitalist mode of production can be comprehended as a mechanism that guarantees the permanent expansion of subsystems of purposive-rational action and thereby overturns the traditionalist 'superiority' of the institutional framework to the forces of production (. . .) to institutionalize self-sustaining economic growth. (Habermas 1968–70)

So potent has the discursive procedure of instrumentality become in the economic sphere of new communications technology that some would argue it has even replaced territorial expansionism:

> Probably some half-consciousness of this fact is at last diverting much of the attention of the statesmen in all parts of the world from territorial expansion to the struggle for relative 'efficiency.' (Gilpin 1980, 248)

A global market for new communications technology is best. Cultural and economic sovereignty should bow to the natural tendency toward growth. New communications technology must be allowed access to the world market. Why? Because only mass production and marketing can make the industry's high cost for prototypes lucrative.

New communications technology is said to play a major role in this 'all essential, all natural' globalization of the economy. Globalization of the economy is made feasible by computer-communications hook-ups that link the centre with the periphery. All of these discourses, both pro and con, operate on the basis of the principle of growth where what is at stake is territorial expansion of the economy made possible by communications networking of marginalized companies, increased efficiency, productivity and control, and by the reduction of risk factors that otherwise accompany decentralization and human intervention (Gilpin 1980, 248; Mackinder 1962).

The assumption that growth of new communications technology links the margins to the centre is somewhat obscured in mainstream discourses on development and new communications technology, while the principle of growth itself is still upheld. The greater the productive capacity of a country the better. 'New communications technology will give wonderful opportunities for developing countries to compete with developed countries.' (Parkhill 1980, 72) The utilitarianism behind technological development is further betrayed by Licklider's somewhat ethnocentric remark that it is not in the interest of the USA to share its new computer-communications know-how unless it gets something of special value in return (1980, 105).

The hierarchy of the value of progress and growth associated with new communications technology is evidenced by the fact that many discourses advocate going ahead with development (and development has gone ahead despite its other economic consequences for which there are neither estimates nor remedies (Russel 1978; Menzies 1981)). Both Russel and Menzies point out the vast disemployment that will arise from new computer-communications technologization of the workplace. Industrial sabotage and the recent Bell employees and air-controllers strikes in the USA indicate the severe socio-psychological implications of new communications technology in the workplace and yet, Gotlieb and Zeeman (1980) and Porat (1978, 3–55) and Bell (1973), all interpret the economic implications of new communications technology in terms of production, efficiency and growth in GNP: instrumentalist values replace humanist values.

In view of these remarks on instrumentalism, Ellul's definition of technology would seem to be extremely pertinent: 'One may define "Technique" as constituted by the ensemble of most efficient means at any given moment.' (1954, 18; 1977, 34, author's translation)

But perhaps even more pertinent is the description of our present technological society by Habermas:

> The institutionalized growth of the forces of production following from scientific and technical progress surpasses all historical proportions. (1968–70, 83)

1.17 Unilaterality of control by knowing subject over object to be known

'The world is my bag.'

> The determination of whether Satellite Technology will take off is a matter of terrestrial manipulation as opposed to one of Natural possibilities – the universe does not forbid its development. (Whittaker/OECD 1980, 39)

Since classical science and pure physics, knowledge has been conceived of as man's unilateral action upon the universe in order to meet his own needs. The essence of technology was born with this linear action upon nature although technologization in the form of machines did not develop historically until the mid-eighteenth century (Kumar 1978). Man acted upon nature in order to control it. And, since other men formed part of that nature, the unilateral control of men, which turned them into uses for the power-seeking, knowing subject, was also quite permissible.

Even Marx begins his *German Ideology* with the statement that man must control nature in order to survive and that a part of his activities, namely work, is directed toward this end. Marx simply disagreed with certain groups of men using other groups as instruments to meet their ends. He disagreed with class exploitation.[18]

Contemporary discourses on new communications technology, such as quoted above, also maintain this subject/object unilaterality whereby nature is to be moulded and controlled by man to serve his needs. Once nature is understood to mean the social environment, control is sought there too. Simon states that we must not deprive ourselves of the computer-communications technology necessary for knowledge of the environment and decision-making in order to preserve our social structure (1980). Licklider, in his hypothetical scenario, happily describes a system of control whereby: 'Social Security and Internal Revenue share the cost of maintaining the National Roster, the ultimate personnel data base with a file for every person, natural or corporate.' (1980, 89) Gassman is yet another who calls for control over the human or worker/consumer environment by means of new communications technology. He expresses the 'need for more precise, relevant data management, administrative and planning processes.' (Gassman/OECD 1980, 11)[19]

Nor have things changed much in this regard. In 1930, an article on electrical communications hook-ups, the forerunners of telematics, also insists greatly on the control motif:

> It is now possible to telephone to a machine, have it begin running or stop running, and have it inform you that it is following your instructions, although it is not quite as simple as the imaginary conversation under the picture. This is the newest method of *controlling* automatic electrical equipment from a distance. ('New Electrical Control,' in *Literary Digest,* 25 October 1930, emphasis added)

Another article, dealing with computerized weaponry, foreshadows the 'remote control' that computer-communications has perfected in today's weaponry.

> The Queen Bee, the British 'robot airplane,' has created quite a sensation on both sides of the Atlantic. On its first flight, with the pilot in the cockpit not touching a single plane or engine control, but merely going along for precautionary measure, it took off, performed every sort of evolution and landed – all under *remote control.* (*Scientific American,* October 1935, 204, emphasis added)

A glance at any of the journals on military technology today further demonstrates an increase in the control/surveillance motivation of new communications technology. The following titles are sufficiently indicative of the central motif:

> 'Spacecraft Survivability Boost Sought' (*Aviation Weekly and Space Technology,* 16 June 1980)
>
> 'Expanded Ocean Surveillance Effort Set' (*Aviation Weekly and Space Technology,* 10 July 1978)
>
> 'Space Surveillance Deemed Inadequate' (*Aviation Weekly and Space Technology,* 16 June 1980)

The desire for control on the part of the military-industrial complex, taken as a society of discourse, or institution, is one of the motivating drives behind the development of new communications technology as these historical examples corroborate.

> The protective functions of government still account for most of it, primarily in code-breaking, with intelligence analysis second, logistics third and command control fourth. (Licklider 1980, 89)

Indeed, one of the first computer-communications hook-ups was the SAGE system designed for the Pentagon.

Hughes and Sasson in their study 'The Usage of International Data Networks in Europe' (OECD 1980, 28–29) constantly argue for increased control. They also indicate that the 'level of security (control) is highest where large financial interests are at stake':

> Control is secured by the companies and because control has not been threatened for companies it never will be for any other social groups (. . .) in each case we found that each national operating company kept its data under secure conditions, and the general reason for processing the information internationally was purely a reason of resources sharing. (Hughes and Sasson/OECD 1980, 31)

Weizenbaum warns of certain developments in computer technology which should be outlawed because their only rationale lies in an increased control over human beings and the elimination of other human activities such as love and interaction. He warns against coupling animals' visual systems with the brain of computers since this cannot be justified in terms of 'any conceivable need of man (that) could be fulfilled by such a device' (1976, 269). And Weizenbaum himself, as the developer of a simulacrum of Rogerian psychoanalysis called 'Eliza,' is clearly in a position of 'authority' to speak on the limitations of such a substitution.

The following remark by Nehru would seem to reconfirm what Stalinism has already borne out, namely that regardless of economic ideology, technology seems to be control-oriented:

> Prime Minister Jawaharlal Nehru of India said today he found the United States and the Soviet Union strikingly alike in 'their worship of technology,' however far apart they might be on political and economic questions. ('Nehru finds the US like Soviets in Part,' *New York Times,* June 1953)

A further derivative of the control motif of technological discourse is the 'technological fix': Yes, there do exist dangers for society due to new technology but society has always had, and will always have, the technological resources to overcome them:

> Should it be found that new risks do indeed arise, and since every risk can be insured at a price – perhaps a completely new field of 'data insurance' may develop. (Gassman 1980, 60)

The unilateral control of the subject upon his world-object is not a new procedure of knowledge. As early as the seventeenth century we find Descartes outlining a form of knowledge, a method, which, if mastered, would make man *'master and owner of the universe.'* It is the knowing, speaking subject that acts upon and changes the universe and never vice versa. Any 'negative factors' may be overcome by technology itself (Dertouzos 1980, 55). Man's discourses of knowlege, his technology, insist that the environment adapt to man's technologies for meeting his needs and even for patching up the disasters of past technology. It is never up to technology or to man to adapt to the environment, at least not according to proponents of the 'technological fix.'

Others, however, argue against the technological fix, and this indicates some *irregularities* and *leaks* in the control-episteme of the discursive procedure of domination. The ecological movement which insists that man and technology must respect, interact with and adapt to the environment is one example of such irregularity. The ecologists doubt the future projections of the 'technological fixers.' They ask what guarantee do we have that technology will always be able to solve its own problems and within what time span will it operate. They tell us to look back into the past rather than blindly into the future, to look at the scars left on the environment by DDT spraying and by technologies which purported to better the quality of life. Such are the questions posed to the control motif of classical discourse by Bateson (1972 and 1979) and Cohn-Bendit and Castoriadis (1981).

That most technology so far controls society and expects society to adapt to it is made clear by Winner's remarks:

> (. . .) with respect to the issue of the autonomy of cultural values as opposed to the autonomy of technology. What we see very often in the social history of technological change is a rescheduling of values. The first time a particular device or system appears in a culture, there exists a certain latitude of choice that is never available again. Once the initial choice is made, the latitude of selectivity among a range of possibilities vanishes. The thing gets set in steel or concrete, and its use becomes an ingrained part of the habits of the culture. (Winner, in *Daedalus* 1980,9)

The control mechanism in discourse about new communications technology may best be summarized by the following two clichés that constantly surround the dialogue/monologue on new communications technology: 'We have to adapt to new things in society, to keep up with progress,' and 'Technology can do it, it all depends on whether society is ready for it or not.' Garret and Wright reiterate these clichés: 'Whether such a system is created is a political decision. The technological means are available.' (1981, 496) Whittaker's point blankly states that it is up to society to adapt to technology: 'What is required is an environment that permits widespread development and application – a need for technical compatibility.' (Whittaker/OECD 1980, 39) And again: 'The only possibility of improving the human utilization of machines lies in the improvement of man himself.' ('Scientist Says Man Outpaces Himself,' *New York Times,* April 1953)

1.18 Confusion of agents: passification of the public's roles

'We'll all be creators (. . .) but first you must be our market.'

In the discussion on new communications technology, technology is set forth as a) desirable, b) desired by the public, and c) by a public that is only ever cast in the role of a grateful consumer. This argument is achieved by the procedure of reducing the public as subject to an object. We call this process 'deactantialization.'

'Actants' or agents (Greimas 1966, 128–33) are generalized semantic categories that refer to the role that a character, actor or other agent plays in relation to other such roles. The *active* roles are those of 'Subject,''Helper,' and 'Sender.' The passive roles are those of 'Object,' 'Opponent,' and 'Receiver.'

Many discourses concerning new communications technology make the claim that this technology will lead to active participation on the part of the citizenry. This type of participation is, of course, the criterion of democracy:

> True participatory democracy in which the town meeting concept is extended via the ubiquitous computer networks to embrace whole nations and even perhaps the world. (Parkhill et al 1980, 70–71)

Obviously if the discourses on new communications technology are to be believed when they declare that new communications technology will encourage public participation at all levels of society, the actantial categories should grant the active roles to the public. This, however, is not always the case. At times, the public is portrayed as active. At other times there is either a confusion of roles or the public is associated with passivity.

If we analyze closely the agents in the Dagenais article, 'L'information á domicile,' we find that the agent, technology, is given most of the active roles. Telematics is the *Sender* in that it 'speaks to us,' 'aims at us,' and 'gives us.' It is a *Helper* in that it is 'economic aid,' 'educational aid,' 'interactive aid,' and 'help to the police for surveillance.' Telematics is also a *Subject* in that it 'allows us to talk back.' Technology is what we might call a 'synchretic agent,' meaning that as an actor it simultaneously fills many roles (Greimas 1966). So far, all of these roles are active.

The citizen, on the other hand, is covered by passive roles. It/he is a *Receiver* in that it receives information from telematics as a sort of 'sought after gift.' It is the *Object* in that the technology industry needs, desires and vies for its support in the form of citizen endorsement and purchasing. The public-consumer is necessary for the take-off of new communications technology. The public is also sometimes described as the *Opponent,* but only if it rejects, fears or criticizes new communications technology, a possibility not included in most futuristic scenarios.

If the public actually desires the Object (technology) it becomes a Subject. Nevertheless, the public may adopt the role of active Subject only through consumerism: 'One of the best illustrations of its possibilities (. . .) is at-home shopping.' (DOC, Telidon publicity, 17) Technology is a Subject hierarchically superior to the consumer – Subject in that one of its tasks is to create the consumer desire in the public sphere: 'to create a public demand.' (Dagenais) Hence, even if the public becomes a Subject that desires to consume, it is a Subject that depends upon the subsuming motivation of yet another subject, be it technology or some other as yet unidentified subject. This unidentified Subject is an occulted third party in the agent structure of the Dagenais text. She speaks of 'making machines,' of 'integrating the new technology,' etc., all of which are actions that some Subject is performing, without identifying this mysterious Subject. Nor is the Object sought by this unidentified subject ever identified.

Moving away briefly from Dagenais' article to Pool's and Solomon's article, it is clear that this major agent is the 'corporation': they constantly speak of legislation of individuals, both personal and corporate. (Pool and Solomon/OECD 1980, 79–139) The mysterious Subject is the 'corporate individual,' a legally recognized entity.

The attribution of the active actantial roles to technology rather than to the public also occurs in Toffler's *The Third Wave* (1980) where technology is the agent of change, of marine and moon development, etc. Also, the technological epoch in *The Third Wave* assumes an active role. The third wave is described as 'crashing down,' 'rushing in' upon passive little us.

The insistence on 'choice' in many of the documents under study does not really activate the public. Fulford describes the public's situation in what might at first sight seem like an active role: 'You sit at the keyboard or hold it on your lap, and use it to call up information on the screen.' (Fulford, in *Saturday Night Review,* September 1980) But the public is not so much a subject here as a Receiver of the message/gift. Elsewhere, Fulford describes technological development as the real Subject, if not Helper. 'The accelerating development of computers makes Telidon's possibilities almost limitless.' (1980) The public is really the Object of the desires of the Subject (technology) or perhaps of destiny, since technology and destiny seem to be collapsed into one another: 'It (new communications technology) may change our lives in ways no one, not even its creators can now imagine.' (Fulford 1980) The actantial configurations point us back to instrumental rationality: technology is the dominating Subject which unilaterally desires and acts upon an Object, be it Nature or human nature.

The free flow debate also relates to this agent relation. It is noteworthy that it is the subjects, the already active, who call for decontrol and free flow. Of course, this simply protects their actantial status. De-regulation or blanket free flow in this case merely reinforces the active roles of corporations and technology as Subjects while de-activating the public as Subject. On the other hand, as Pool's and Solomon's recommendations indicate, these same subjects call for control and regulation/legislation whenever the public challenges existing actantial roles. For example, decontrol is replaced by legislation when the public threatens, through piracy or other means, not to recognize the supreme possession of information by the corporate subjects. The corporate subjects call for control of intellectual proprietary rights to exclude the public from the role of Sender and Subject of information. The controversy and court cases surrounding photocopying are a case in point.[20]

The various roles that structure the debate on new communications technology also shed some light on the development debate. We are all developing nations in that we are all in the process of acquiring more technology which will become the Subject of our economy and of our organization.

> We are all 'developing' nations in the new information society and solutions to these new problems cannot always be found in tradition or precedent. (Faulkner quoted in OECD 1979, 59)

However, apart from development, where technology is becoming an ever stronger Subject, there is also a question of dependence (Cardoso and Faletto 1979). Cardoso and Faletto define dependence as the interiorization by one nation or group of the interests of another, dominant nation or group. To rephrase in terms of agents, what occurs in cases of dependence is that the Subject of desire of a weaker nation or group becomes the Receiver of the desires of a dominant Subject-nation or group. Consequently, the dominated subject becomes either a Helper or an Object for the desires of the dominant Subject, because the former internalizes and serves the desires of the latter. Dependence is a case of actantial struggle and domination:

> the problem of transnational data flows has created the potential of growing dependence, rather than interdependence, and with it the danger of loss of legitimate access to vital information and the danger that individual and social development will largely be governed by the decisions of interest groups. (Faulkner quoted in OECD 1979, 57)

> The paradox of Canada's dependence is that her use of modern telecommunications links of all kinds is greater than that of any nation on earth (. . .) Despite the technical head starts which Canada constantly provides herself (. . .) she has found that most of the content of her media consists of American material. Many Canadians treat the phenomenon today as a kind of running national crisis. No other country in the world, probably, is more completely committed to the practice of free flow in its

culture and no country is so completely its victim. (Smith 1980, 53–4)

But, if dependence is defined by a Subject being hierarchically dominated by virtue of internalizing the desires of another Subject, what exactly are these internalized desires? Indonesia, a country where only a very small percentage of the population has even a radio receiver was sold a prohibitively expensive satellite system. Herbert Schiller (1976) argues that the dominating countries – Subjects – manage to make the Third World internalize a desire for technology in such a way as to ensure that they suffer trade deficits regardless of their large exports of raw materials. One thing is certain, the developing or dependent countries do not always have the right to produce, develop, or use technology to serve other more pertinent desires. For example, several Pacific rim countries wished to use their satellite networks to conduct an interactive, international debate on the ban of nuclear testing in the Pacific. By some strange sleight of hand, some omnipotent Subject pulled the plug on the satellite connection.

The same actantial dominance operates at the level of the debate on disemployment and new communications technology in the workplace. The worker is to internalize a desire for the Object (technology). But once implemented, it often occurs, and is the cause of much labor unrest, that the worker is reduced to the status of Helper to the Subject (technology), which does everything from set agendas to survey worker efficiency. This confusion of actantial structures whereby one wonders whether man or technology is the Subject of desire appears as early as 1940 in an article entitled 'Men and Machines and TNEC' (*Business Week,* 4 May 1940), a debate about whether technology serves man or vice versa.

One exception to this deactantialization of the worker subject is the negotiated consensus in West Germany between government, industry, and workers regarding the degree of actantial dominance that technology will be allowed to assume in industry (Friedrichs/UNESCO 1981, 274–84).

'We'll all be creators,' is another of the clichés that circulate within discourses on new communications technology. Still, the public is never told what it should do to ensure these creative possibilities. It should just wait passively and patiently until the home computer arrives on its doorstep. It is the Subject (institution) that 'stimulates' and 'activates.' Even this 'creativity' betrays the presence of some other force acting upon the public rather than any supremacy of the public. Subject: 'A premium will be placed on creativity, which being more risky will give rise to concern over the psychological health of executives.' (Russel 1978, 41)

Elsewhere, it is suggested that if new communications technology is to thrive as a mediun, hence becoming the Subject of its desire for vast public acceptance, man must play the role of Helper by learning skills such as how to work a keyboard. These skills are the Object of the desire, not of the public but of technology, and are stimulated by the Sender (corporation) as the following quotation reveals:

> Every new communication medium needs new creative skills of writing and display. Prestel is no exception. We are stimulating the development of these skills by allowing free commercial competition between information providers (Reid/OECD 1980, 41–42).

At this stage of our analysis of the agent structures of discourses on new communications technology we have to doubt the thematic claims that computer-communications services are fully participative and interactive. Participation and interaction would presuppose a symmetry of agent roles by all participants, something which is clearly not the case. The exclusion of the public from dominant actantial, active roles also belies the claims to democratic participation made by these same discourses. The public as Subject remains a highly passive, Receiver, Object of desire, and at times a Helper or Opposer in relation to either the technological Subject or the more occulted corporate Subject. Sometimes the public is even omitted altogether as in the following description of a balanced communicational system: 'a communications infrastructure in which the traditional carriers, the cable companies and broadcasters would work together with better productivity.' (CRAB/DOC 1979, 13) The exclusion of the public as a voting member of such information providers – Senders – as VISPAC and INFOMART in Canada is a further example of the exclusion and deactantialization of the public.[21]

Where the public is allowed the role of desiring Subject it is usually exclusively within the sphere of consumption, and as such the public is serving the desire for profit of a subsuming, hierarchically superior corporate-subject.

Thus, one must be very wary of the participatory, interactive capacity claimed by discourses on computer-communications networking. Such a claim is a semantically evident ploy, used during the mechanical age, and intended to make people believe that democratic participation exists. If one looks closely at the agent roles, however, the public is excluded from assuming dominant roles of Sender and Subject, just as it was in the broadcasting model of communications:

> Of course to *feel* less a cog is not necessarily to *be* any less a cog: the home terminal might function as a subtle form of social control, damping down dissent by contenting people with an illusory sense of political participation. (Boden 1981, 433–450)

Calls for balance and free enterprise, the liberalist slogans of the past, do not seem to have restored symmetry to the participative roles. This applies at least to the discourse on new communications technology which constantly suggest that man is in control:

> 'Man is in control,' Fishman concluded, 'and layman must not abdicate that control to an aristocracy of experts. It is our responsibility to keep our hands on the button, because we are going to get what we deserve.' (Brow, 'The Manners of a Machine,' in *Saturday Night,* September 1980)

What Fishman turns his back on is that the button is in someone else's court, even as concerns discourses on new communications technology, not to mention new communications technology as discourse. Before making wild claims of participatory democracy, we need a theory of democracy in terms both of political theory and of a fully developed theory of what participatory communications entails. (cf chapter four) One such entailment is certainly a more symmetrical agent structure than has been manifest so far. Cameras in courtrooms do not necessarily make for public justice. Nor does two-way cable for symmetrical participation.

1.19 Usurpation, discrediting, stereotyping of the public(s)' voice(s)

'The public is ignorant, plebean, lumpen, and paranoid. Therefore any public response beyond consumerism should not be heeded.'

> Quentin R. Smith wrenches the top off another beer, hunches over the console of his home communicator and spins the television dial through its circle of channels (. . .) Ah, that's more like it – a *Bonanza* rerun on the entertainment band, Quentin leans back, sucks his beer, sighs with contentment. (Stewart, in *Maclean's,* December 1968, 1)

> Scenario II: Millions of men in their undershirts, beer cans in one hand, digital response pads in the other, tuning out Monday Night Football. . .
> Scenario III: Similar throngs eagerly spending \$25–\$100 to purchase a video-cassette. . . (Berkman, 'The Video Revolution')

These portraits of the public do not exactly encourage confidence in the public's capacity to make rational decisions about the design and function of new communications technology. One might describe these portraits as a sort of *disenfranchizing* of the public in decision-making by virtue of a *derision of the public* – Mr Everyman. We are presented with pictures of a stupid, lazy housewife, the porno-lusting adolescent, the blasé penny pincher, the boozing sports fan, and so on, a constant series of deriding portraits of the citizenry. The public response to new communications technology is often portrayed as completely childish as the conclusion to Stewart's article shows, a quotation which also *naturalizes* (Barthes 1957) new technology by taking it as necessary and for granted:

> Once, when my son Craig was about four years old, he asked me if they had television when I was a little boy, and I had to tell him no as a matter of fact, they did not. 'Gee, Daddy,' he asked, 'did they have houses?' (Stewart, in *Maclean's,* December 1968)

Many theoretical studies adopt this derisory attitude toward the public. Ellul, in the same vein as Luhman's exclusive decisionists (1971), argues that the public is too 'backward' or 'retarded' to participate in any decision-making about highly complex issues as new computer-communications or technology in general:

> The public is a half-century behind the times and does not understand anything about the real problems that are posed. At the very most, popular decisions if applicable would only serve to stunt technical growth, to disrupt the system and to provide a socio-economic regression which this same public is not at all prepared to accept. (Ellul 1977, 23, author's translation)

If the public is not attacked for being too far behind the times, it is derided as being unprepared for the complexity of the debate which must be reserved for the technical specialists (Luhman 1968). Or else, the public is too fragmented and uncertain about what it wants to be able to reach a consensus upon which to base legislation: 'the degree of consensus on information about large and powerful organizations is likely to be less than about privacy,' and therefore, unilateral action by the institutions themselves is advisable (Pool and Solomon 1980, 97). Generally, such a system where only technical specialists and bureaucrats make decisions at the exclusion of the public is known as a 'technocracy.'

If the receiver-public is not degraded it is *stereotyped* into pat roles which make its responses to communications technology pre- and over-determined. Discourses on new communications technology rarely recognize plurality. They say, to paraphrase: 'You are a consumer and therefore you will welcome and buy this videotext system which in turn will facilitate your consumption of other things'; or 'You are a lazy housewife who hates to work and needs company, so look what we have invented for you, smart toasters that talk.' Or again, you could be a nationalist requiring supremacy in technology, as the title of a recent article in *Le Monde* (December 1981) entitled 'L'intelligence artificelle, nouvel enjeu du défi japonais' makes clear.

These stereotyped, ungratifying portraits of the receiver speak to him in a condescending tone which credit the receiver with no personal reflective capacity or critical intelligence. He must rely on the author of the article. Berkman jovially empathizes with his reader on a lowest common denominator: 'A majority of that vast majority of *us* who are arithmetic illiterates' (Berkman, emphasis added).

Dogma refers to the pragmatic presupposition of common sense which could be appealed to in order to reinforce the rhetorical strength of one's argument. In the case of these texts on new communications technology dogma is a construction of the author as well as of past theories of the media that make of the receiver what Gustav Le Bon's archaic socio-psychology had long ago portrayed, a 'manipulatable, dumb mass.' The authors presuppose the stupidity, the lack of education, and the vulgarity of the public in order to bar the way to any legitimate call for informed public input into the discussion. It is a pragmatic discursive strategy of exclusion.

It is interesting to note that one of the few times the public was actually consulted about new communications technology only 32 per cent of the groups surveyed indicated a desire or a need to know about or have access to new communications technologies (*Community Communications in Five Ontario Cities,* DOC 1972, 49)

Such stereotyping of the public is crucial in the strategy of interpreting for the public its own needs, which in the case of the Headingley Telidon trials, have surfaced primarily as the need to consume (Foidart/DOC 1980). In this case one has to wonder how much stereotyping pre-determined the formation of the questionnaires and the replies subsequently given. It is quite obvious that one of the major concerns of the developers of videotext is whether and for how much people will buy it. This concern cannot help but bias research about public attitudes and choices.

Of significant importance here is the fact that most of the documents under scrutiny appear in journals and media that are addressed to an educated upper middle class elite, be it the pseudo-leftist liberals of *Le Nouvel Observateur* or the conservatives of *Time*. Although a specific sampling of a corpus of more populist literature on this subject would be required to corroborate this hypothesis, it seems that discourses on new communications technology are addressed to and appeal either to corporate executives or to the upper middle class, always as potential consumers. What is surprising is how successfully jargon and cliché seem to prevail in a discourse addressed to relatively well-educated receivers:

> some 15–20% of the population made up of more educated individuals in higher socio-economic categories, with well-

> defined information-seeking habits, with a pattern of using disposable income to buy information to accomplish specific ends (. . .) The information-poor include the rest of the population who have less developed information-seeking patterns, whether because they *reject* the information culture, are *indifferent* to it, or do not have adequate opportunities or abilities. (Guité 1977, 57–58)

Further marketing studies would be required to determine the success of these discourses in the extra-discursive context of the market place. And indeed, the studies have already been carried out. The newsprint industry has conducted many international studies to see if videotext will replace newspapers. The answer is 'No.' In Montreal, the Tamec Report, carried out for the cable companies (Vidéotron), also stipulated the conditions under which these companies could seek to monopolize the incasting business.

One way to explain the public gullability is Ellul's argument that the more information we seek and are exposed to, the more we are 'pre-propagandized' by the habit itself. But this hypothesis also risks stereotyping the receiver. A better explanation of the survival of the cliché text on new communications technology is that the public has so far been provided with a scarcity of fora in which to reply and to dialogue with these derogatory texts. Thus the texts continue to reproduce themselves regardless of how out of touch they may be with the public. Indeed, studies in journalism indicate this to be a general case: journalists tend to write for their sources (corporate and other) rather than for their public, about whom they have little idea (Darnton 1975).

Strawman adversaries or *puppet interlocutors* are the easiest to disarm. Apart from thematically denigrating or stereotyping the receiver, another way of denigrating anything that he may have to say about new communications technology is to usurp his voice, substituting a half-baked and watered-down version of his objections in order to squelch them more readily. In much the same way that it was a little too easy for Socrates to win over the dialogue partners that Plato created for him, it might be more difficult for a discourse to contend with its opponent if it allowed that opponent to speak for itself rather than quoting it directly or indirectly. There would be a *dialogue* instead of a *monologue* where each discourse would have the right to speak for itself. It is primarily out of fidelity to this principle that this study is so cumbersomely laden with direct quotations.

'Nur Idioten gegen technischen Fortschritt' – 'Only idiots against technical progress' (*Die Zeit,* June 1981). Here in a nutshell is a common example of the way in which adversaries are set up as strawmen and easily knocked down in discourse. Their positions are stated for them in a simplified and vulgarized way. Their voice is usurped. They cannot initiate a discussion or change the level on which it is carried out. They are transformed into push-over adversaries.

Dave Berkman is another who, through his rhetorical technique of fictive narrative scenarios, constructs flimsy possibilities of new communications technology in order to dismiss them more easily. Anyone who opposes computer-communications technology is dismissed by the technical and bureaucratic specialists as 'paranoid,' 'irrational,' and 'influenced by rumor.' Licklider's rebuttal of his creation of the paranoid public adversary is a good example of this enunciative bullying:

> Once in a while there is a flurry of rumours about international espionage within the multinet, but widespread belief in them is precluded by the fact that the entrances to and exits from national networks are under national control, access control is so highly developed, and complete audit trails are recorded. More frequently, a national government is accused of monitoring more assiduously than the law allows, but in the US at least, no one has yet come up with audit trails to prove it. (1980, 95)

Denicoff, in his article 'Sophisticated Software: The Road to Science and Utopia,' yet another 'Gee Whiz' text, speaks of 'near paranoia about the inherent dangers that networking represents to efforts to prevent the nightmare of 1984 society.' (1980, 390) What counts here is not so much that Licklider semantically assures us that fears about the vulnerability of stored information are unfounded or that Pool says computer spying is unfeasible, rather it is the enunciative modality whereby these authors do not debate with their adversaries but quote or paraphrase an enfeebled version of their argument, then defeat it, and leave no room for reply. If Licklider et al are convinced and convincing enough about their argument then they should have the courage to debate or dialogue with Blumenthal and Parker (Blumenthal, 'Electronic Fraud Accompanies Tellerless Banking, in *New York Times,* 6 March 1978, 1; Parker 1976). A true dialogue would not permit the reporting voice to control the voice and the counter-argument that it is reporting.

To say the least, these discursive practices reflect a very a-symmetrical dialogue, so a-symmetrical that it is in fact a monologue. Why should one believe the claims of these discourses that new communications technology will lead to dialogical, democratic participation when they themselves are extremely monological? Many of the problems resulting from dependence in communications are also linked to such a usurpation of the voice of the participants. Very often, for example, Latin American countries are forced by their prestige newspapers to rely on the voice of foreign news agencies for reporting issues, including new communications technology issues, related directly to Latin America. Although access to hardware is not only determinant of this voice control, it certainly plays a decisive role in prohibiting certain voices from speaking for themselves (Guité 1977).

Lévi-Strauss, writing on primitive societies, demonstrates also how the control of writing by chiefs yielded control of the sender/writer over his receivers who could not write (1970, 288–94). As Roland Barthes (1978) and many others have made quite evident, he who has control over the right to speak of the other has enormous control over that other, be it in the personal, social or political sphere of interaction.

The choice is said to be society's and yet, society is not allowed to speak for itself nor is it listened to in any coherent way. Discourses speak of the great importance of decisions regarding new communications technology and yet, no place is created for the voice of the public sphere.

The whole *content-carrier debate* could be seen as a discursive strategy to allow the participation of government, and content and carrier companies at the expense of the public sphere. Major participants in the recent CRTC hearings on the licensing of pay tv (1981–82) were companies which were not representative of the public sphere. Such policies as the 'arm's length policy' and the separation of content and carrier (CRAB/DOC 1979, 13), are snow jobs of public exclusion. The reservations that Ouimet (1979) and Babe (1975) have about the efficacy of this policy to legislate socially responsible communications consolidate such an argument. A glance at the two following quotations about future choices is yet a further indication of the exclusion of the public sphere in decision-making:

> Information technology can contribute to an open or a secret society. *The top elites in governments and major corporations are making policies* today that will cumulatively determine the outcome. No new invention or institution was ever more directly related to the prospects of an open or a secret society. (Lowi 1981, 454)

> The future is full of promise, but there are many uncertainties about the final shape and direction of coming systems and application (. . .) recommendations must take the form of broad and flexible policy guidelines. (*Green Paper*/DOC 1973, 6)

1.20 Claim to authority by subjects of enunciation: institutionalized will to truth

'Listen to the gentleman, he's an expert.'

While all of these discursive procedures that constitute the episteme within which traditional debates on new communications technology take place will tend to ensure the authority of the speaker, these discourses also make explicit claims to authority. The main procedure used to claim authority is that of reference to a technical specialist or corporate head. The important individual is 'in the *field of truth* and has the right to say the truth' (Foucault 1971).

The following is an introduction to a speaker in an article entitled 'NATO Defence Technology Outlined':

> Dr George Heilmeier's technical background and experience in the office of the director of defence research and engineering (DDR&E) and the Defence Advanced Research Projects Agency (DARPA) have given him unusual insights into the problems of the North Atlantic Treaty Organization. His penetrating comments at a recent Air Force Avionics Laboratory award dinner distill this experience (. . .) He is now vice-president for corporate research and engineering at Texas

Instruments Inc. (*Aviation Weekly and Space Technology*, 17 July 1978, 62)

What better credentials could one have than this bureaucratic, military-industrial grooming? The authority is invested exactly where the procedures of knowledge have said it shall remain for the rest of the 'technocratic' age.

TV ads about new communications technology also rely on the use of a voice for authority. As opposed to soap ads where we might find a dialogue between two housewives (although often even their conclusions are ratified by some stern, handsome male figure such as Mr Glad), ads for new communications technology are narrated by an often visually present, middle-aged, authoritative male figure – the episteme of the responsible businessman. The Bendix ad featured the paternalistic narrative of Cliff Robertson who tells us what 'we know.' Another familiar guest in our living rooms is the orchestrating, apple-crunching, crew-cut Mr CNCP speaking in his undeniable, clipped phrases about how efficient your office will be made by his expert team.

Professionalism, the fetishism of the technical expert, is one discursive trait that lays claim to an exclusive power to say the truth. Almost all of the texts that we examined were infiltrated with references to engineers, bureaucratic officials or corporate heads: 'Gordon Thompson, a researcher and sort of one-man think-tank at the Northern Electric Company of Canada.' (Stewart, in *Maclean's,* December 1968) 'Sacred Electronics' asserts its authority by quoting T. Watson, president of IBM. Mumford's book on technology is filled with inventors' names which lend 'credence' to the status of what Mumford himself is saying about technology. Bernard Ostry speaks for the Canadian Department of Communications, and so on.

In most cases, including the two anthologies published by MIT (Forester 1981; Moses and Dertouzos 1980), the authors form an interdiscursive consolidation of the discourse of 'experts' drawn from the military, engineering and bureaucratic sectors of discursive societies. The discourses constantly quote each other, refer to each other, and exclude non-members from their discursive 'cartel.' Winograd quotes Simon who is quoted by Boden, and so on.

Where professionalism or technical speciality is not the criterion of authority it is replaced by State representation (UNESCO 1981) and connection with revered institutions of learning. In an article entitled 'Oppenheimer Calls Computer Just as Vital to Science as Invention of Microscope' (*New York Times,* April 1953), the credentials of authority are first served up on the institutional tray:

> Dr Oppenheimer, director of the Institute for advanced Study in Princeton, New Jersey, was head of the Manhattan Project in World War II.

But, if the author must constantly quote other sources to lend authority to his statements, how does he derive his own authority? In texts addressed to an audience larger than a handful of specialists, the speaker's authority seems to derive from his capacity to mediate between at least two spheres of discourse, the technically specialized and the common public sphere. The lexicon of many texts indicates such mobility by virtue of simultaneous juxtaposition of highly technical – 'cathode ray tube' – and highly vulgarized – 'tv' – terminology. These Janus-faced texts delimit two spheres of discourse, one that the public has access to and one that it cannot broach without the authority of a translator, i.e., the author of the text. Such mediation is usually called scientific vulgarization. The Baer 'state of the art' text (in Robinson 1978) as well as articles in *Ms.* and *Le Devoir* all resort to this technique, though the champion of scientific vulgarization is undoubtedly *Gutenberg Two* (Godfrey and Parkhill 1980).

As the preceding section demonstrated, authors also claim a certain degree of authority by claiming to represent public opinion. Dertouzos (1980, Preface) admits that he chose not to canvas the public but is confident that the public shares the points of view expressed by the authors he has selected.

Another consequence of this shifting from specialized and exclusive to familiar and inclusive spheres of discourse is to ease, disarm and familiarize the reader, hence reducing the risk of alienation. Slippage back and forth between spheres also attempts to deal with the new and to relate it to the old.

The mobility between spheres of discourse is a certain claim to authority, and most of this claim derives from an interdiscursive hierarchy between various types of discourses. The field of technical expertise seems to come out on top, a fact borne out by President Reagan's eagerness to be photographed on site at the

NASA landing strip of the Columbia space shuttle, with the astronauts and the shuttle forming part of the backdrop (July 1982). He won more publicity points for this than did NASA. The hierarchy of technical specialization is so firm that we even find journalists abdicating their right to speak 'objectively' and 'scientifically' about new communications technology:

> As this article unfolds, I seem to be drifting into a pessimistic bias, the bias of a non-technical person lost in a technical jungle (Stewart, in *Maclean's,* December 1968).

1.20.1 The controlling voice: reporting vs reported discourse

If you simply quote a discourse, leaving it intact within its quotation marks, there is a limited degree of control that you can exercise over it, with the exception of a slight recontextualization. However, if you take someone else's discourse out of the quotation marks, reporting it directly within and adding elements of your own discourse at the same time, then you have a great deal of control over the meaning of the reported discourse. The voice of the reporting discourse infiltrates, subsumes and modifies the subsumed, reported discourse to suit its own purposes (Bakhtin 1977, 161–220).

For example, 'Sacred Electronics' uses the third person, omniscient narrator and 'indirect speech' to quote other voices while infiltrating them with the narrator's own values and positions: 'man put his electronic brains to work for the glory of God.' Here, the narrator, perhaps with a touch of irony, is infusing his 'objective' reporting with both the Christian values affirmed by the discourse of the priests and with the work ethic and progressivism affirmed by the discourse of American know-how. The constant references to *1984* and to *Brave New World,* which very few texts seem able to resist, is clearly a case of controlling the discourse over these two novels. Most of the host discourses do not seem to have read these texts and yet quote them in order to insert the concept of paranoia and pessimism which they will attack in the same breath while leaving no room for the reported discourses to reply: 'The brave new world of television has decided that the Sacco-Vanzetti case, now thirty-six years old, is too controversial for present-day viewing.' ('Our Gray, 21-Inch Lives,' *The Nation,* December 1956) Often we read about those 'defunct predictions of another *1984*-type scenario,' references which insert the concept purely in order to defuse it as trite, fictional and paranoid. The authorial voice claims authority and exercises it over another discourse by virtue of the way in which that discourse is allowed to exist within the reporting discourse.

Advertising has also used this discursive infiltration technique in order to control values and concept. As early as 1919 an ad for Benjamin electrical industrial lighting (*Scientific American,* 12 July 1919) uses this technique:

> Benjamin Industrial Lighting Helps to Make Happier Healthier Workmen.
> This Scientifically correct method of factory illumination reduces the physical and mental fatigue that results from eye strain. It keeps workmen in better spirits. It makes them better husbands and better fathers as well as better workmen.

The two discourses being infiltrated here are that of the factory owner who requires better productivity and the middle American view of the family which seeks material and physical well-being. The voice that incorporates these other voices is obviously that of the lighting salesman. By indirectly citing these other discourses, he manages to portray the message that science, profitability, and well-being are united in the embodiment of new technology, which should not be feared but purchased. Taking the analysis a step further, however, we can ask if the voices cited are the same as those of the ad's addressees. It is quite unlikely that *Scientific American* catered to an audience of blue collar workers. If the ad were addressed to executives, a more specifically-targeted advertising campaign would have been used, one sending salesmen directly into the industry. The reader of *Scientific American* is the upper middle class American, with above normal education. He is Mr Public Opinion. The admixture of discourses here addressed to Mr Middle America is not intended to sell anything but rather to have the pragmatic effect of creating a 'good feeling in the minds of Americans about technology in society.'

The text of a more recent advertisement for Fujitsu in *Japan Echo* adds another dimension to this good feeling. Once again, it is unlikely that the readers have the means to install computer-communications systems. Thus they do not have the status of potential customers. However, the readers are addressed as persons who may feel new technology to be a threat that could take away their jobs or render their skills obsolete. Hence the good feeling that Fujitsu creates is one infiltrated by this fear and a more authoritative voice that allays it. In this case, the discourse reported is based on the *presupposition* of this anxiety:

> Not so long ago, about the only person who could operate a computer was a highly trained expert. Today, operating a computer is quickly approaching the level of child's play (. . .) We at Fujitsu spend most of our time in tomorrow. It's an exciting place, an untapped country of heretofore undreamed potential.

A discourse may also obtain authority and power over the receiver by means of gratification. He is presented with images that appeal to the right hemisphere. He is told a story. He is asked to perform a simple puzzle like reading difficultly spaced print or answering a riddle, and then receives a pat on the back. The one sure tactic of the voice, however, is to avoid the past and to reveal very little, if any, new information. This would be to make his authority accountable and to oblige him to share it. All discourses are characterized by interdiscursive control techniques. No discourse is uncontrolled or unregulated. The question is rather one of which societies of discourse are in control at the expense of which others. Which societies of discourse control the right to be right? And it is here that discursive control and social power as well as the speaker's authority reside.

1.21 Trivialization of new communications technology and its functions – depoliticization

'Vive homo ludens.'

To paraphrase Oscar Wilde, there is nothing so essential as the trivial, nor so trivial as the essential. And nothing more trivialized than new communications technology. Examples of trivialization could be given almost endlessly. News coverage of the multi-million dollar video game phenomenon reiterates the pitch. 'Gronk. Flash. Zap. Video Games are Blitzing the World.' (*Time,* 18 January 1982)

The trivialization of new communications technology is achieved by means of a certain intensified hype about all of technology's uses and flashing forms. A certain familiarization of the new technology is prevalent also. There is an aura of the banal in some of the texts on new communications technology. As mentioned in the section dealing with fictive narrative scenario, a 'new object' is recounted within a familiar surrounding as a sort of disarming tactic, one which allays fears and stimulates appetite. Dagenais describes new communications technology as 'not very spectacular,' 'not very mysterious,' and 'like your tv set.'

The new gadget is always defined and described in relation to something familiar. In 1940 the calculating machine was known but not the electric one, hence 'The Electric Calculating Machine' (*Science News Letter,* 14 September 1940). In the 1950s new techniques of assembling television sets are referred to as 'Tinker Toy TV' (*Newsweek* 1954).

Ms. suggests that we will be able to talk to our toasters and our oven doors. Toffler suggests that technology will bestow upon mankind a second stomach so that he can eat grass. Machines will be taught to recognize voice so that we can play 'out-loud scrabble.' A recent observation of the activities of an Apple computer club has convinced us that most of the members were especially keen on computer games.[22] Artificial intelligence is building machines that think like humans in elementary decision-making processes, though of course no one questions the need for such machines in light of the fact that humans are perfectly capable of carrying out these elementary decision-making processes themselves. TV ads inform us that our new RCA converter will allow us to receive over 100 channels (CTV, Winter 1982). We can watch hundreds of situation comedies but only one at a time. A boy uses the new videotext to read *Fanny Hill* on his display screen. Computerization in industry makes made-to-measure alligator shoes. You can send a letter electronically and receive your junk mail in similar fashion. These are just a few of the magnificent, awe-inspiring things that we can do with this new technology. The technologically sophisticated is ceaselessly reinscribed within the banal and the mundane. The purpose of this trivialization appears to be to achieve depoliticization, to remove the object from the sphere of public debate by showing that it is both already familiar and already everywhere.

CHAPTER 2

New communications technology as discourse: techne

> Man's very perception of the world about him is programmed by the language he speaks, just as a computer is programmed. Like the computer, man's mind will register and structure external reality only in accordance with the program. Since two languages often program the same class of events quite differently, no belief or philosophical system should be considered apart from language. (Hall 1966, 1–2)

Unlike madness in the seventeenth century, technology is not reduced to silence. It speaks its own reasoning.

Michel Foucault describes Bentham's panopticon as a technology which is, simultaneously, a design for a building, the concretization of that building, an institution, a form of social organization, and finally a pattern for scientific discourse:

> at the periphery, an annular building; at the centre, a tower; this tower is pierced with wide windows that open onto the inner side of the ring; the peripheric building is divided into cells, each of which extends the whole width of the building; they have two windows, one on the inside, corresponding to the windows of the tower; the other, on the outside, allows the light to cross the cell from one end to the other. All that is needed, then, is to place a supervisor in a central tower and to shut up in each cell a madman, a patient, a condemned man, a worker or a schoolboy. By the effect of backlighting, one can observe from the tower, standing out precisely against the light, the small captive shadows in the cells of the periphery. They are like so many cages, so many small theatres, in which each actor is alone, perfectly individualized and constantly visible. (Foucault, in Lentricchia 1982, 65)

Telematics is an electronic networked computer-communications system architecturally designed for individual users to negotiate their information with and from a central data banking facility which maintains a record of their information use. The individual, or individualized receiver, has no link with other users, and enjoys only partial knowledge of what is in the central facility. Governments which have invested vast amounts into the development of telematics systems such as Telidon, Antiope, Prestel, and HI-OVIS hope that this architectural design for telecommunications hook-ups will be incorporated into homes, schools, hospitals, farms, police forces, in short, into all of society's institutions.

Telematics networking patterns look somewhat like the architectural design for the panopticon:

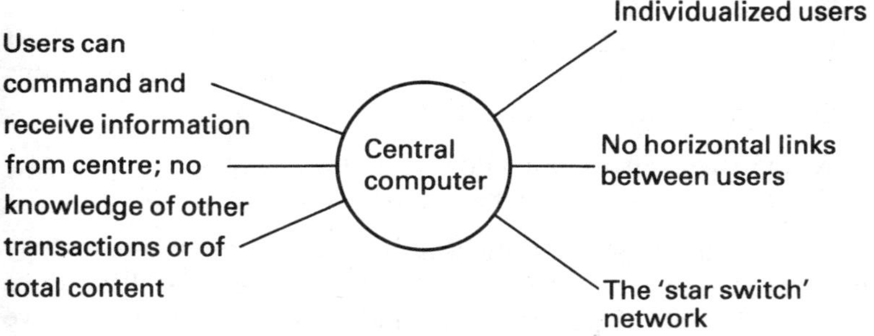

Is the panopticon, then, an architectural design, an institutional system, a form of social organization, a discourse of science, a kind of information theory, an electronics design, or all of these at once?

If we now think of new communications technology as a form of discourse rather than simply as some kind of machine-object, then we can put the same questions to it as we put to the discourses on new communications technology in the preceding chapter.

We can no longer be content to say that technology is inherently neutral or inherently valorized. We must examine the contextually specific practices of new communications technology and the generalized patterns that arise from them. By taking new communications technology itself as a discourse and by examining its procedures, we will try to discover if there are consistencies and differences between the procedures of new communications technology as discourse and the procedures of discourses about new communications technology. In short, do the discourses *on* new communications technology and new communications technology *as* discourse share the same procedures? And, consequently, do they belong to the same episteme?

Weizenbaum, for one, recognizes that the design/structure of technology works with the context for which it was intended and, consequently, that it cannot simply adopt another set of practices or uses once it has been designed:

> A tool gains its power from the fact that it permits certain actions and not others. For example, a hammer has to be rigid. It can, therefore, not be used as a rope. There can be no such thing as a general-purpose tool, just as there can be no general-purpose ends. (1976, 37)

He also draws an analogy between words and technology, which is very pertinent to our treatment of both new communications technology and texts about it as discourses:

> A language is also a tool, and it also, indeed, very strongly, shapes what the world looks like to its user. (1976, 102)

In order to know what effect new communications technology will have on society, whether it will lead to the centralization or decentralization of power, what kind of knowledge it will give society, how it will affect our perceptual and cognitive patterns, etc., we will have to begin by looking at the procedures or rules of discourse that characterize actual, specific discourse practices of new communications technology as manifest in society.

By discourse practices of new communications technology, we mean the procedures of both hardware and software, though we recognize the increasing difficulty in distinguishing between the two:

> I think it becomes increasingly difficult today to talk about technology as hardware because the boundary between hardware and software is becoming fuzzier and fuzzier. By software I mean not only software in the conventional sense that is used in connection with computers, but also management science, operations research, organization theory and applications of those things. (Brooks, in *Daedalus* 1980, 10–11)

The single most fundamental assumption of this discourse analysis is that communications technology is determinate in its nature as contextualized forms of knowledge and communications and that these forms can be expressed as discursive procedures. The eventual pertinence of this investigation lies in the subsequent attempt to demonstrate how social impact and uses are inherent within these discursive procedures and how intervention to change the latter would have to address itself to the former.

Many people ask what a socially beneficial videotex system would be or suggest that videotex systems will increase our quality of life by providing educational services, housing regulation, employment systems, etc.[1] In what follows several of these new communications technology 'services' will be explored in terms of the discursive procedures that make them possible. The main question that must be put to them concerns whether or not their discursive procedures, i.e., the conditions of possibility of their services, really increase, philanthropically, the quality of our lives or simply reinforce the power, legitimation, and control of groups that are less than publically-spirited.

One other question needs clarification. Objections might be raised because we have not singled out and described in great detail any particular new communicational object or machine. Our decision not to organize this chapter along the lines of particular machines, including a section on home videotex systems, satellite, word processors, and so on, might also seem contentious. Our reasons, however, are quite precise. To begin with, it would be a reversion to a referentialist epistemology which would accord importance to the 'reality' of the 'machine-object.' Secondly, we can only insist once more that what interests us here are the discursive procedures of an episteme. These procedures cut across the boundaries not only of disciplines but also of particular machines. For these reasons, we have chosen our examples from a wide gamut of new communications technologies and services.

2.0 Surveillance: the 'hybrid' net closes in – increased memory and cross-referencing capacity/order

The form of new communications technology known as 'telematics' is a hybrid of telecommunications hook-ups and computer technology. Among the many intended functions of computers, data banking or extensive record-keeping is only one. Governments and credit agencies alike tend to use computers designed primarily for this purpose. The data banking capacity of computers allow these institutions to maintain ever growing pools of information, usually about their citizens and clients. Such data banking or memory and record-keeping functions have been a source of concern to citizen groups for over twenty years, though, with the exception of the Scandinavian countries and Germany's 'Land Hessen Akt,' there exists little protective legislation.[2]

2.1 Compu-nications

If to this data banking capacity is added the telecommunicational aspect, as airlines (SABRE) and the Pentagon (SAGE) did over thirty years ago, then we have not only a great storage capacity but also a tremendous cross-referencing capacity. This hybrid of computer and telecommunications has been termed 'télématique' (Nora and Minc 1980) or 'compu-nications.' Not only can departments of public health maintain vast amounts of data concerning citizens but so can credit agencies as well as insurance or advertising companies. Furthermore, by means of telecommunications hook-ups, the potential exists for these various data banks to be cross-referenced or even stored together in a common bank. Qualitatively, cross-referencing has always existed as regards credit ratings or the use of educational records by prospective employers, etc. Quantitatively, however, the potential for cross-referencing increases one hundred-fold with the increasing telematization of society. For example, it is not uncommon, upon receiving a credit card, to also receive a computer print-out informing one that use of the card signifies acceptance of having any information concerning one's financial dealings made available to other credit agencies.

The possibility naturally arises of the same institutions having access to a wide variety of information about individual citizens and social groups. Such wide access would perhaps entail a qualitative change in the very nature of information. Information changes qualitatively when to isolated bits are substituted cross-referenced banks of information concerning one's health, financial dealings, education, work habits, leisure activities, information consumption, etc. Herein lies the tightening of the web.

Our main concern here lies not necessarily with the individual and his privacy, even though our society clearly values this sacred cow above all others, but with a whole social configuration in which certain societies of discourse are allowed to cross-reference extensively information concerning the activities of other discursive societies. Not only does it become increasingly difficult for the individual to escape the web of surveillance but we also find two classes of society arising: that which yields information about itself, and that which gathers and associates the information. The essential question here is not addressed merely to the individual but rather to a form of 'public management' made possible by the existence of vast interconnected reservoirs of knowledge which can be made to yield group profiles. One might wish to refer to this information as a new form of control of social reproductive forces, as 'Bureaucratic Capital.' Futhermore, it is precisely the individualist perspective that disarms the class view by stating that 'each man is an atom unto himself and that if he has nothing to hide he should not fear increased surveillance capabilities.'[3] One thing is certain, however, as Foucault has shown of all philanthropic enterprises since the seventeenth century: while telematics services such as instant electronic banking and remote area health services may appear philanthropic, they inevitably increase the amount and extent of information that institutions have about citizens.

2.2 Discursive cartels: totalizing integration of spheres of discourse

The specificity of the computer-communications hybrid lies in its capacity to integrate or link various spheres of discourse, to form *cartels* or data banks. If we were to consider government census information as one sphere of discourse, and information concerning educational performance, medical health records, commercial consumption, etc., as other spheres, then computer-communications has the networking capacity to transcend the boundaries of these spheres and to integrate all of them within one 'Multinet' information network. On the other hand, the opposite of integration also occurs. Just as certain discourses are integrated into a totalizing network, so too are others excluded and outcast from accepted spheres of discourse. Solidarity and exclusivity: two parallel routes to social control?

Where power may be defined as the 'codification of a continuous surveillance' (Foucault 1980, 104), it becomes obvious that telematics, precisely by increasing that codification, is playing into the hands of power for certain discourse users.

As the following quotations evidence, much new communications technology combines computers and telecommunications which are specifically designed for increased surveillance via cross-referenced record-keeping. Weizenbaum suggests that there is no rationale other than surveillance for the creation of a computerized speech-recognition device:

> Granted that a speech-recognition machine is bound to be enormously expensive and that only government and possibly very few large corporations will therefore be able to afford it, what will they use it for? (. . .) But such a listening machine, could it be made, will make monitoring of voice communications very much easier than it is now. Perhaps the only reason that there is very little government surveillance of telephone conversations in many countries of the world is that such surveillance takes so much manpower (. . .) why should a talented computer technologist lend his support to such a project? (. .) why should my government spend 2.5 billion dollars per year on this project? (1976, 271–2)

A further example is provided when the federal Minister of Communications proudly announces the federal government's contribution to the development of computer-communications hook-ups for police cars so that instant information about drivers can be obtained in situ (Fox/DOC 1979-80, 14).

Lowi points out that much computerized software is surveillance-oriented in its design and that it is also very costly and time-consuming to maintain:

> Surveillance pre-determined by Software Information must be kept up to date and credible, and credibility and currency require continual surveillance, occasional house-cleaning, and regular viability checks. (1981, 456)

What is especially interesting in Lowi's remarks is that surveillance is said to require and to breed yet more surveillance. The increasing need for more self-surveillance has long been recognized in psychoanalysis as one of the defining characteristics of neurosis. Perhaps surveillance-oriented and propagating technology will give birth to neurosis on a social and State level. In any case, there can be no doubt that research and development on computer-communications technology for surveillance purposes continues.

This cross-referenced surveillance capacity ties up with the previously mentioned discursive procedures of panopticism, analysis and order. Order, because cross-referencing is continuously involved in listing, organizing, and associating various givens with other givens, which have been atomized and compartmentalized for cross-referencing by the procedure of analysis.

2.3 Analysis/atomism

Dreyfus' main critique of the limitations of knowledge representation by artificial intelligence computer technology is fundamentally a critique of the classical discursive procedure of analysis, that is to say of the belief that knowledge of the world can be acquired by breaking it down into its elementary particles. Most computer programs reduce understanding tasks to a listing of their so-called elementary, atomistic concepts. Dreyfus quoted Simon's description of the SHRDLU program as ample evidence of the analytical knowledge procedures of computer technology:

> The SHRDLU system deals with problems in a single blocks world, with a fixed representation. When it is instructed to 'pick up a big red block,' it needs only to associate the term 'pick up' with a procedure for carrying out that process; identify, by applying appropriate tests associated with 'big,' 'red,' and 'block,' the argument for the procedure and use its problem-solving capacities to carry out the procedure. (Simon, in Dreyfus 1972–79, 12, note 23)

The fundamental problem with analysis or atomism as a discursive procedure is that it ignores the larger context within which it is considered to be a procedure of knowledge. Atomism is an abstraction from larger social contexts of knowledge and is, hence, an a priori, absolute principle of knowledge which operates as a first principle that can only be accepted on faith. It ignores many aspects of knowledge processes as Kuhn also suggests:

> Why is the *concrete* scientific achievement, as a locus of professional commitment, prior to the various concepts, laws, theories, and points of view that may be abstracted from it? In what sense is the shared paradigm a fundamental unit for the student of scientific development, a unit that *cannot* be fully reduced to logically *atomic components* which might function in its stead? (1970, 11)

2.4 Cordoning off pathways: constraints of accessing and association patterns – order

> There is great unawareness as regards the actual constraints of software systems. It is time to think about this, to expose these constraints and to suggest how they can be circumvented. (Steele 1980, 47)

The following are some of the ways in which computer-communications currently allow information to be accessed, and according to which information is collected and stored: 'keyword searches,' 'tree structure,' 'master index,' 'host computer central,' 'possible windows,' and 'third party handshakes.' Handshakes occur when magnetic tape is transferred from one computer to another: it is also known as a 'dump' (Kurchak 1981, 9). In the tree logic system the information provider determines the paths and correlations of information accessing that the user may follow, and excludes others. As anyone doing computerized bibliographical research can attest, it is more restrictive of the paths and associations it enables than is a library cataloguing system (Kurchak 1981, 10). Master index and host computer central are also accessing systems entirely within the hands of the information provider and programmer.

It is certainly the case, perhaps with the exception of code breaking ('hacking') and computer fraud, that he who controls the accessing patterns of the software can indeed control all of the information entered, received and released by the system. While it is also the case that very nuanced and flexible patterns of accessing may be made available, including windows to other data banks, the major information providers in Canada, Infomart and Vispac, have tended to concentrate on a strategic choice of information and on patterns of accessing that are mass-marketable, and consequently, very rudimentary. As Arrow explains, even though the information may be in the data banks, without a more sophisticated accessing system the seeker of anything more complicated than grain prices or the yellow pages will be helpless:

> Thus, being in the lobby of the Library of Congress, a few feet away from much of the world's written knowledge, is of almost no value to us unless we also have methods of locating and collecting all of the information relevant to our purpose. The same is true of a large information network, whose utility will largely be limited by the evolution of the technology for accessing information. (1963, 317)

Recent publicity for the Grass Roots Telidon program further exemplifies the pre-programmed rigidity of accessing. Not only accessing patterns but 'content management' as well enforces preconceived, rigid information gathering restrictions. Content management is predetermined by centralized technologically – and commercially – oriented interests in the choice and mode of presentation of information, a sort of 'electronic gate-keeping,' as Lewin used this term to speak of newspaper editors (White 1950, 383–90). As the contributors to Infomart and Vispac already indicate, the types of information provision and accessing technology are beginning to lend themselves far more to commercially interested presentation of stock reports and commodities pricing than to furnishing facsimiles of Middle Ages manuscripts. The VISTA *Content Applications Plan* (6 March 1980) and Rex Winsbury's study, *The Electronic Bookstall* (1979) point this out quite clearly. A listing of the members of Canada's information provision organizations also demonstrates a predominance of commercially interested organizations and the traditional news providers, e.g., Torstar and Southam. From this it is difficult to conceive of how our information accessing is going to provide us with anything more sophisticated than the yellow pages and newspapers already offer. It is not that stock reports are any less valuable than the Library of Congress but that, at present, there is no choice for information users since access is given to the former but not to the latter.

Decisions concerning the strategic choice of data to be banked and of patterns of access are known as 'content management.' 'Content management' is a form of discursive constraint deciding not only what may and what may not be said or accessed but also who may and who may not access it. It can choose to include or exclude whole fields of discourse. From the descriptions of the functions of content management below it is easy to see how rules of exclusivity and claims to truth function in direct relation to rules and designs of accessing patterns:

> Content management involves attracting information providers and determining training policies; setting terminal allocation policies and choosing between different types of user terminals; deciding upon index structure and design and formulating policy for formating pages; and devising a policy or plan for content. (Kurchak 1981, 39)

It is quite certain that not only intra- and inter- but also extra-discursive relations enter into play in the game of access and exclusivity, given that, for example, the fee to become a voting member of Vispac ($1000) is prohibitively higher than to be a non-voting member (Kurchak 1981).

Given that the strategic choice of content and the paths of accessing are so inextricably linked, the separation between content and carrier which was the central element of past policy-making, is no longer tenable. OECD has also contended that policy can no longer rely on this separation, even in definition (1979 and 1980). Given the technical impossibility of this separation, it must be admitted that any information carrier is also a provider and vice versa. If policy wishes to exercise some control over content in, for example, the name of Canadian content regulations, it will first and foremost have to legislate the very design of the accessing structure of computer-communications.

One might close this section with two questions: who has never experienced the joy of finding something for which they were not looking simply by browsing through the shelves in a part of the library not indicated by the master card index? This would become quite impossible with electronic accessing. As computerized bibliographical searches indicate, one usually acquires a huge quantity of information which is too vast to be usable and must subsequently be broken down 'manually.' One often misses out on vital documents because one's subject associations do not fit the tree accessing pattern and one often ends up browsing through the shelves anyway. Secondly, would Einstein have discovered the theory of relativity had he had to rely on some centralized text book data banking company for all his information on physics? As Mattelart had documented so well, this is precisely what is occurring today. Publishing houses such as Rand and Macmillan first monopolized the educational publishing industry and are now investing huge sums in computerized electronic publishing (Mattelart 79, 277–85).

2.5 Closed/static language vs open-ended/dynamic language: how much can a computer move?

Computer-communications will lead to more dynamic communication because information will have no material form but will be electronic and will thus be updatable. Such is Gassman's hypothesis:

> New information infrastructures which will permit the transition from the present essentially *static* information (such as represented in books, journals, printed matter) to *dynamic* information in electronic form, which is increasingly needed to master the intricacies of the modern world and the rapid changes with which we are confronted today. (1980, 11–12)

Our reply to Gassman is, once again, to call for contextual specificity. Exactly what aspect or component of information is rendered more dynamic by new communications technology? Over-generalized terms such as 'static,' 'dynamic,' 'centralized,' and 'decentralized,' mean nothing out of the context of concrete communications practices.

There is no question that the graphics and updating capacities of videotex are more 'dynamic' than those of written texts. Videotex information has a significant potential for alterations and for actual physical mobility. However, whether this is actually used remains to be seen. Data banking could remain as static as our present-day yellow pages if the public accepted it and the corporations found it to be financially convenient. Any number of reasons may prevent data banking systems being designed with built-in multi-channelled update systems.

But there is also a further consideration. Dynamics refers not only to chronology and graphics. It can also qualify the accessing pathways, the types of content, the filters and funnels through which information passes, in short, the discursive procedures themselves. The question of dynamism or statis becomes pertinent when it refers to whether or not new communications technology actually dynamizes the discursive procedures of communication. Will the accessing pathways, the possibilities for combination and association of content, the cognitive assumptions of perspective, voice of enunciation, and themes of mediation become more elastic or will they harden? The degree of dynamism of discursive procedures depends a great deal on the degree to which the technical design of systems is centralized. Where the medium of information storage and retrieval is always the same and always controlled by the same system, as in 'telematics,' great redundancy not only of content but also of access, storage, association and retrieval patterns and pathways is a risk because of the *monovalent* or 'monophonic' nature of the sources and perspectives involved in programming and designing. Where programs and data are mass-produced by large companies, we can expect not only content but also various tree accessing routes and association patterns to be static.

Earlier we argued against a reliance upon the categories and models used by mass media analysts for discussing new communications technology. Still, if television is any indicator at all of what electronic media designed for mass distribution do to communication, we should expect the telematization of all communications transactions in an infrastructure of major monopolistic data processing, storing and networking institutions to lead to what Raymond Williams has called a 'hardening of formula,' i.e., static communication:

> In actual content and presentation now, the formulas seem to be hardening: 'the masses' – crime, sex, sports, personalities, entertainment, pictures, 'the minority' – traditional politics, traditional arts, briefings on popular trends. It is then a matter for argument whether 'the masses' and 'the minority' are

> inevitable social facts, or whether they are communication models which, in part, create and reinforce the situation they apparently describe. (1974–78, 103)

A glance at some of the Telidon publicity certainly reveals some hardened formulas, i.e., commercialism and surveillance as primary discursive procedures. But, then again, the sum of the very discursive procedures discussed in this chapter also amounts to what we consider to be a sort of hardening of formulas although we would have to reconfirm the degree of regularity and rigidity of these 'hardened' procedures over a period of time.

A further hardened category has to be the 'recreational' function of new communications technology, so much of it being designed and touted as a frivolous pastime. But, as we will discuss in greater detail, perhaps the most hardened and recurrent formula of all is the military rationale behind the design and development of new communications technology: surveillance, control, and destruction. Also, it remains to be seen whether or not a finalized version of videotex will continue to maintain the hardened formulas of 'entertainment,' 'international politics,' 'life styles,' etc, such as are found in the press. Williams gives us a clue here in suggesting that in broadcasting there is:

> an increasingly close connection between the methods and content of advertising and editorial material, and second the marked division of material into classes, which then normally keep to their own world. (1974–78, 99)

What we do know from Carey's study on the professional communicator is that, since the coming of the wire services and the increasing centralization of both government and news chains, sources have acquired increasing control over the dissemination of information, leading to the hardening of formulas. Innis would call this an effect of 'space-binding' interference of new communications technology such as the telegraph. And Carey adds:

> the net effect of the press conference, the background interview, the rules governing anonymous disclosure and attribution of sources, and particularly the growing use of public information officers within government is to routinize the reporter's function and to grant to the sources exceptional control over news dissemination. (1969, 33)

It has, in turn, been suggested that videotex amounts to a public information utility whose degree of centralization will determine the degree of control over dissemination patterns and over formula of content (Bakcsy 1979, 13).

As it now stands, we have seen that the discourses on new communications technology tend to be very redundant, to 'over-code' their message by relying on cliché and slogan: 'We'll all be creators.' What remains to be seen is whether videotex systems will turn all teachers, doctors, social workers, etc., into nothing more than what Carey calls information brokers and purveyors of redundant formulas.

In Dreyfus' discussion of new computer technology as well as in the quotations from artifical intelligence experts, a tremendous accent is placed on *memory* and *script programming.* This is nothing more than filling the computer memory with hardened formulas, be they single terms, such as the definition of a triangle, or whole narrative scenarios, such as what to do when you go to a restaurant. Even the treatment of context relies upon filling the computer bank with a collection of givens about the background, which the computer does not manage to change or manipulate in a dynamic, relativist or event-specific way.

> But Schank claims that he can use this approach to understand stories about *actual* restaurant-going – that in effect he can treat the sub-world of restaurant-going as if it were an isolated micro-world. To do this, however, he must artificially limit the possibilities; for, as one might suspect, no matter how *stereotyped,* going to the restaurant is not a self-contained but a highly variable set of behaviors which open out into the rest of human activity. What 'normally' happens when one goes to a restaurant can be reselected and formalized by the programmer as default assignments, but the background has been left out so that a program using *such a script* cannot be said to understand going to a restaurant at all. This can easily be seen by imagining a situation that *deviates from the norm.* (Dreyfus 1972–79, 42, emphasis added)

In other words, the computer is too static or over-coded in its interactive communications patterns to improvise, even in the most rudimentary everyday human situations.

Weizenbaum as well has admitted that the Eliza program is nothing more than a pattern for pat responses going no further than a mirroring of what the patient has just stated. For Weizanbaum, this is a fixed stereotype, in no way approximating the dynamics and compassion required for real, humanly interactive psychotherapy.

Yet a further encouragement of static formulas is the reaction of computer-communications to any deviance from the standard procedures of commands, accessing and associating information. Deviation is simply unfeasible because the computer refuses to function. Its obstinacy is responsible for what has been termed 'computer Angst.'[4] Computer Angst furthermore has been responsible for the trend towards the design of 'user-friendly' systems such as Apple's Lisa and MacIntosh. Deviance becomes that which goes against some conventionally established technical categories which decide what is right and what is wrong. The ambiguous, the third term, is the excluded middle. 'Wrong' is the computer's blurted out social surveillance, its sanction against anything that does not play according to its own discursive rules. Interestingly enough, any successful deviance, such as code-breaking and computer fraud, is quite threatening to the entire realm of computer-communications. So dangerous, in fact, that millions of dollars are being spent to counter it.

Another type of stasis in computer-communications involves that of the participants of the social sphere in which it operates. As Rosenblith stated 'the more technology we have the more levels of technology we have, for example, in the medical profession.' (*Daedalus* 1980, 20) With assembly lines came the division of labor. In the medical profession we now have x number of different types of technicians. New communications technology creates a demand, not yet completely filled by training, for any number of types of technical specialists whose job it is to manipulate one type of machine or improve one particular program. Kiddler's book describes such compartmentalized specialization in his portrayal of computer engineers (1982). This overspecialization in both the building and use of computers gives way to a static form of labor interaction whereby a worker is trained to function as a very specialized technician with few other skills and little reflection required. He has become one specialized function within a technical machine. His roles in work interaction and eventually in the public sphere are limited both by his training and by society's demands on him.

2.6 On-line-real-time mediation: ersatz reality

As the above title indicates, one of the claims of new communications technology is that it sets the world at our feet 'immediately.' Satellites beam instant pictures of 'how it is,' 'here and now,' onto our television screens. Airline reservations systems confirm our tickets immediately. Videotex will bring vital commercial information immediately into offices and vital news and community information immediately into our homes. Computer guided weaponry has made the destruction of war immediate.

But there is another perspective to be taken on the question of immediacy. Besides shortening the time-span between an event and its electronic representation, communications technology does still mediate an event in the sense that without it our perception of the event would be quite different or nonexistent. New communications technology interposes itself between our perception and the world. And still we are constantly being persuaded of the immediacy of it all.

New communications technology, by threatening to usurp the place of many face-to-face transactions, such as shopping, and by interposing a new language (other than everyday language) between participants is indeed an increasingly ubiquitous mediator. Weizenbaum describes this mediation by new computer-communications technology as creating a psychological distance between people and between people and events. Experiments have shown that people are far more prone to acts of inhuman cruelty when the consequences of their acts are mediated and distanced. This is the type of psychological distance that Weizenbaum suggests must be reduced by reducing the ubiquity of mediation by new communications technology:

> The scientist and technologist must, by acts of will and of imagination, actively strive to reduce such psychological distances, to counter the forces that tend to remove him from the consequences of his actions. (1976, 276)

Interestingly enough, as Galtung (1970, 259ff) has shown with regard to the mediations of the news media, in a strange epistemological and ontological twist of fate, it is possible to conceive of new communications technology as actually 'producing' the realities which they claim to represent immediately. For example, during the 1980 Iranian hostage crisis the presence of cameras might be said to have actually provoked – co-constituted – the 'fanatical' demonstrations. But what happens if mediation and its accompanying myth of immediacy become so ubiquitous that distinguishing between immediate and mediate is no longer possible? A deadening of one's awareness of the impact of one's own actions upon others seems to be the result. Mattelart suggests that that is precisely what has happened in the use of telecommunications in weaponry. The weapons bear names such as 'Popeye,' already a mediation through the media, and their consequences are far removed from their use. But they are not less devastating because of it.

The question of mediation has been a social and epistemological problem throughout the history of Western thought. It began perhaps in *The Cratylus* with the questioning of the adequacy of our representations of the world. In his discourse on mass media, Walter Lippman, one of the fathers of American mainstream communication theory, spoke of the 'ersatz reality' that replaced immediate contact with reality. It does seem to be the case that new communications technology makes us rely progressively on mediation, though it be an almost instantaneous one. Nevertheless, Lippman's very critique of 'ersatz reality' presupposes the possibility of a direct and immediate knowledge of reality, something that an a-referentialist epistemology would not allow. The jargon of immediacy is itself no more than a mediated convention. It is a contradiction in terms to speak of new communications *media* and of how they render the world *im*mediate to us. All we have is an increase in the speed of mediation with a simultaneous complexification of the mediating apparatus. It is necessary to bear in mind that the claims of immediacy by and of new communications technology merely derive from an increase in their time- and space-binding capacity (Innis) which nevertheless cannot altogether eliminate electronic time lapses or collapse Peking into New York. Shortening a distance or a time span is not equivalent to identifying the opposite poles. The problems of referentiality are more obfuscated than solved by the on-line-real-time of new communications technology.

2.7 Control of the source is control of access and association

Those who know (episteme) vs those who know how (techne) may become the principal class distinction of the new communications society.

It is generally agreed that the computer software at the basis of new communications technology depends on a systems model theory of knowledge which has little to do with the procedures of traditional, everyday human knowledge, even though information theory has a great deal to do with the procedures of classical instrumentalist reason. This makes the common user of computer-communications, armed only with his traditional procedures of knowledge, both frustrated by and vulnerable to a system which functions according to other discursive procedures of knowledge. Few will understand the rationale behind the programming and Weizenbaum even suggests that programmers themselves do not really understand computer knowledge processes but only know how to function with a few rules. Kidder's description of the debugging procedures of newly built computers also insists that the system is too complex for any single man to understand or debug.

> The days of the debugging wore on. In March, West said, referring strictly to the debugging, 'Most of the fear is gone now.' He was speaking only for himself, however. The team has passed through the first sharp fear. But they had designed the machine too fast for prudence. It had features that none of the group had dealt with before. At this stage none of them dared claim to understand in detail how all the parts worked and fit together. Sufficient cause for worry about debugging remained. (1982, 151)

The user, and even the debugger of computer-communications, often resembles the child who, through trial, error and memory, manages to put the square block into the square hole without the slightest understanding of geometry. It may or may not work but few will know why or how to fix it if it stops. In this case, the source of computer-communications technology will be a very restricted society of programmers and designers who can only partially understand and change the procedures of computation. The rest will have to be content with using it, with no control over the rules of the game, i.e., over the discursive procedures.

The discrepancy between knowing and knowing how reduces knowledge to an automatic model with 'bugs that are to be eliminated.' It also leads to some erroneous and somewhat dangerous metaphorical parallels between computer and human intelligence. Weizanbaum, for example, was amazed at how seriously users took his Eliza program, since, for him, it was merely a poor *simulacrum* of human interaction.

As we will argue in the next chapter, although it is likely that no one single person or elite group controls all discourse, various societies which dictate the rules of discourse and understand its functioning are in a position to exercise knowledge and power over those who do not. In this case, societies such as content managers, program designers and systems analysts are certainly in a superior position as compared to those who simply *use* the program.

Even home computer systems like Apple and home computer link-ups like Telidon provide ready-made programs on the assumption that their users will not be able to program them. The programs come with instructions for text editing, financial accounting, and so on, which the user simply follows without needing to understand the knowledge procedures in action. Only the designer does, along with a few computer wizards who have managed feats of code-breaking and computer theft such as beguile even the designers. Marvin Denicoff reconfirms this vulnerability of a discourse user to a discourse whose procedures he does not understand while, like Kidder, suggesting that even computer builders suffer from this malaise:

> Unfortunately, today software is not based on the theoretical foundations and practical disciplines that are traditional in the established branches of engineering. No doubt, many of the problems that occur during the software development, acquisition, operation and maintenance cycles are in part symptomatic of this lack of knowledge of the fundamental nature of computer software and its life-cycle process. (1980, 384–5)

A discourse out of control? The parallels with Frankenstein are perhaps not too futuristically spectacular. They may be a forewarning of a phenomenon so current that no futurisms are needed to produce a science fiction effect.

If it is agreed that the designers and programmers, i.e., the sources of new communications and computer technology, have a certain advantage in dictating the rules of the knowledge- and language-game, does it follow that home terminals and other smaller information providers decentralize and diversify this control? Not necessarily. So far, as Kurchak's study indicates, home terminals and small information provision companies obtain their information and programs from *brokers,* which are the equivalent of major broadcasting networks in the field of computer-communications. One reason for this trend is the comparatively high cost of page creation and updating, as compared to that of mere storage (Kurchak 1981, 40). Furthermore, given the high cost of buying computer memory, the trend is for smaller information providers to store their information in a larger broker's bank, hence once again, making accessing, association, and organization (analysis and order) dependent upon the discursive patterns, interests, and perspectives of larger, centralized brokers.

Furthermore, the patterns of knowledge and access in computer-communcations software are likely to remain invariant for long periods of time. Because of the cost differential between storage and creation, even large information providers will tend to use old, commercially profitable, centrally produced, and widely distributed material as opposed to diverse, original programs and contents, of limited distribution potential. They will be the Double Days of electronic information provision. The dreams of more horizontal and diversified programming, catering to all tastes, as opposed to a reduction to the lowest common denominator, seems to be just that: a dream. Therefore, not only is there an inequity between the users and the designers in terms of their control over discursive procedures, but there is also a decrease in the more diversified and publicly involved exercise of control over discursive procedures. Here there does seem to remain a one-to-many ratio common to the transmission or broadcasting model.

2.8 'Closed user groups': the electronic incarnation of the procedure of exclusivity

We have already seen one type of control over discursive procedures enjoyed by the designers of software programs. There exists yet another type of discursive control based entirely on exclusive rights to accessing.

For example, there is already a growing trend toward the establishment of what are called 'closed user group' accessing rights. These amount to certain types of materially manifest exclusions of the public from certain spheres of discourse. This exclusion is insisted upon by traditional groups or disciplines whose power and profits depend upon exclusive access to certain types of information. For example, travel agents, notaries, and to a degree, doctors, lawyers, and stock brokers, also derive their special status as members of professions and disciplines by virtue of access to a discourse that is private and closed to the general public. These 'initiates' are very concerned with maintaining their exclusivity in and through the design of telematics networking (Kurchak 1981, 10). A Maritimes-based experiment has shown that, ideally, the closed user group could be opened up by telematics as was done for the notarizing of house deeds. That did not catch on, however, as the income of notaries depended upon exclusive access. Consequently, *encryption safeguards* are constantly being devised to ensure closure so that the entire social structure of disciplines and professions does not suffer a substantial revision. Ideally, the system could be open, but specifically, technically, and contextually, it remains exclusive and closed as is indicated by the fact that electronic publishers such as Viewtron of Viewdata Corporation (USA) are highly selective of which companies or information providers are allowed access to its information during field trials.

Here we come full circle back to Foucault's theory of discourse: discursive procedures are restrictions and constraints operating on a principle of exclusivity of access to discourses of knowledge which legitimate the knowledge and power of certain societies of discourse over others. 'Closed user groups' would seem to reconfirm the existence of a strong alliance between discourse, knowledge, and power.

The procedures of exclusion named by Foucault in 'The

Discourse on Language' (in 1972) seem to describe quite perfectly the rationale behind the 'closed user group.' *Prohibition* amounts to taboos placed upon certain discourse users to speak the truth, to know and, consequently, to govern. A patient is subject to a taboo on self-diagnosis by the medical profession. *Rarefaction* is a division and rejection of certain information at the expense of other information, i.e., a strategic choice to store some 'information' but not other information, with information meaning not only data but also programs. For example, travel agents are widely concerned that they will lose their customers if the public can access the same fare and on-line-real-time reservation systems as the professionals. This information would therefore have to be 'rarefied' from the public information bank. The 'will to truth' manifests itself when the members of the 'closed user group' want to be in the know at the expense of those outside the group. Otherwise, knowledge gives no special status.

Being in the field of truth in order to say the truth (Foucault in 1972) is another procedure relied upon by those who defend the maintenance of 'closed user groups.' The 'closed user group' is akin to a knowledge class, which grants users access to certain information which places them in a certain field of discourse and knowledge thus permitting them to say the truth about certain things, such as disease or law. Were this field opened up by ending the closure of the group's discourse, then the exclusive field of truth would no longer function quite so exclusively. 'Closed user groups' are so vital to maintaining the status quo of various professions and disciplines, including the military, that millions of dollars are being spent on the research and development of encryption devices to ensure that closed discourses remain impenetrable to would-be invaders. Efficacious encryption devices are relatively expensive and it is usually only the traditionally superior societies of discourse such as the military and corporate societies which can afford to ensure the exclusivity of their banks, whereas public groups have little funding for such assurances.

'Closed user group' telematics technology is equivalent to a private language, a secret discursive society of the initiated. Control of encryption and of the networking system of accessing ensure the privacy of language, but at a cost. Curiously enough, the 'closed user groups' which insist most on their exclusivity are precisely the same as the societies of discourse that enjoy exclusive status and prestige of both a social and a financial nature today: doctors, notaries, brokers, the military, banking and other financial institutions:

> Computerized access to vast information pools is far more likley to enhance the capacity of old groups to persist than of new groups to form (. . .) given a sufficient amount of technology private groups can in effect become 'private governments.' (Lowi 1981, 462–63)

Videotex systems designed for 'closed user groups' seem to be repeating the paradigm of discourse elaborated by Foucault and according to which the control of a discourse is not only exclusive or private, but also yields control over the subject without access and knowledge to those who have them:

> Videotex sits at the cross-over point between (. . .) *two competing private languages*. One, the computer language, speaks of a data base maintained by manual or automatic update, with operators keying in, or bulk input by magnetic transfer or computer to computer handshake, with data reformatted to fit the Prestel display mode and generate the indexes. The other, the publishing language, speaks of a magazine or information service published on videotex with up-to-date by editors and presented on Prestel pages with suitable layout and cross-references, using editing keyboards. (Winsbury, in Kurchak 1981, 9, emphasis added)

All of the above considerations suggest that with new communications technology the source rigidly determines the seeker's accessing patterns and possibilities. Not only must the receiver adopt the association schema and patterns devised by the source, not only must he learn another's language in order to express what he wants, but the source may arbitrarily, in favor of some other user, decide that he shall not have access at all.

The scant elasticity of association and of the search for information that is still permitted by discourse is fast vanishing. In electronic information provision, the system's discourse replaces browsing, intuition, and imaginative association.

> Excess and fuzziness are incompatible with the medium and an IP must know how to define and route his or her information

> very clearly. Similarly, because Telidon is not a browsing medium it assumes the user knows more or less exactly what he or she wants. (Kurchak 1981, 64)

Woe betide the researcher in the days when computer-communications systems will have fully replaced the more elastic archival and library systems of today. This is not to say, of course, that library and cataloguing procedures do not also determine access and restrict associations. They do, but in these older systems, it seems to be the very lacunae that allow one to skip, ferret amongst certain unprescribed paths, and profit from the element of chance that still provides a source of joy for the information seeker. Try to skip or invent a step or a path in computerized accessing and the main reply blurted out is : 'ERROR - TRY AGAIN.' The constraints of tree logic are perhaps not qualitatively different from traditional discursive constraints, simply less avoidable, more water tight. If one does not follow them it is impossible to function within that medium, whereas one can always walk around a bookshelf. Perhaps the net has simply been tightened.

2.9 Reduction of knowledge: data base as knowledge base

In their SHRDLU program, Papert and Minsky try to build a knowledge base that would constitute the micro-world that a child would need to know in order to understand the following conversation fragment: 'Janet: "That isn't a very good ball you have. Give it to me and I'll give you my lollipop".' They go on to suggest that the computer memory must contain knowledge of what talking, social relations, playing, owning, eating, liking, living, etc., are (Papert and Minsky, in Dreyfus 1972–79, 10–11, note 18). Dreyfus criticizes this conception of knowledge although it is still the one used by most programs, with the exception of the Winograd work on fifth generation computers. According to Dreyfus, this representation of knowledge is fundamentally atomistic (analytical) in that it conceives of a knowledge base as a collection of itemized or atomized pieces of information which are separable from each other as well as from the larger context of knowledge of the world possessed by the average intelligent human being. It is also divorced from thought processes or critical learning capacity such as how to determine which situation is in play and which contextual meaning applies in which case. Hence, in most cases when computer specialists speak of the computer's 'capacity to learn,' what is meant is a capacity to absorb and categorize increasing amounts of data and not necessarily to become more adaptable in the handling or pertinent data and the exclusions of non-pertinent data. Even Bell admits of the difference between knowledge base and data base and of the confusions which result from trying to consider the computer's construction of a data base and the use of that base as a knowledge base.

> One is the intellectual question of the distinction between programming a data base and constructing a program for use as a knowledge base. Retrieving some census items from a data base is a simple matter; but finding kindred and analogous conceptual terms – the handling of ideas – raises all the problems that were first encountered and never successfully solved, in the effort to achieve sophisticated machine translation of languages. (Bell 1981, 529)

What does this schism between knowledge base and data base mean for the social implications of new communications technology? There has been much talk about the information explosion in society, about the complexity of society because of the exponentially increasing quantities of information that unassisted human beings simply cannot handle. Lukasiewicz in his article 'The Ignorance Explosion' (1972, 373–91) continues this trend of remarks by suggesting that the information explosion is really an 'ignorance explosion' in that the proportion of knowledge/information extant far surpasses the proportion absorbable by human beings, hence the ratio of what needs to be known or exists to be known, to what can be known, is increasing constantly. This line of reasoning is of course a perfect rationale for the installation of computer-communications in every walk of life since the computer memory would give us more 'knowledge,' where knowledge is taken as information and as data.

However, what Lukasiewicz ignores, as do those specialists who have conflated knowledge base and data base, is that knowledge has, throughout the ages been conceived of as much more than a collection of tidbits of information. Knowledge is especially a set of varying capacities manifest as a discursive capacity to handle information in a contextually pertinent way and to be able to process information critically, as opposed to just memorizing, analyzing – breaking down – and categorizing it. The fact that knowledge is more than data collection is corroborated by the failure of computers in AI experiments to understand circumstances presented to them with a background constructed only of data. As Dreyfus points out, the computer is unable to answer such simple questions as 'do we eat with our ear or our mouth?' without having this information placed in its banks. Therefore, the 'restaurant' program cannot really be said to 'understand' what going to a restaurant means. Futhermore, as Dreyfus' 1979 preface to his book illustrates, since 1972 many AI experts themselves have had to concur. Not all computer specialists have conflated knowledge and data bases. The best-known and most praiseworthy of these is Terry Winograd who has sought to devise a far more holistic conception of what understanding and knowledge are and of the functions of which computers would have to be capable in order to be knowledge practitioners. But first, a closer look at how traditional computer-communications models conceive of discourse of knowledge in relation to knowledge of the world, i.e., of context.

2.10 Unilateral control of voice and perspective

As mentioned earlier, the one conclusive notion that contemporary theories of discourse have taught us is that there is no 'zero degree' of information (Barthes 1953). That is to say that data are never completely neutral and objective; they never come from nowhere, from no point of view, from no source of bias. These same French theories (Genette 1972) have also taught us that voice – source – and perspective – mode – are language's revelation of value, belief, and bias.

Now it has been a habit of computer scientists and AI experts to claim absolute objectivity and neutrality for their science and programs. Their information has often been treated as context-free and hence neutral.

> Yet Winston's program works only if the 'student' is saved the trouble of what Charles Sanders Peirce called abduction by being 'told' a set of context-free features and relations – in this case a list of possible spatial relationships of blocks such as 'left of,' 'standing,' 'above,' and 'supported by' – from which to build up a description of an arch. (Dreyfus 1972–79, 21)

However, as even the AI experts themselves admit, AI research has remained somewhat stagnant in the last decade due to the computer's inability to handle context, value, and presupposed beliefs. This is because information always comes to us from a particular source and a particular point of view, even though discourses may try to occult these in order to claim complete objectivity. Indeed, the claim to objectivity is itself a value judgment, a certain epistemological perspective which claims power for the speaking source. It is, among other things, a refusal of other points of view and an interdiction of science to intervene in practical problems of value. Foucault, referring to Port Royal and Descartes, calls such a claim to objectivity and neutrality the classical, scientifist perspective.

It is impossible to divorce data from the perspectives and sources of their groups and originators and controllers. The claim to neutrality is nothing but pretense. Yet the perspectives and sources imbedded in discourses of new communications technology still remain to be uncovered and studied in detail. For example, when your credit card statement gives you a print-out message that you have been a good card holder and consequently that your credit limit is to be raised, you might detect not only the rationalist perspective whereby efficiency and maximum financial gain are at stake but also a materialistic perspective of punishment and reward. One must be very wary about allowing monopolization of computer-communications services precisely for the way in which bias is naturalized. One of the saviors of 'free' society has been the presence of many perspectives in the information available to citizens. Were all information to come to us from a central telematics system one would face the possibility that a single homologous voice and perspective would underpin all of our information. Perspectives would risk being entirely monopolistic and reflective of corporate, bureaucratic or commercial interests in the same way that the discursive procedure of rationalism in discourse on new communications technology was seen to reflect these same interests.

Telematics has also been suggested for computerized medical treatment. It is touted for its professional objectivity. In what follows we will quote an extensive passage from Weizenbaum's Eliza program designed to *imitate* Rogerian psychotherapy. The computer's responses are in capitals.

> Men are all alike.
> IN WHAT WAY?
> They're always bugging us about something or other.
> CAN YOU THINK OF A SPECIFIC EXAMPLE.
> Well, my boyfriend made me come here.
> YOUR BOYFRIEND MADE YOU COME HERE.
> He says I'm depressed much of the time.
> I'M SORRY TO HEAR YOU ARE DEPRESSED.
> (. . .)
> You don't argue with me.
> WHY DO YOU THINK I DON'T ARGUE WITH YOU.
> You are afraid of me.
> DOES IT PLEASE YOU TO BELIEVE I AM AFRAID OF YOU.
> My father is afraid of everybody.
> WHAT ELSE COMES TO MIND WHEN YOU THINK OF YOUR FATHER.
> Bullies.
> DOES THAT HAVE ANYTHING TO DO WITH THE FACT THAT YOUR BOYFRIEND MADE YOU COME HERE.
> (Weizenbaum 1976, 3–4)

This program, which Weizenbaum developed in order to show its limitations rather than as a genuinely proposed psychoanalytical method, met with wide enthusiasm and even with suggestions for its implementation from the psychiatric profession itself (Weizenbaum 1976, 4–5). As regards perspective, however, this computer dialogue is not an objective tool or discourse. Weizenbaum admits of two perspectives that went into the design of the program. First of all, the Rogerian perspective on analysis and on the human psyche which holds that in therapy humans merely need an interrogative reflection of their own perspectives. The underlying perspective then concerns man's needs for psychological therapy and restoration. Secondly, it follows that the perspective is also that of the patient to a degree, minus the interrogative mode. The program reflects back to the client his own perspective. Thus, the Eliza program contains at least two perspectives, a meta-perspective about what Rogerian analysis should do to restore health and a specific perspective, that of the reflected patient.

Coupled with the computer-communications world's occultation of perspective, this is often an occultation of a shift in perspective. If you do not acknowledge perspective in the first place you are not likely to acknowledge a shift in it. Before computer-communications networking, a data bank would store information about a certain topic or from a certain perspective. For example, a data bank on health care would look at patient records from the perspective of medical care. However, once medical data banks and employment record data banks, not to mention financial records banks, are cross-referenced, what we have is a shift of perspective on the information considered. Medical information is now viewable from a potential employer's perspective and as may easily be imagined, it takes on an entirely different meaning and function than when viewed from a medical perspective.

Policy-making will have to answer some weighty questions about this perspective shifting. For example, they will have to decide the boundaries within which this shifting may and may not occur. Once again, the potential for social control of the play of perspectives, both exclusively and inclusively, is significant. Not only do certain institutions have the right and the technical capacity to store large amounts of information about citizens but they now also have the telecommunicational capacity to transfer this data into other contexts or perspectives which completely alter its original significance and use.

Furthermore, this capacity to control source and perspective is not separable from the exercise of power. As anyone involved in an argument can vouch, the capacity to take another's words and reposition them from one's own perspective is an extremely potent rhetorical and argumentative tool. So is the capacity to control access to what information your opponent/partner may have and how he or she may get it. Doug Seeley, for one, concurs with the position that power arises from the very structuring of discourse in new communications technology, from the control of software and hardware processes such as accumulating, filtering, storing, structuring, networking, and placing into perspective any type of information whatsoever:

> In fact, in the overall interlapping circuits of information flow, wherever there are nodes that 1) filter and 2) structure and bias, 3) accumulate and 4) control access and provision, you have nodes of information power. (1980, 48)

2.11 Context-free vs context-bound discourses of knowledge

Early makers of computer programs which were used to informatize society, such as translation and interpretation programs, were confident that they could store in the computer data bank a sufficient body of 'denotative' or context-free meaning to facilitate certain understanding and translation patterns by the computer. Realizing the ridiculousness of a context-free discourse, where computers could not interpret shifters such as 'it,' or equivocal terms such as metaphors, later programs, such as those for chess (MYCIN and CHESS 4.5), successfully operated within contexts which were highly restricted and watertight. Chess programs could consider all possible moves by means of a tree search since only the board's coordinates had to be dealt with. Winograd himself acknowledges this contextual limitedness:

> The AI programs of the late sixties and early seventies are much too literal. They deal with meaning as if it were a structure to be built up of the bricks and mortar provided by the words, rather than as a design to be created on the sketches and hints actually present in the input. This gives then a 'brittle' character, able to deal well with tightly specified areas of meaning in an artifically formal conversation. They are correspondingly weak in dealing with natural utterances, full of bits of fragments, continual (unnoticed) metaphor, and references to much less easily formalizable areas of knowledge. (in Dreyfus 1972–79, 15)

Given that most programs cannot deal at all with context or else only with a very restricted context, it follows that computer-communications' treatment of more social information, which the 'communications society' is supposed to herald, would amount to a *contextual distortion* of information about society. Information could be restricted to watertight contexts or transferred from one context to another with no links between them, amounting to a distorting de- or re-contextualizing control over information. What is more, given that such de/re-contextualized information is used for very weighty decision-making, such as employment or even legal proceedings the importance of this contextual weakness on the part of computer-communications becomes even greater. For example, one area where there is much optimism about the implications of computer communication is in education (Godfrey and Parkhill, 1980). Not only would computerized testing completely ignore or restrict the context of each particular student's answers but the results of the testing would also be de/re-contextually interpreted. The same applies to health information, consumerist trends, and employment and financial records.

Once again, the limited contexts built into the computer-communications systems are done at the instigation of programmers under instruction from the institutions that order the design (Kidder 1981). The public sphere has little say in the design of government, educational, corporate, or employment data banking procedures, and hence exercises little control over the contextualization of information about itself. Minc expresses concern over this problem of context from a cultural angle. He suggests that the limited contextualizations of French history by the information provider *Time/Life* or *Washington Post,* would threaten cultural genocide of peoples of other lands restricted to these sources or banks of information (Minc/OECD/ 1980).

But, does hope for a solution to the contextual problem lie in the success of Winograd's programs or those of the fifth generation computers to deal with context in a far more holistic, more reflective, less atomistic and less restricted way? It is certain that Winograd's projects epistemologically revolutionize the conception of computer knowledge. Winograd begins with the assumption that knowledge and meaning cannot be context-free or even context-restricted, and that computer knowledge representation must place any word interpreted in the most holistic context possible, i.e., the context of the whole field of knowledge and critical association and exclusion patterns of social interactional beings.

> Rather, learning of specific details takes place on a background of shared practices which seem to be picked up in everyday interactions not as facts and beliefs but as bodily skills for coping with the world (. . .) A more plausible, even if in the last analysis perhaps no more promising, approach would be to use the new theoretical power of frames or stereotypes to dispense with the need to preanalyze everyday situations in terms of a set of primitive features whose *relevance is independent of context.* This approach starts with the recognition that in everyday communication 'Meaning is multi-dimensional, formalizable only in terms of the entire complex of goals and knowledge (of the world) being applied by both the producer and the understander.' (112) This knowledge, of course is assumed to be 'A body of beliefs specifically (expressed as symbol structures. . .) making up the person's model of the world.' (113) Given these assumptions Terry Winograd and his co-workers are developing a new knowledge representation language (KRL) which they hope will enable programmers to capture these beliefs in symbolic descriptions of multidimensional prototypical objects *whose relevant aspects are a function of their context.* (Dreyfus 1972–79, 47)

Once again, KRL has had a degree of success in a limited context of experimentation, namely a small field with geometrical figures to be manoeuvred. However, the main problem of the program is in developing the capacity of the computer not only to contextualize but to choose the *pertinent* context. So far, Winograd comes further in his theory than in his practice in recognizing not only contextuality but contextual pertinence and specificity:

> The results of human reasoning are *context-dependent,* the structure of memory includes not only the long-term storage organization (what do I know?) but also a current context (what is in focus at the moment?). We believe that this is an important feature of human thought, not an inconvenient limitation. (in Dreyfus 1972–79, 51–52)

As Dreyfus suggests, the way is not barred for Winograd's project but it is a tremendously difficult and complex task:

> In the meantime everyone interested in the philosophical

> project of cognitive science will be watching to see if Winograd and company can produce a moodless, disembodied, concernless, already adult surrogate for our slowly acquired situated understanding. (1972–79, 55)

A context-sensitive computer-communications system would conceivably be a more socially equitable system. However, as our own experiments in developing an explicitly discursive theory of context to deal with complex communications such as irony have shown, and as Peirce stated years beforehand, context is an ever-expanding sign-field – phaneron – which would have to be considered all at once but which cannot possibly be, thus making the task of interpretation an infinite one, where the degree of error is always as great as the degree of certainty.[5]

2.12 The elusiveness of data processing to monitoring: unaccountability

Data processing has been seen to operate according to a set of discursive constraints such as rules of accessing and association, as well as to categorizing patterns engraved and determined in the very design of the technology. But there is yet another design-inherent trait. This trait involves the impossibility of monitoring either the contents or the transfer patterns of information once within the system. It amounts to an occultation of discursive procedures by the very procedures themselves. To the degree that monitoring of discursive procedures is possible it is too costly and time consuming.

These very procedures may themselves, after all, be surveillance procedures operative in the social as opposed to the corporate or technical spheres. Indeed, coding or encryption is itself a technical design inherently structured to avoid monitoring and hence to evade accountability be it to government, social or public agencies:

> Experience with prohibitions on export of unlicensed personal data is that such rules are hard to enforce. (Pool & Solomon/OECD 1980, 118)

> It is unlikely that any cyphers will be deposited with authorities.

> Once text is loosed within a computer network, outside of its owner's files, it is virtually impossible to police it, to know who has used it, how often and when (. . .) protection of intellectual property is going to be difficult. (Pool & Solomon/OECD 1980, 117–20)

> It is impossible that digitalized voice and data traffic may be optimally handled in a single mixed bit-stream, particularly in expanded corporate telecommunications networks which will flow over national boundaries. Any regulations applying separately to those flows, would then require the application of some form of content control of the flow (raising) even more dangerous issues of invasion of communications privacy than those frequently raised regarding transborder access to personal data files. Simply stated, in the proximate future, the only method of determining whether data is being transported across national boundaries in contravention of some national restrictions will be some form of wire-tapping and for full control, that wire-tapping will have to be on a gross scale covering all transnational telecommunications traffic. So issues of standards are fraught with possibilities for restricting data flow. (Pool & Solomon/OECD 1980, 100–1)

While we acknowledge the accuracy of the above description as regards the factual status quo of existing constraints on the potential for monitoring data flow, the one difference of opinion we have with Pool concerns the degree of acceptance of this technology of the 'factual' status quo in terms of other possible rules of discourse. The technology does seem to have been structured to elude monitoring of transborder, transnational corporate communications. But, it is possible, despite these difficulties, to restructure networking so that countries are not left terribly vulnerable. Networking could conceivably have built-in decentralization and blocking mechanisms to curtail free rein data flow so that some degree of national autonomy might be preserved. However, these policy suggestions aside for the moment, the fact still does remain, as Pool and Solomon describe, that the present structure of new communications technology does not include procedures which might allay fears over the increasing corporate, and decreasing government and public sphere, control over communications. A citizen's every act of consumption must be monitored 'for billing purposes,' while the vast scale of transborder corporate communications networks are immune to demands for accountability.

2.13 Valorization of control for speed, growth, efficiency and productivity

One of the major tasks for which videotex has been developed and designed, and is already being used, is EFTS, Electronic Funds Transfers. Here, payment for goods is carried out through electronic transfer and record-keeping without the necessity of any actual cash transaction. Since this system is already used by many banks, the only new development will be the degree and range over which transfers are done electronically.

This capacity for electronic funds transfer has been partly responsible for the obsession in many discourses on new communications technology with the issue of billing. First of all, this throws us back into the procedure of discourse whereby all communication is a means-ends logic serving the purpose of maximizing production and profits through efficiency. As Simon reports, electronic records of funds transfer and billing for information use serve 'to gain economies of scale, to coordinate interdependent activities and to control lower level activities in the interest of higher level goals.' (1980, 212, emphasis added)

Electronic funds transfer is both an issue and a technique designed to maintain present property relations regarding the control of goods and information industry, It does not exist for any philanthropic reason.

> The most agonizing problem about the Hell of Administrative Boredom is that all information is proprietary, open access to lower members and nonmembers too readily deranges the authority structure and the delicate balance of interdependence of all parts of a well run organization. (Lowi 1981, 468)

The link of EFTS to control, hence to instrumental reason, is also recognized by researchers for OECD:

> With the expansion of various trees of information flow, for example, credit wealth, insurance, etc., there is a trend towards

> more and more information collection about individuals. (Gassman/OECD 1980, 11–12)

EFTS not only reaffirms liberalist proprietary values but also vastly increases the capacity for citizen control because records of all individual and social information-consumption must be kept in order to bill the users. The need for horrendously extensive record-keeping poses no problem to the immense capacity of modern data banks. A researcher for the Canadian Institute for Public Policy Research testifies:

> EFTS creates records of payments that can easily be used to trace the actions of individuals and to monitor buying habits. The risk to protection of individual privacy is obvious (. . .) purchases would be cleared immediately through the banking institution, rather than a day or two later (. . .) consumers might find it difficult to keep track of their expenditures: and banking institutions might decide over the long term, to accept debit payments only from established households. (Gotlieb 1978, 15)

Lowi expresses clearly this admixture of concern for control and economic efficiency and productivity in the instrumentalist deployment and design of communications technology:

> The governing proposition is that, as a result of information technology, man's power over his environment will increase greatly and his susceptibility to manipulation will rise proportionately. (1981, 454)

Electronic funds transfer is a case in point of information technology being designed to control the environment and not to understand or interact with it.

2.14 Quantitative over qualitative evaluation: the 'pig principle' revisited

One of the claims made to justify the implementation of new communications technology was that it would increase information choices. Multi-channelled cable, made possible by fibre optics and coaxial cable, purports to increase 'choice' in the sphere of communications. This claim follows a discursive procedure of substantive rationality, namely that of quantificational calculation of effects and variables, a partial off-shoot of American reception of Vienna School positivism. A current illustration of the quantificational bent is Warner Corporation's QUBE system which provides over thirty channels of programming for the community. Although some of the channels are used for community and information services most are still devoted to 'sitcoms' and Hollywood movie entertainment. Indeed, most requests by users turned out to be for more pornographic channels. The choice available here is quantitative, between already existing information services. Situation comedies and movies already exist on normal broadcasting systems as well as in the cinema. The choice is simply for more of the same. Qualitative choice would involve far more software development especially designed for the videotex service, something that, as of 1982, is barely underway in Canada. As most literature on Telidon field trials, as well as on cable tv in Canada indicates, there is a penury of resources being devoted to the creations of alternative and original programming, content and software, whereas a great deal is being devoted to hardware facilities to increase available channels.[6] What will fill them? Pay tv's most innovative response so far seems to be that slightly aged box office hits will (CRTC 1980, 49–116). One notable exception to this qualitative lack of choice is Manitoba's Telidon – Ida *Grassroots* system whereby farmers actually do have access to very contextually pertinent information that fulfills their needs, though it be mostly commercial.

2.15 Télématique or privatique: centralization or decentralization

The specific type of networking instituted will clearly influence whether or not new communications technology leads to a binding of the periphery to the centre – space binding (Innis 1951–77, 92–131) – or to a decentralization of communications and control patterns.

Some computer-communications specialists argue a priori that these systems will lead to centralization and that this is not necessarily a negative factor; a rather apologetic stance. Simon contends that computers will indeed lead to centralization in society but that common beliefs about centralization and decentralization are not necessarily true:

> Decentralization is commonly equated with autonomy, self-determination, or even self-actualization. Centralization is equated with bureaucracy (in the pejorative sense of the term) or with authoritarianism and is often named as a prime force in the dehumanization of organizations and the alienation of their members. (1980, 212)

The point that has to be made here is that even the apologists for computer-communications appear to agree that computer-communications is leading to an increasing centralization of social and economic institutions.[7]

Centralization is likely to occur, however, because of the corporate 'need' – desire – for centralized control of the margins which drive the development of new communications technology and because of the very structure of design of the hardware, more specifically of the patterns of networking (cf tables I and II).

> As the total number of installed computer systems expands, there is also a growing need to link these systems with each other and to remote terminals via telecommunications systems. (Gassman/OECD 1980, 11)

Indeed, computers may be seen to have actually required telecommunications hook-ups in the first place in order to centralize data banking from remote areas. There was a need to draw in towards the centre, via linkage, all of the dispersed or peripheral systems. Licklider, a researcher for the American National Defence Council suggests the desirability of developing a totally centralized computer-communications network as a function of the need for social control:

> An international network of digital computer communication networks serves as the main and essential medium of information interaction for governments, institutions, corporations and individuals. The multinet, as it is called, is hierarchical; some of the component networks are themselves networks of networks. And many of the top-level networks are national networks. The many sub-networks are electronically compatible and physically interconnected. (Licklider, 1979, 91–2)

We are not speaking of an a-contextual, over-generalized kind of centralization. From the above quotation, it is clear that centralization is both made feasible and encouraged by specific technical factors such as networking patterns and electronic compatibility. For example, it has recently been announced that the USA has accepted the Canadian Telidon videotex system as the standard electronic hook-up for its own videotex industry (Fox/ACC June 1982).[8] This is a tremendous boost for the hardware industry in Canada. However, it also means electronic compatibility which will facilitate the transfer of data and software from the USA to Canada, thereby threatening once again the autonomy of Canada's communications.

Simon is unequivocal about the technical foundations for centralization in computer communications:

> (1) Facilitating the construction and use of systems models that can incorporate expert knowledge about system structure and system behavior or
> (2) Permitting the assembly of expert knowledge in large data banks that can be consulted readily from any organization location provided with a terminal. (1980, 226)

Further on in this same text, Simon reconfirms that centralization and decentralization are consequences linked to the actual programming of computer-communications technology: 'depending on how we program computers they will lead to centralization or decentralization.' (1980, 226) He then suggests that the extant programming technology has led to centralization: 'Computers encourage centralization of data and decision-making. (They) lead to greater increase and efficiency of strategic planning and instrumental action.' (1980, 221)

Thus, Simon once again points to instrumentalist discursive procedures as the drive behind the development of a technology which itself in turn manifests the instrumentalist procedures of control, efficiency and problem-solving models. Just a glance at treatments of this technology in *Business Week* and *Fortune* reveals what this technology is designed for in the business place is to reduce risk factors and increase control over the periphery by means of centralizing information treatment capacity.[9]

If knowledge is power, then computerized knowledge is more

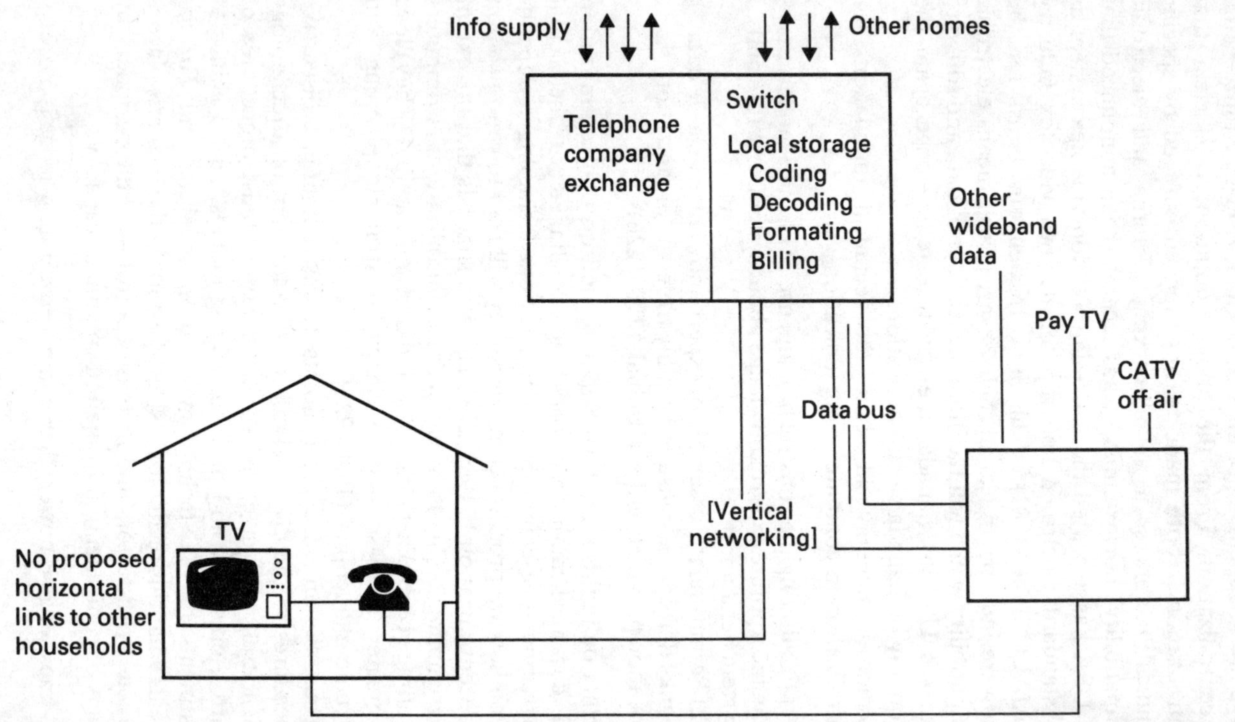

from: *Gutenberg II*, p. 78, Parkhill

[] — our additions

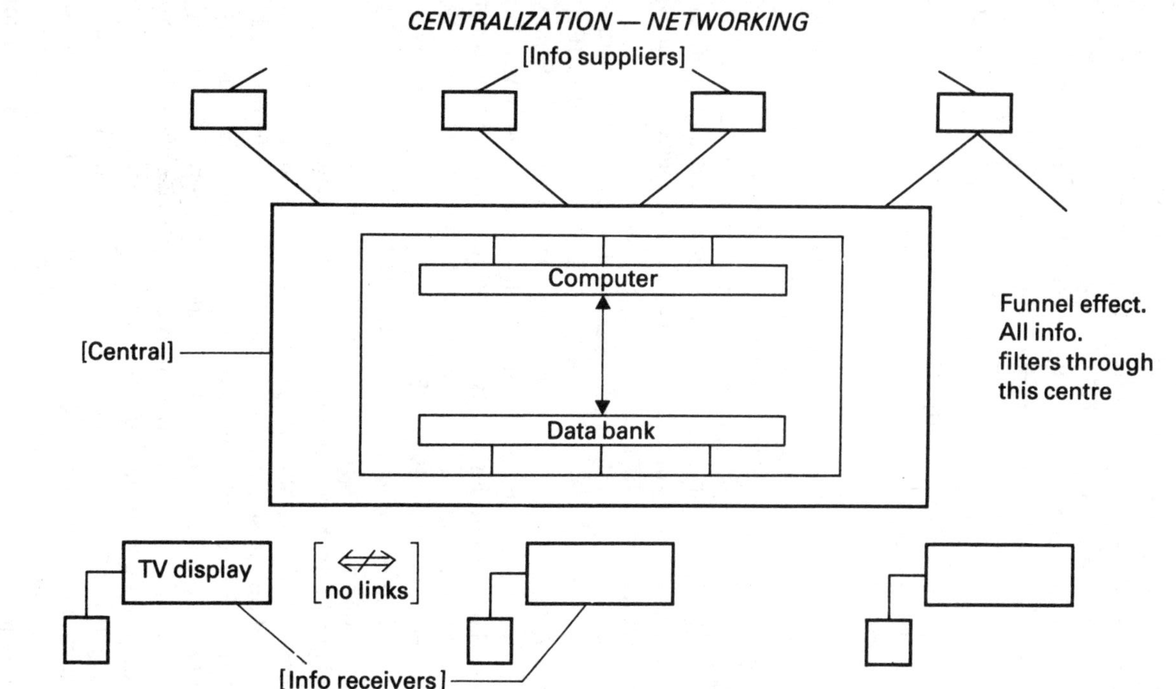

from: Madden *Gutenberg Two*, p.77

[] — our additions

> power. Knowledge is power and control, provided it is timely, ample, and relevant (. . .) Today's advanced electronic computer enables a man to control his business and to assess its environment with incomparable effectiveness because it enables him both to lay hands on relevant facts swiftly and to understand their changing relationships. Besides supplying him with facts in historical time, or after they have happened, it supplies him with facts 'on line,' i.e., as soon as they are born, and in 'real time,' i.e., promptly and abundantly enough to control the circumstances they describe while these circumstances are developing. (Burck, in *Fortune,* April 1964, 141)

In videotex trials, the predominance of a few centralized information providers and of the centralized interpretation of citizen feedback, reconfirms this trend toward centralization.[10] Nevertheless, citizens and communications specialists tend habitually to think in terms of centralist models. This tendency is due to the broadcasting (one-to-many) theoretical model which dominates both our broadcasting and narrowcasting media.

But this debate on the impending centralization or decentralization due to computer-communications technology is still further contextually specific and sophisticated. It involves a debate between whether the dominant model of this technology will be the networking system known as 'télématique,' a term coined by Nora and Minc (1980) or 'privatique,' its counterpart coined by Lussato and Bounine (1979). Both types of networking exist today. 'Télématique' emphasized centralized, vertical networking patterns whereas 'privatique' accentuates horizontal, limited, localized person-to-person terminal links of interaction.

There is little doubt that the 'télématique' scenario tends toward a centralization of social groups and institutions since all are hooked up to a central computer terminal. There is at least a possibility that 'privatique' could lead to a more 'dialogical,' horizontally interactional communications structure whereby various interest groups or individuals communicate with each other, by-passing any central terminal. Office partners or members of the gay community, for example, would interact, thus favoring regional, local or topical alliances over blanket, centralized ones. The computer-communications systems contained within a single office and the self-contained home computer are also cases of the 'privatique' scenario. 'Privatique' could lead to interaction amongst members on the margins without necessarily binding them to the centre:

> *Télématique* as described in the Nora-Minc report envisages a vast electronic highway or utility accessible to anyone anywhere in the world for any conceivable information-based purpose. The term *privatique* is taken from the notion put forward by two dissenters to the Nora-Minc report (Lussato and Bounine 1979). They maintain that the integration of computer and communications technology will not necessarily lead to a single global network, but could produce a diffusion of many independent informatics loci, each with its own computer power and having relatively little need for communications with central computers of giant data bases in remote locations. Hence the term *privatique*. The *télématique* scenario developed in the Nora-Minc report arose from a focus on the enabling technology itself. By extension, it is possible that the *télématique* approach will triumph where corporate planners concentrate on and try to exploit the full capacities of the new technology
> (. . .)
>
> There are profound implications to each of these two different approaches. The *privatique* approach suggests that informatics will be implemented in a diffused manner with each corporation and corporate department relatively free to design its own system. The *télématique* approach seems linked to a more centralized process, with a master plan being implemented either from the top down or from the bottom up. The suggested scope and pace of change differ as well. Under the *privatique* approach, the traditional office seems to remain, although its support equipment and job functions are transformed to the electronic model – for instance, through the purchase of stand-alone word processing equipment. On the other hand, the *télématique* orientation would favor automating information-related office functions and integrating them within a multi-functional decision-making support system, held together by a company wide, electronic communications network. It is also possible that the presence of a strong communications network such as that which already exists in Canada could itself encourage a *télématique* approach emphasizing infrastructure

> rather than content. This possibility would become even more probable if the specialization of this country's communications capacity became a matter of policy – in an attempt to ensure Canada's stake in the office of the future. (Menzies 1981, 19–21) (Nora and Minc 1980) (Lussato and Bounine 1979)[11]

What should become apparent here is that the choice between a centralizing and a decentralizing technology is not simply one of use but also crucially one of discursive procedures, pathways of access and content, procedures designed and built into the technology as networking patterns, structures of consoles and extensiveness of data banking and computer facilities.

In this light, no credence should be lent the policy recommendations of Pool and Solomon for the OECD according to which only the uses and abuses of technology are to be regulated, and not the technology or its development and design.

Harold Innis' insight that new communications technology can lead to decentralization or centralization (space-binding) is now fathomable (1950–72). However, in view of the present day trends toward the 'télématique' or centralized model, it is also understandable that Innis' studies revealed a steady trend toward space-binding centralization (the centralization and consolidation of power), as a result of new communications technology (Innis, in Salter and Melody et al. 1981, 193–207).

While in Canada the CRAB report (1979, 15) states the government's intention to maintain a dispersion of data banks, the current information providers and computer services providers tend to be centralized in the hands of companies such as QUBE (USA), Southam and Torstar. Telidon itself conforms to the 'télématique' model. Indeed, as the events that lead up to the Kent Commission of press monopoly in Canada, as well as the commission's actual report indicate, Torstar and Southam have sought nothing short of a press monopoly (Kent 1981). Is there any reason, based on past experience, to expect that they will not also seek centralized control in the electronic publishing endeavors? But, as the TAMEC consulting report carried out for the Montreal cable companies indicates, there will be competition in the quest for monopoly (Ouimet 1980, 145).

One way to ensure centralization is to ensure standardization in a company's hook-ups. At present, the English Prestel and the French Antiope systems are not compatible with the Canadian

Telidon system making an exchange or centralization of their data banks unfeasible. The huge distribution of IBM computer-communications hook-ups across the globe has led Nora and Minc to suggest that two separate spheres of discourse will arise. The first will be the globalized computer-communications of IBM whose wide distribution and huge economies of scale have allowed it this monopoly. The second will be smaller networks of computer-communications such as Telidon. Minc projects that as far as program compatibility is concerned there will be two territories, IBM territory and that of the other computer-communications companies. Futhermore, with regard to concerns of national sovereignty, Minc suggests that there will be a curious bi-hemispheric centralization delineated along company rather than national lines:

> Virtual frontiers will be traced between territories of the various manufacturers: communications from one to another will therefore be impossible. (There will be a) possible emergence of teleprocessing and electronic frontiers. (Minc/OECD 1980, 155)

The fact that none of the big American computer-communications companies has developed a home videotex system as sophisticated as Telidon, though AT & T is now close, may indicate that they are concentrating more on corporate, military and governmental centralized computer hook-ups. This contrast risks creating certain transnational institutionally delimited spheres of discourse, just as societies of discourse known as professions and disciplines have done in the past.

The whole issue of national sovereignty is almost a moot question given the contemporary degree and form of development of computer-communications technology. Canadian and American home computer users seem likely to share American software, for the most part, on Canadian Telidon hardware. The hardware/software trade-off seems to be in the making as the recent American purchase of Telidon for field trials would indicate. In Brazil, the third world country which has most protected its microcomputer market, companies rely on foreign software and even on foreign-built chips. Furthermore, there is a major thrust in the Congress, led by senator Roberto Campus, to deregulate the whole industry (Summer 1984). As Sweden's emphasis on the development of its own pertinent software suggests, no country dependent on another country's software or data banks in an environment of expanding use of electronic communication could expect to maintain its communicational, cultural or political sovereignty (Gotlieb and Zeeman 1980). And yet, as Pool and Solomon describe, networking systems and software packages are already in place which preclude 'customs' control over national boundaries. This is a case of technologically built-in transnational centralization:

> In a packet-switched international network the individual packets are relatively meaningless and travel through the system in random routes. Any attempt by one country to restrict the code that was allowed to pass through its nodes would be unenforceable, or if an attempt was made to enforce it, would prevent the operation of the packet net in that country. Thus, from a technical point of view, freedom of encryption has to be allowed on such a net. (Pool and Solomon/OECD 1980, 101)

All of this to say that the actual technical structuring of new communications technology including networking and software packages as well as the design of the hardware and its compatibility with other hardware, all combine to reinforce a discursive procedure of transnational, corporate-oriented centralization beyond the control of nationally or regionally autonomous groups.

However, this is not merely to accept the technical status quo. As the 'privatique' scenario indicates, the technico-discursive structure of new communications technology can be designed in ways less conducive to centralization. It is to this possibility that critics of the factual status quo must address themselves in the realm of policy.

One final note must be added concerning the centralization versus decentralization debate. Should it be the case that a 'privatique' scenario becomes dominant, this would not exclude the exercise of power through control of discourses of knowledge. As will be discussed in the next chapter, power may be exercised at the most minute and dispersed levels of discourse in society. It is quite possible for monopolistic discursive constraints to be programmed into 'non-centralized' equipment, just as they are into centralized types. For example, the 'floppy disc' programs which can be purchased for separate, un-networked personal computers are nevertheless redundant sets of constraints or discursive procedures of knowledge designed into the technology. Buying the script processing program 'Wordstar' is tantamount to buying into a whole set of discursive procedures for text composition.

2.16 Networked privatization: the individualization of subjects of enunciation

While many discourses on new communications technology say it will lead to a more fully participatory society, much of the networking capacity of these machines seems rather to stress a form of individualist or privatized communication. Home networking, especially for consumerism, is one of the main thrusts of videotex development. 'You won't have to enter the store, just punch in your shopping wishes: they will be delivered to your doorstep and you'll be electronically billed.' Indeed, one of the initial inspirations of this investigation was a perceived quantitative over-emphasis of home-based consumerism by videotex publicity.

One of the main objections on the part of the computer operators has been the great isolation involved in working on video display terminal (VDT) due to minimalized contact with fellow workers. There is much talk of a return to the cottage industry structure (Toffler 1980) in which workers stay home all day, simply connecting with the company's central computer in order to carry out their administrative and executive functions. Futurist prediction? Perhaps, however, present-day networking patterns would make such a plan feasible. The so-called 'home computer revolution' might be expressed in terms of a relocation of many communication functions within the locus of the sacred a-social molecule – the family. The result is a further atomization and retrenchment of the social unit into the nutshell of the individualist-oriented group. Although technologically-oriented interaction such as shopping will take place as mediated by this networking, any a-technological interaction that accompanies such tasks in face-to-face communication tends to be eliminated.

Furthermore, the privatization is uni-directional. While the users may only have contact with central data banks as opposed to contact with other groups in horizontal patterns, the computer-communications data storage and retrieving central has access to communication with many individuals and social groups. This enables it to form regional, group and social profiles, such as audience- and buyer-surveys. The members of these groups follow isolated communications patterns, believing themselves individuals – islands unto themselves – while they are really having their information needs catered to on the basis of what the senders know about the group profile. The 'individualist' paradox is somewhat like the car ad that suggests owning such and such a car sets one off from one's neighbors. The ad agency is meanwhile capitalizing on the marketing information that the group of people belonging to the upper middle class share the common trait of wanting to be different from their neighbors. The above example may seem labyrinthine but this is precisely because of the paradox between an information system that gives voice to individuals and one that addresses 'masses,' i.e., specific groups, based on their privileged position of being able to generalize the reception of millions of 'individual' communication inputs. Marketing companies have been among the first to take advantage of this double-edged form of communication, as Mattelart's chapter 'When Advertising Becomes Politics' tends to prove quite statistically (1979, 235–85).

Once again we might refer to these resources as Bureaucratic Capital which favors 'public management.' The apologists of mainstream communication models for mass media would run up against a brick wall in trying to allay misgivings about the capacity for manipulation of home networking systems. Lazarsfield et al. (1948) insisted that the mass media did not have any great influence on people because of the 'two-step' buffer zone of interhuman contact which seemed to have more 'effect' than purely mass media input. Should the home networking patterns universalize the present trend of hooking up isolated individuals to central computer facilities – the télèmatique scenario – then this interhuman buffer zone clung to by the Columbia School in order to disprove the exercise of mass media power and manipulation would seem to shrink away. The social sphere of public interaction would be unequivocally threatened by a one-to-many, closed circuit electronic sphere.

2.17 Unilateral participation: hardware and software for more than feedback

In the area of social services the trump card of new computer-communications technology is that it supposedly allows for 'real' two-way communication and hence for a more participatory democratic society, and all of this because of two-way channel capacity.

First of all, as Pergler reports, this two-way capacity is not so sophisticated. 'Existing methods (of two-way incasting) are clumsy and incomplete.' (1980, xv–xiv) Noam Lemelhstrich, one of the most respected engineers in information technology concurs with the finding that the two-way capacity is vastly limited (1974). Despite multi-channelled fibre optics and coaxial cable, the major capacity for user feedback is a keypad on which 'yes' and 'no' and numeric answers on a scale from 'one to ten' are permitted. Such are the possibilities of the *Votaphone* developed by Bell Northern and in use in South Eastern Ontario and of the key pad used in the 'Grassroots' trials. Users can neither initiate subjects of concern nor reply with any significant degree of nuance and complexity let alone produce visual images of themselves. Consequently, the enunciative capacity for symmetrical dialogue between central base and home user is non-existent. As Barthes (1978) and Beneveniste (1966), amongst others, have stated the one who initiates the discursive practice usually manages to make the respondent play by his or her rules. This would seem to be the case to date given the degree of response allowed by two-way communications technology. The complexity of the feedback, so far, is quite simply very constrained.

What is perhaps more dangerous than the limitations of feedback capacity built into the system so far are the claims made for this capacity. When electronic polling and 'yes/no' or '1 to 10' voting are claimed to be communicational procedures of a democratic society, regardless of the nuances of a situation or the degree of information required for informed answers, there is always the danger of a plebiscitary-type rule whereby citizens are told, on the one hand, that they have choice but where, on the other hand, they can only respond within the parameters – the discursive procedures – set up for them by the initiators of the technological 'dialogue.'

Hegel once said that just because a man had a head did not mean he could philosophize just as we would not claim that any man could make shoes just because he was given the tools. We might apply this remark to the question of participatory communication and democracy: just because one has a channel does not necessarily mean that one can communicate. While many documents have been concerned with equitable access to hardware as the precondition of a democratic participatory communications society, it is our belief that one must distinguish between the purveyor of access to hardware and the purveyors of communications, interactive capacity. For example, networking would have to be devised so as to link not only a periphery with a centre but also the peripheries amongst themselves in order to provide a public forum, a public sphere, for interactive participation. But what is more, education in communicational procedures of knowledge, such as excluded middle, authority of the subject of enunciation, perspective and mode, and others mentioned throughout this study, would be required in order for symmetrical dialogue to occur. As it stands, even if everyone had access to this new technological hardware, those with advantages in traditional communicational skills, such as political rhetoric, would still have a discursive edge in legitimacy and power to say the 'truth.'

As we have seen earlier in the debates on new communications technology, there were certain redundant, quite traditional and classical discursive procedures (referentiality, order, exclusivity, enunciative a-symmetry, etc.) which had to be mastered, followed and recognized in order to license the subject of discourse to say the 'truth' authoritatively about the object. These meta-communicational procedures must be recognized as the parameters of procedures of technology itself, and new communications technology as a discourse has been seen to share many of them. This is perhaps what Gotlieb means when he suggests that 'Anyone can get into the electronic publishing business (. . .) the barriers (. . .) are not technological but rather social, political and economic.' (1978, 17–18) While we would argue that there are indeed technological barriers in the form of discursive constraints built into new communications technology, we agree here with Gotlieb that barriers in other spheres of discourse also play a major role in allowing for or prohibiting participation. In the end, it is the episteme which circumscribes the conditions of possibility for participation: no less than the total field of discursive procedures recognized by a particular society as generative of true practices or statements.

These same constraints apply to the utopian claims that new communications technology will 'make us all creators.' Technically, the cost of electronic page creation is so formidable that, so far, private small information providers are prohibited from becoming active, whereas large corporations, with their advantageous economies of scale, are commandeering the market: 'A particular interest was whether costs and constraints might inhibit small or non-profit organizations for participating.' (Kurchak 1981, iv) Furthermore, there are major concerns of distribution that remain to be solved, such as the incentives, if any, to encourage users to order up private publications on their screens. This inhibition is further enhanced by trends in funding for 'R&D.' For example, the Canadian government is concentrating on seeding the hardware industry where concern is more for elaborate and colorful graphics than for software packages to enhance public and small groups' participation in information provision:

> Some information providers believe field trial operators are interested only in testing the technology and argue that, if enough thinking isn't done about content, then permanent precedents may be set. (Kurhchak 1981, v)

Before pronouncing any final word on the participatory capacity of new communications technology, what is required, is a communicational and political theory of participation in relation to democracy. Purely in terms of interaction, what criteria must be met for participatory democracy in what Karl Popper calls 'open society'? From then on we can begin to ask: what procedures of technological discourse would have to be implemented for these criteria to be met?

2.18 Systems theory: the ideology of the information society

In the preceding we have taken new communications technology to mean the practices of electronic information processing and transmission – the procedures built into and practised by computer-communications equipment. We have seen that there is a substantial overlap between the procedures of discourses *on* new communications technology and those of discourses *of* new communications technology.

What explains this overlap? This can be answered by stepping back from the discourses on and of new communications technology. We have insisted time and again throughout this book that discursive procedures are 'conditions of possibility.' They themselves also have other 'conditions of possibility.' What makes the emergence and eventual dominance of some procedures possible and not of others? We have already pointed to a number of factors: the class interests which surround the emergence and propagation of discourse, forms of reasoning and rationality such as classical scientifist episteme, the constellation of other inter-, intra-, and extra-discursive relations, and so on.

In the most immediate sense, however, one might wish to point to systems or information theory as the principle 'condition of possibility' of discourses on and of new communications technology.

Systems or information theory is itself a result of the classical scientifist episteme and contains and propagates all of the procedures we have so far identified in discourses on and of new communications technology. Systems or information theory is, in a very real sense, the 'ideology,' the metaphysical justification and practical belief system for the 'information revolution.'

Irvin Laszlo, author of *Introduction to Systems Philosophy: towards a new paradigm of contemporary thought* (1972) provides a succinct résumé of the traits of the systems paradigm. First of all, he leaves no doubt as to the role of the 'information society' in the practice of the systems paradigm:

> The present situation in the field of theory is characterized by the fruits of the scientific 'information explosion' on the one hand and by highly sophisticated methodologies and conceptual analyses on the other. The obvious task is to bring them together. (1972, ix)

The discursive procedures of referentiality and order are two underlying assumptions of the systems episteme:

> Its basic conceptual assumption (general systems synthesis) is

> that the first-order models refer to some common core termed 'reality,' and that this core is generally ordered. (1972, 32)

The role of systems knowledge practices is to find order in the world, to find universal laws that circumscribe the system, and never to yield to the 'appearance' of chaos, which is never anything more than illusion:

> Extended into general *systems philosophy,* this instrument can polarize the contemporary theoretical scene as a magnet polarizes a field of charge particles: by ordering the formerly random segments into a meaningful pattern (. . .) The world is, at least in some respects, intelligibly ordered (open to rational enquiry). (1972, 8)

The metaphor of the Newtonian world machine seems appropriate even though systems theorists often try to distinguish themselves from Newtonian physics. What they derive from the Newtonian model is a confidence in the existence of laws and a vision of a system which changes yet remains stable:

> the concept of a dynamic, self-sustaining 'system' discriminated against the background of a changing natural environment. (1972, viii)

Furthermore, the differentiation which Laszlo establishes between systems theory and Newtonian physics is a question of degree and level of the application of the procedures of classical knowledge. Systems laws are different from Newtonian laws not as laws per se but in that the former apply to all science, at a sort of meta-level whereby all sciences may be globally qualified by the same set of laws:

> These creatively postulated invariances are not laws of physics as these laws have been hitherto understood, nor are they laws of biology or of any other special enquiry. They are general laws of natural organization cutting across disciplinary boundaries and applying to organized entities in the microhierarchy at each of its many levels. (1972, 32)

The liberalist and Newtonian notion of a naturally achieved equilibrium in nature is also a postulate of systems theory:

> We have shown that ordered wholes, i.e., systems with calculable fixed forces, tend to return to stationary states following perturbation introduced in their surroundings. (1972, 48)

The tendency to define knowledge in terms of the procedure of ordering, taxonomy and hierarchization also shines through in systems theory. Heinz Werner gives the following systems account of developments:

> Wherever development occurs it proceeds from a state of relative globality and lack of differentiation to a state of increasing differentiation, articulation and hierarchic order. (1957, in Laszlo 1972, 48)

Moreover, where systems theory and classical scientific rationalism are most inextricably related is in the explanation of these procedures of hierarchization and division into units on the basis that they best serve the ends of efficiency. The following systems theory explanation for hierarchical analysis could also be read as the instrumentalist rationale behind the division of labor, for example:

> Phrased qualitatively, it is 'simpler' for systems to cooperatively constitute higher systems than to do the job of complexification alone. And here simplicity is equal to efficiency and is measured by the time required for the process. (Laszlo 1972, 48)

The emphasis on quantitative criteria of measurement is yet another procedure borrowed from instrumentalist rationality.

But Laszlo, to be fair, does insist on the 'newness' of the systems paradigm, as the subtitle of his book, 'towards a new paradigm of contemporary thought,' indicates. He claims theory to be more open-ended than models of classical science. Interestingly enough, the example furnished to demonstrate this is taken from cybernetics rather than from informatics (in the case of new communications technology the informatics models of Shannon and Weaver still predominate above and beyond even the most simplistic cybernetic models):

> It is in this wider sense that 'cybernetics' will be used here, to wit, as systems-cybernetics, understanding by 'system' an

> ordered whole in relation to its relevant environment (hence one actually or potentially open.) (1972, 38)

Indeed, we do find in the ecological, cybernetic theories of someone like Bateson, such an open-endedness and perhaps even in more sophisticated computer programming, such as Winograd's models for context treatment. This openness as a form of interaction with the environment as opposed to an action upon the environment is indeed something quite different from what we have previously seen to qualify instrumentalist reason. In the Conclusion, we will discuss briefly the potential for such open-ended interaction on the part of new communications technology at which point it will be more feasible to suggest whether or not this technology does deviate substantially from the instrumentalism of systems thought paradigms.

The other area in which Laszlo wishes to distinguish thought from classical science is in that of epistemology. The author suggests that systems theory, by viewing the world as what Bateson calls an 'ecosystem,' i.e., a series of systems where every system is interrelated with the others, deviates from the theory of knowledge where the knowing subject is separate from the object to be known. In other words, the classical traits of the immunity, objectivity and positivism of the known subject is not a premise of systems knowledge. The knowing subject is part of the system it seeks to study.

> Human beings are environmentally transacting open systems; and they do not perceive and cognitize something just because it is there, or necessarily the *way* it is there. (Laszlo 1972, 197)

Once again, we would suggest that Laszlo is referring here to a more philosophical and interactional tradition of systems theory, one more akin to Bateson than to the informatics of Shannon and Weaver. Furthermore, Laszlo is not entirely consequent here in 'de-objectivizing' knowledge since a bit further on the example that he gives of difficulties of subject-involvement in knowledge reverts to a confidence in an objective structure of communication, the equivalent of Chomsky's generative grammar or of Saussure's 'language.' While referring specifically to a new communications technology, Laszlo is suggesting here again that there are real structures in the world that must and can be known in full before particular sensations can be interpreted in experience.

> It was originally thought, for example, that a machine functioning like a voice-typewriter could be easily built, analyzing the frequency and intensity of the sound-waves produced by a speaker and translating them into a printout proved to be impossible however. The reason is that the actual sound waves are remarkably undetermined by the speaker, and also show great variation depending on context (. . .) *The full structure ('grammar' or 'set of rules') of the pattern of which the perceived situation is a part must be known before the situation can be understood, or even perceived.* (1972, 197, emphasis added)

Furthermore, while Laszlo does recognize some active role for the perceptual system of the knowing subject in the act of knowing, he does revert to rationalism with the uncanny trick of suggesting that we can know man's perceptual, nervous system and via this knowledge judge the validity of the knowledge that it produces: 'all men have similar nervous systems, hence similar perceptual frameworks.' (1972, 210) Following a brief moment of doubt about the non-objectivity of knowledge and of the interaction of knowing subject and object to the known then, Laszlo does revert to a fully referentialist theory of knowledge when he states that: 'Nature supplies the limits of the interpretable perceptual patterns, and science supplies their interpretation.' (1972, 210)

Finally, Laszlo is absolutizing even the systems episteme as the only epistemology possible when he says that it is the notion of feedback that correctly adjust paradigms of knowledge to be adequate to the world as revealed and confirmed through experiment:

> *all* confirmation involves 'experimentation' in some sense and that all successful experimentation is self-stabilizing through negative feedback (. . .) He (the scientist) has found a 'paradigm' and confirmed it by experiment. (1972, 197, emphasis added)

Judging from Laszlo's description of systems theory, it does seem that for the most part there is nothing new about it in relation to the previous discursive procedures of knowledge of classical, liberal instrumentalist science. There were instances when Laszlo seemed to be associating interactionism and open-endedness with the systems episteme. Where this is the case we would suggest the systems episteme distinguishes itself strikingly from classical science. There did, however, seem to be some procrastination on the question of the interactive relationship of knowing subject and known object where interaction was initially recognized but subsequently suppressed in order to reclaim confidence in the capacity of the subject to know the real order of the universe.

Therefore, we would suggest that the 'systems' episteme is only a refinement of the instrumentalist, classical rationalistic episteme. The only difference may be in the success and degree of the exclusivity of these procedures whereby, today, all *other* forms of knowledge are socially and politically disenfranchized.

CHAPTER 3

Powermatics: focusing in on power and social control

> Certainly in the West we have not 'got used to the fact' that nobody rules or that the administration is open to every non-illiterate. What remains of the liberal hope is that at least within the confines of particular blocs, living well together will be achieved by administration. As the secularised, eschatological framework disappears because men can no longer hope for it what remains in the practice of modern politics is the purified conception of technical rationality. Whatever general or particular criticisms should properly be levelled against the thought of Max Weber, he laid before us what 'rationality' means in the souls of those who are most often influential in the realm of modern politics. (Grant 1973, 191)

In the introduction to this book we expressed a dissatisfaction with the lack of focus exhibited by many texts on new communications technology with regard to the question of social impact. In this chapter, we will attempt to show how all of the previously isolated discursive procedures are linked in the end to a certain order of social control and domination. The question of social impact focuses necessarily on relations of control and power.

3.0

We have seen that discourses on and of new communications technology are not only produced and accumulated but also function and circulate according to certain procedures. It is important to move beyond the descriptive and towards the interpretative in order to provide a critique of these procedures as they function in society. Traditionally, descriptive analysis has been distinguished from interpretative or critical analysis. The time has come to give a socio-political critical focus to the procedures which have controlled, selected, organized, and distributed in society discourses on and of new communications technology. Such an interpretation implies nothing short of a full fledged 'critique' (in the sense of critical philosophy)[1] of discourses on new communications technology.

Our critique of the social functioning of new communications technology will draw upon several theorists of communication and of its relationship to knowledge and power. Foucault, Weber, and Innis have each described *and* criticized the procedures of communication and of knowledge from the perspective of power and social control. Their critique is quite simple: control of procedures of communication/discourse leads to control of knowledge which in turn legitimates the exercise of power and social control. Apart from this simple argument, however, they all define and situate power in very nuanced and a-deterministic ways. This socio-political critique aims to show that constraints on discourse and knowledge result in social constraints and in the exercise of domination and power.

> In society such as ours, but basically in any society, there are manifest relations of power which permeate, characterize, and constitute the social body, and these relations of power cannot themselves be established, consolidated or implemented without the production, accumulation, circulation and functioning of a discourse. (Foucault 1980, 200)

3.1 Lip service

Today, almost any piece of writing on new communications technology, writing that could be called post-Orwellian, pays some sort of lip service to the 'future conditional' dangers of social control associated with this new technology. But this lip service rarely goes beyond a passing reference to the threats to liberalist individualism as portrayed in *1984*. Although the potential for social control via new communications technology is frequently mentioned, the futurological perspective coupled with the 'utopian' reliance upon the technical fix of instrumental reason interposes qualifications such as: 'it's really quite inconceivable,' 'Just futurist science fiction,' or 'Man will find a technical way out before disaster ensues.' Thus, while the power motif does receive at least some slight recognition, a serious acknowledgment and critique of its presence is effected by historicist, technological optimism. This sort of tokenism is a form of occultation.

3.2 Occultation

For the most part, the texts examined in the preceding chapters claim to argue in favor of democratic participation while themselves using procedures which precisely exclude such participation.

Sometimes, discourses which wish to occult the power focus nonetheless let it slip through. For a brief instant, it is held out as an unlikely possibility:

> Essentially, either the citizen will control the television set, and influence the world around him through its incasting capability, or the television set will condition and control the citizen and through it those in power will reach and influence the citizen. (Pergler 1980, xiii)

Still, certain contemporary discourses have explicitly argued that there exists a very great potential for the exercise of power and social control via discourses of new communications technology. While H. Vandeberghe, writing for the OECD, confuses information and communication, he does nevertheless point out power focus:

> the balance of power in a democratic society is the problem of the relationship between those who have information and those who do not, i.e., between the executive and parliament, between public authorities and industrial and consumer organization. (Vandeberghe/OECD 1979, 249)

Such theorists as Mumford, who are not usually critical of technology, do nevertheless point to the role of power in the technological/social configuration. The most familiar indication may be Mumford's discussion of the power that the clock gave to the clergy which initially controlled it and of the power claimed by the clergy by virtue of its exclusive ability to produce and circulate manuscripts at a time when the printing press had not yet superseded scribing.

3.3 Historical examples of an awareness of control

The recognition of social control as a central axis of the procedures of discourses of knowledge is not something entirely contemporary.[2] Although an explicit critique of the control motif in discourse and knowledge was never an integral part of the mainstream, classical thought, this motif was recognized in discourses of the seventeenth, eighteenth, and nineteenth centuries.

Auguste Comte fittingly made this link clear when he stated that the role of science was to 'prévoir pour pouvoir.' He advocated that science had the power and the authority to promote and accelerate the progress of man. In the eighteenth century, Saint-Simon had already begun to write about the consequences of the industrial society and of rationalized production in terms of social control. Bacon praised the linking of discourses of knowledge to the exercise of (State) power. In *New Atlantis* he suggests that the 'progressive application of science' would consist of 'a body of scientists constituting the supreme body within the State (. . .) the spirit of science would infuse itself into the thought and behaviour of all citizens of the new order.' (in Kumar 1978, 25)

The discourses of the Enlightenment make it quite clear that the exercise of power by science is strictly linked to another fundamental discursive procedure, the knowing subject's 'will to truth.' Condorcet said of Bacon:

> The love of truth assembles there the men whom the sacrifice of ordinary passions has rendered worthy of her; and Enlightenment nations, aware of all that she can do for the happiness of the human species, lavish upon *Genius* the means of unfolding its *activity* and *strength*. (in Kumar 1978, 25)

Foucault has also amply demonstrated the connection, ever since the seventeenth century, between power, discourse and knowledge in his analyses of texts from various diciplines dating from that period. For Foucault, the Age of Reason was the age of truth's exclusivity and of a claim to power based on a claim to truth. One example of the power of this exclusivity was the science of madness, psychiatry, which managed to exclude by confinement the 'unreasonable' from society. Discourses of reason defined and legitimated themselves by calling other discourses 'madness' and by excluding them. The difference between the historical acknowledgment of power and Foucault's analyses of power is the difference between recognition and critique. Sometimes these historical documents hide power entirely. But when it is acknowledged, it is done positively or with a sort of a-critical, mythical naturalization. It was considered 'natural' for science to rule the world and for science to follow the correct discursive procedures of knowledge, thereby excluding all other procedures.

3.4 Two axes of a socio-political critique of technology

If we wish to argue that the discursive relationships of exclusivity and hierarchical control have a parallel in social exclusivity and domination then two demonstrations are required. Firstly, we must explain how discursive procedures actualized by any communicative act affect our perception and our conception of the world and also how our knowledge frameworks affect the possibilities for action in society. In other words, rather than simply assuming a hegemonic centre of domination, our critique should build its demonstration of domination by showing the relationship of discursive domination to scientific domination and then to politico-social domination. We must interpret the functioning of rules of discourse at the levels of communication, science, and political practice.

Secondly, a socio-political critique of discourse must also identify the *sites* of power and domination. Rather than simply assuming that multinational corporations, multi-millionaire capitalists or socialist State bureaucracies hold the reins of power, we must identify the possible sites of power by seeking discursive confirmation of those which are inhabited by which groups or societies of discourse.

These two preoccupations of a socio-political critique of power in discourses may be rephrased as two questions which will provide the main framework for an exposition of the traits of power in new communications technology and the social environment:

> A// How will new communications technology affect our perception and our cognitive grids which filter our knowledge of the world and of ourselves? Or, in Foucaldian terms: 'What rules of right are implemented by relations of power in the production of discourses of Truth?' (Foucault 1980, 93)
> B// Which institutions and groups will consolidate or accumulate power via these discourses of knowledge, thereby securing a 'monopoly of knowledge' or a position of the initiated in the 'dominant episteme' under their control to the exclusion of others? Again, in Foucaldian terms: 'What type of power is susceptible of producing discourses of truth that in a society such as ours are endowed with such potent effects?' (Foucault 1980, 93)

3.5 A re-examination of the characteristics of power

One way of beginning to criticize power is to avoid the preconceptions about it made by classical discourse as well as by political science's theories of the State. Most critics of power assume that it is seated in the State or lies in the hands of capital. Most, including Marcuse, assume that power is purely repressive and is hence a threat to liberalist conceptions of freedom as the absence of constraints. Many assume that power manifests itself only in important or high institutions. Others insist that power is always dependent upon some form of base structure, such as the economic or the technological.

We would prefer to develop an alternative critique of power. One that is more nuanced and less dependent upon *a priori* theories of the State, linear causality, or historicity. Rather than seeing power as a unilateral determination or as a simple, symptomatic causality, we prefer to see it as a complex of changing, interactive hierarchical discursive relations.[3]

> to reverse the mode of analysis followed by the entire discourse of right from the time of the Middle Ages. My aim, therefore, was to invert it, to give due weight, that is, to the fact of domination, to expose both its latent nature and its brutality. I then wanted to show not only how right is, in a general way, the instrument of this domination – which scarcely needs saying –

> but also to show the extent to which, and the form in which, right (not simply the laws but the whole complex of apparatuses, institutions and regulations responsible for their application), transmits and puts in motion relations that are not relations of sovereignty, but of domination. (Foucault 1980, 95)

It is in and through such an alternative description and critique of power that we may arrive at grounds for an evaluation of power in the 'information' society and for the postulation of an alternative. Thus a whole set of questions concerning the role of knowledge and power in society will be re-examined:

- Does power always make itself semantically explicit?
- Is power situated only in the State, in the corporate structure, or may it be sewn throughout the social fabric?
- Is power only repressive or is it also conducive?
- Is power always hegemonic or centralized – State sovereignty – or may it be decentralized as dispersed nodes?
- To change power relations does it suffice to change a deep structure of society, for example, the economic structure, or must many other aspects of society be changed, and which are they?
- Does power legitimate itself solely through repression and the allocation of resources necessary for needs fulfillment or does power legitimate itself in and through discourses of exclusive knowledge?
- Is there anything wrong with rationality's exercise of power except that it places constraints upon otherwise free individuals?

3.6 How does discourse affect knowledge: or how does changing the way we communicate change the way we perceive and conceive?

> The limits of my language are the limits of my universe. (Wittgenstein, *Tractatus Logico Philosophicus*)

> A language is also a tool, and it also, indeed, very strongly shapes what the world looks like to its user. (Weizenbaum 1976, 102)

Procedures of discourse circumscribe our knowledge of what is exclusively considered to be truth. Our participation in the world depends on the knowledge that either we have or claim to have. Discourse of science and technology not only reflect but also constitute social organization.

During the 1980 Iranian seizure of members of the American embassy, an interesting phenomenon appeared in the news, both international and local. The seizure of the Americans by the Iranians was called the 'Iranian hostage-taking crisis.' What is perhaps more interesting is that around that time, any minor kidnapping incidents were reported lengthily on the national news broadcast and the term 'hostage-taking crisis' was constantly used to refer to them.

Although the above example has little to do with the latest communications technology, it does show how the molds of our language affect very much what will be perceived as pertinent and how it will be described. The hardened formula 'hostage-taking crisis' had infiltrated the cognitive news filters not only of every Western news gatherer but also of every newswatcher. Of course, one might refer to historical examples in order to show the same principle at work with regard to terms such as 'reform,' 'revolution,' 'democracy,' etc. Robert Darnton, in his article 'Writing News and Telling Stories,' shows how these discursive formulae, such as 'Jack the Ripper,' actually serve as filters for picking out news stories, i.e., for knowing and interpreting the world (Darnton 1975). Darnton calls these phenomena the 'cookie cutters' of news stories. Of course, there is nothing new in suggesting that our discourse influences our knowledge processes. The *Cratylus* might be cited as an earlier exposition of this position. Still, positivism and classical science did seem to have forgotten Socrates' lessons and to have postulated discourse as a transparent mediator between knowledge and the world. This faith in discourse's referential transparency was shaken by thinkers such as Heisenberg whose experiments in quantum mechanics could only be explained by suggesting that the instruments themselves, which are the equivalent of the scientific discursive apparatus, along with the cognitive discourse of the scientist were 'perturbing' the results. Quantum mechanics terminology such as 'perturbation effect,' 'indeterminacy,' and 'potentia' all refer to the impossibility of prying discourse and knowledge loose from the world.[4] Discourse, interaction with/about the world, modifies, indeed co-produces, our cognition of the world. The focus on the relationship of discourse to knowledge raises the issue of what the consequence of the social discursive procedures of new communications technology will be for *social knowledge*. Minc worries about an impoverishment of our natural language at the hands of a system where most of our communication for successive generations of school children is on throughways compatible with the computer. He suggests a consequent social inequality between those who can fully master both technical and non-technical types of communication and those who can only reply in some form of abbreviated key punching. We agree with Minc as regards the strong links between communication, knowledge, and values, on the one hand and social power relations on the other. However, his contention that the discursive procedures of the 'communications' society will be qualitatively different remains to be proven.

Lowi is another who does not hesitate to call a spade a spade: 'The new technology offers still newer and more potent means of social control through "manipulated consensus".' (1981, 462) What Lowi is stating here is the possibility of using new communications technology as a means of control of language and subsequently of the knowledge filters and value systems of other men. By imposing a certain type of communication upon the social sphere, certain societies of power manage to legitimate themselves through *manipulated consensus,* that is to say consensus concerning a) the rules of communication and, consequently, b) the content of knowledge. Because this consensus depends upon an imposition of discursive constraints and thus upon an interiorization of certain values, it is a managed consensus as opposed to a freely formed consensus.

> The earliest and most direct impact of the information revolution is probably on the individual's conceptual apparatus – his way of thinking (. . .) he must become adept at analysis and abstract thought. He must use this valuable commodity in special forms and through special kinds of procedures and technology. Already we see successive generations of school children brought up on thoughtways compatible with the computer (. . .) graduate students whose analytical powers have increased but whose theoretical perspectives are largely unquestioned commitments of the persons who designed the canned programs. (Lowi 1981, 457)

In short, procedures of technology are discursive procedures of knowledge which organize our social consciousness – our 'imagery' – which organize our image of the 'real' and so cannot fail to 'influence' our action in 'the real' (Castoriadis 1975, 25).

3.7 Panopticism: social control via the internalization of discursive constraints on knowledge

Destutt de Tracy once wrote that a true politician binds his subject most strongly not by real chains but by the chains of his own ideas. The anchor at the end of this chain is the discourse of reason. Fundamentally, the control of consciousness via the control of discourse is a vast generalized case of interiorized surveillance, what we earlier referred to as the discursive procedure of surveillance or panopticism. Panopticism, here, is not merely a discursive procedure. It is a socio-political goal. Coercion, rather than operating by means of external pressures such as physical restraints, functions by means of the internalization of procedures of discourse within the consciousness of the 'controlled' subjects. For example, Foucault showed how the discourses of penal and psychiatric institutions took the chains off criminals and madmen only when they could be assured that the criminal or patient had internalized the socially normative verbal and behavioral rules of discourse. They removed the chains once they were assured that their charges would 'watch themselves.' One of those internalized rules was the feeling of guilt over one's actions; another, the compulsive striving for piety in one's every thought and act:

> This mode of surveillance was to induce in the inmate a state of consciousness and permanent visability that assured the automatic functioning of power (. . .) he who is subjected to a field of visibility, and who knows it, assumes responsibility for the constraints of power (. . .) he inscribes in himself the power relation in which he simultaneously plays both roles: he becomes the principle of his subjection. (Foucault 1979, 202)

With new communications technology, there is an internalization of such discursive procedures as means/end logic, a problem-solving conception of knowledge, and certain patterns of information and ordering. Once internalized, all of these procedures have a surveillance capacity over the subject's social interaction. Since such procedures characterize both discourses on and of new communications technology, it is not far-fetched to suggest that such an internalization is occurring in the 'information society.' In other words, the users of new communications technology have become the principal operators of their own subjection by assuming and interiorizing the discursive procedures on and of new communications technology.

Panopticism is a seminal concept for understanding the relationship of communications procedures to social control. It is far too costly and complicated for a State or an institution to maintain control over its citizens by means of external, physical pressure. Panoptic discursive control has three advantages: (1) it allocates power at a very low cost; (2) it provides for maximum breadth of social control, especially as discourse is mass-diffused; (3) the output of discursive formations may be directly linked to the economic growth of power.

Just as Carey (1969) has argued that the mass media have been a most efficient way of creating national and social cohesion which certain institutions require in order to remain intact, new communications technology, by becoming ever more ubiquitous and more standardized, also congeals discourse into fixed, internalizable patterns of knowledge and, consequently, of social behavior.

In that new communications technology sets up certain new forms or procedures of knowledge in ever more sites, its panoptic function may be described as a 'multiplication of the effects of power.' But the effects of power are not just manifest in and through State-controlled data banks. They are also to be found in the knowledge of how to behave in the kitchen, at school, in the body, etc. Its effects are not merely repressive but also compulsive. They compel us to think, eat, drink, move, sleep, and talk in certain ways.

> It is a double process: an epistemological 'thaw' through a refinement of power relations; a multiplication of the effect of power through the formation and accumulation of new forms of knowledge. (Foucault 1979, 221)

A theory of power based on a theory of the discursive constitution of the cognitive grip as opposed to, for example, economic theories of media imperialism, is an explanation for how control operates at many levels of society without any empirically evident physical means of coercion. It is important to realize that, for Foucault, these systems of communication and cognition operate by means of a multi-faceted internalization of control and surveillance – 'panopticism' – rather than by means of class brutality. Lowi, while not beginning with the discursive, concurs here:

> The new integrated information system presents new potentials, in the extent to which it can provide the individual with an altered set of fact premises on which to base his opinions; or it can provide him with new and well-developed sets of value premises that could alter his opinions as well as his treatment of the facts. Once such generalized views are internalized in a people the influence is within rather than *on* the individual. (1981, 462)

It is interesting to note that even while trying to defend the 'neutrality' of computer/communications technology, Herbert Simon is also forced to admit that computer-communications will be discourses of internalized control or opinion engineering, a sort of collective internalization of conceptions and values concerning the world:

> If the computer has any implications for the effectiveness of democratic institutions they have to do with processes *for forming and informing opinion* rather than the process of recording it. (1980, 225, emphasis added)

While we would argue that the capacity for recording or surveying public opinion is directly linked to the capacity for controlling it, and that neither will lead to an increase in democratization, Simon does at least, if inadvertently, point out the capacity for internalized control of cognition via control of computer discourse. He reinforces his statement with the following one which explicitly draws the link between cognitive and behavioral control: 'It will help us only to the extent that we have valid scientific theories of the systems *whose behavior we are trying to model.*' (1980, 225-6, emphasis added)

Nor is the main thrust of governments' communications research completely divorced from directive communications systems of behavioral control as this résumé of research by the Canadian Minister of Communication amply illustrates:

> A comprehensive plan for behavioral research was prepared based on available information about office communications systems. Research planned for next year will assist with terminal design and implementation of new office systems, as well as the analysis of taste and measurement of performance in offices. During 1979/80 the department started an attitudinal study to identify reactions of potential users toward the technology. (Fox/DOC/*Blue Paper* 1979–80, 15)

The control envisioned here is micro-behavioral. We are not now making the delirious claims of the gurus of the information age that new communications technology will radically alter macro-economic behavior but rather that it penetrates our every minor gesture, if allowed, the sum of which is then tantemount to social control. In short, panopticism ensures: 'this fixed conception of a population that reproduces itself in the proper way, composed of people who marry in the proper way and behave in the proper way, according to precisely determined norms.' (Foucault 1980, 124)

3.8 A de-centred non-causalist theory of control

Power emanates neither strictly from the State nor strictly from capital. The theory panopticism illustrates is that relationships of power and discursive control have throughout history been more nuanced and more dispersed than suggested by the causalist claim that knowledge belongs to a class which rules by virtue of that knowledge and which has knowledge by virtue of its rule, as classically formulated by Marx:

> The ideas of the ruling class in every epoch the ruling ideas, i.e., the class which is the ruling *material* force of society, is at the same time its ruling *intellectual* force. The class which has the means of mental production, so that thereby, generally speaking, the ideas of those who lack the means of mental production are subject to it. (1959–1974, 64)

Weber, for one, had a more contextually flexible view of the power/discourse relationship. In his sociologically and anthropologically sensitive epistemology, Weber tried to show how various rules of reason characterized various types of society although many different rules could exist. Each society was particularized by a dominant form of reason. Weber was quick to see the exclusivities and tensions that arose between the interests of certain forms of reason and certain modern forms of rationality. This conflict arose within various classes or groups of individuals. For Weber, power and knowledge relationships are not confined to the lines of economics and class (Weber 1927–64, 62). Weber demonstrates how there are many types of value-relations that link up with forms of knowledge.

These elective affinities or interests of knowledge may be quite complex for Weber. And, these epistemological elective affinities, these pluralistic interests of knowledge, are each strictly related to rules governing conduct, i.e., to power.

> He (Weber) constructs social dynamics in terms of a pluralistic analysis of factors, which may be isolated and gauged in terms of their respective causal weights. He does this by comparative analyses of comparable units, which are found in different cultural settings. (C. Wright Mills, in Weber 1927–64, Introduction, 65)

Foucault's work also deals with the cognitive relationships of discourse in a very nuanced way. In a series of historical studies of various discursive artefacts he shows the discursive procedures followed and how they became accepted as constraints upon knowledge and behavior, i.e., as scientific discourse that legitimates social action. While Foucault (1970) shows that these procedures are the same for various disciplines within the same spatial and temporal coordinates – grammar, biology, economics – he stipulates that these procedures change within the same discipline from epoch to epoch.

Accordingly that which was scientific discourse on madness in the middle ages was not necessarily a dominant discourse in the renaissance or in the age of reason. Foucault describes the exercise of power by discourses of knowledge as a dynamic, changing, and localized operation: 'a discursive practice exercises its power by transforming a hitherto passive and unreflected segment of the social system into a formal object amenable to scientific technique.' (1980, 22)

For Foucault, power cannot simply be assumed in the abstract. Power is not an object. It is a complex set of relations that have to be practised in discourse in order to be meaningful: 'Power is employed and exercised through a net-like organization.' (1980, 98) 'In the substantive sense, "le" pouvoir doesn't exist. In reality power means relations, a more-or-less organized, hierarchical, coordinated cluster of relations.' (1980, 198)

3.9 Power as the hub of reason

While power relations have no linear causal relationships and are nuanced and complex, there is no question that these power relations are the central axis of the will to truth, or reason. While Weber and Foucault both relate communication to a control of knowledge procedures or cognitive filters in a nuanced and un-determinist way, both unequivocally insist that the central focus of classical rationality is *power:* hence 'Power/Knowledge' (Foucault 1980).

> In the last analysis, the processes of economic development are *struggles for power*. Our ultimate yardstick of values is 'reason of state' and this is also the yardstick for our economic reflections. (Weber 1921, in C. Wright Mills/Weber 1927–64, 36)

Weber defines power in terms of capacity for willed action within a social context. This capacity is enabled or legitimated by knowledge and communicational relations rather than constrained by physical force:

> 'Power' (Macht) is the probability that one actor within a social relationship will be in a position to carry out his own will despite resistance, regardless of the basis on which probability rests. 'Imperative control' (Herrschaft) is the probability that a command with a given specific content will be obeyed by a given group of persons. (1947–64, 369)

Control for Weber is therefore the ability to impose a certain behavior on others. His definition of 'discipline' is very close to Foucault's description of power as conventional constraints upon users of discourses, constraints based on habit, i.e., customary epistemics-discursive procedures:

> 'Discipline' is the probability that by virtue of habituation a command will receive prompt and automatic obedience in stereotyped forms, on the part of a given group of persons. (Weber 1947–64, 369)

The interrelation between discipline, in the sense of an academic of scientific discipline, and in the sense of a behavioral compulsion, i.e., to discipline someone, is essential to the theories of how rules of might become rules of right for both Weber and Foucault. Scientific communities of discourse reserve for themselves the right to command actions:

> Bureaucratic administration means fundamentally the exercise of control on the basis of knowledge. This is the feature of it which makes it specifically rational. This consists on the one hand in technical knowledge, which, by itself, is sufficient to ensure it a position of extraordinary power. (Weber 1947–64, 339)

The above remark not only reconfirms the interdependence of knowledge and power, it also leaves no doubt as to the role that technics plays in the power/knowledge relationship today. Technical knowledge for Weber refers to the mode in which administration is carried out, an extent of effective jurisdiction by authority. Organizational knowledge is operative in many disciplines, scientific, economic, political, and military. In all cases, technical knowledge operates as an appeal to knowledge procedures for the legitimation of action.

3.10 Power is not only repressive but also conducive

Power not only forbids us to behave in certain ways, it also compels certain types of behavior. One point concerning power that Foucault makes and which sets his analysis of power off from Marcuse's studies, is the contention that power is not only repressive but also conducive. Foucault's conducive view of power is derived from Nietzsche. Power can *compel* us to do certain things: it compels the body to behave in certain ways, the mouth to speak in certain ways, etc.

Applying a theory of power as compulsive to the question of new communications technology could illustrate how this technology operates as a discourse of knowledge and power not only by forbidding certain actions and relationships but also by compelling others. For example, many discourses on new communications technology suggest that consumption is a much prized behavior in new communications society. Indeed, as Telidon advertising for retailers indicates, new communications technology could conceivably compel consumerism. A still greater compulsion in the realm of new communications technology is simply the compulsion to follow the discursive procedures elaborated in chapter two. New communications technology as discourse compels its users to code, classify, order, survey, etc., along the same pathways of networking and accessing that these new systems oblige us to follow. Extant networking, for example, was shown to be a kind of bottleneck of both compelled and forbidden interdiscursive relationships, of included and excluded acts.

3.10.1 Compulsive issue-raising

It will be recalled that one of the main working hypotheses of this work is that various discourses enabled, indeed pre-conditioned, the raising of specific issues in specific ways. At this point it is also clear that discursive procedures actually compel the raising of certain socio-political issues in certain ways. The discursive procedures which function as the apex of control and domination compel the issues of privacy, free flow, individual freedom from constraint, billing, etc., to be raised to the exclusion of certain other excluded or taboo issues, such as those pertaining to the erosion of community structures or to public management.

The compulsive aspect of discourse may be responsible for the very insistence upon an information revolution. It is what ‘prepares the way’.

> If information is less a new force coming into the world than a powerful and abiding socio-economic factor reaching a new and determining position in the evolution of society – how much of

> its revolutionary potential is due to intrinsic causes and how much is due to the fact that *the way has been prepared.* (MacClean 1981, 17, emphasis added)

All of the discursive procedures that have been isolated and described in the previous two chapters, regardless of their 'philanthropic' content, raise issues in certain ways which re-inscribe power relations at every level of society and of the individual. The discursive procedures dictate the laws of truth and the legitimacy of power in all of the sites in which Foucault situates the operation of political power:

> The role of political power, on this (Clausewitz's) hypothesis is perpetually to reinscribe this relation through a form of unspoken warfare; to reinscibe it in social institutions, in economic inequalities, in language, in the bodies themselves of each and every one of us. (Foucault 1980, 89–90)

It would be feasible though tedious to isolate systematically each discursive procedure that has been pointed out in the preceding chapters and to show how it compels a certain action or attitude towards social relations. One need only think of 'gadgetphilia' as a case of compelled fascination with the machine, a fascination that has been seen to operate not only in the Pentagon, in theoretical works and research projects, but also in the imaginations and body gyrations of Atari users.

On the other hand, if discursive procedures no longer seek to occult the power focus then certain other issues would of necessity be raised. For example, in this book the power focus raises the major issues of how the various communicational procedures of new communications technology will affect certain hierarchical social alliances. Issues of individualism would have to yield to the compulsion to raise issues of social autonomy and identity as well as issues of social participation in democratic decision-making once the focus of social domination is no longer occulted.

3.11 External procedures of discourse

In *The Discourse on Language,* Foucault says that there are both internal and external procedures of discourse. The external procedures subsume and suffuse the internal procedures by which means they function. Chapters one and two uncovered internal procedures of discourses on new communications technology. While there is a degree of critique in the uncovering of the internal contradictions of these procedures, our most severe critique of new communications technology arises when we elaborate upon the external procedures of discourse and contrast them to some of the overt semantic claims of those same discourses.

To conclude this section we might say that the internal discursive procedures seen in the second part of this book have as their conditions of possibility several external discursive procedures which all serve to reinforce relations of social control, domination, and exclusive power. The external procedures are:
(a) the will to truth as the force behind discourse;
(b) the exclusivity of access to truth;
(c) the legitimation of power by truth;
(d) the exclusivity of domination by virtue of the exclusivity of access to truth.

3.12 Control of whom?/power to whom?

What institutions or groups will consolidate or accumulate power via these discourses of knowledge? Nora and Minc seem to concur with the Foucauldian view that a change in the procedures of communication will yield a profound social change in relationships of power:

> questions obviously arise in terms of social organization and the influence exerted by various social groups. Since I lay no claim to forecasting long-term events and do not believe in futurology I shall go no further. (Minc/OECD 1980, 159)

Do we dare go further? Can we do so without falling back into unacceptable futurological projections? Lowi, for one, does dare to forecast who will be endowed with power as a result of new communications technology. The powerful are the dictators of the procedures of knowledge, the few designers and programmers who understand the procedures of the system.

> Many will know how, yet few will know. Those who know will be setting the agenda for those who know how. This means a

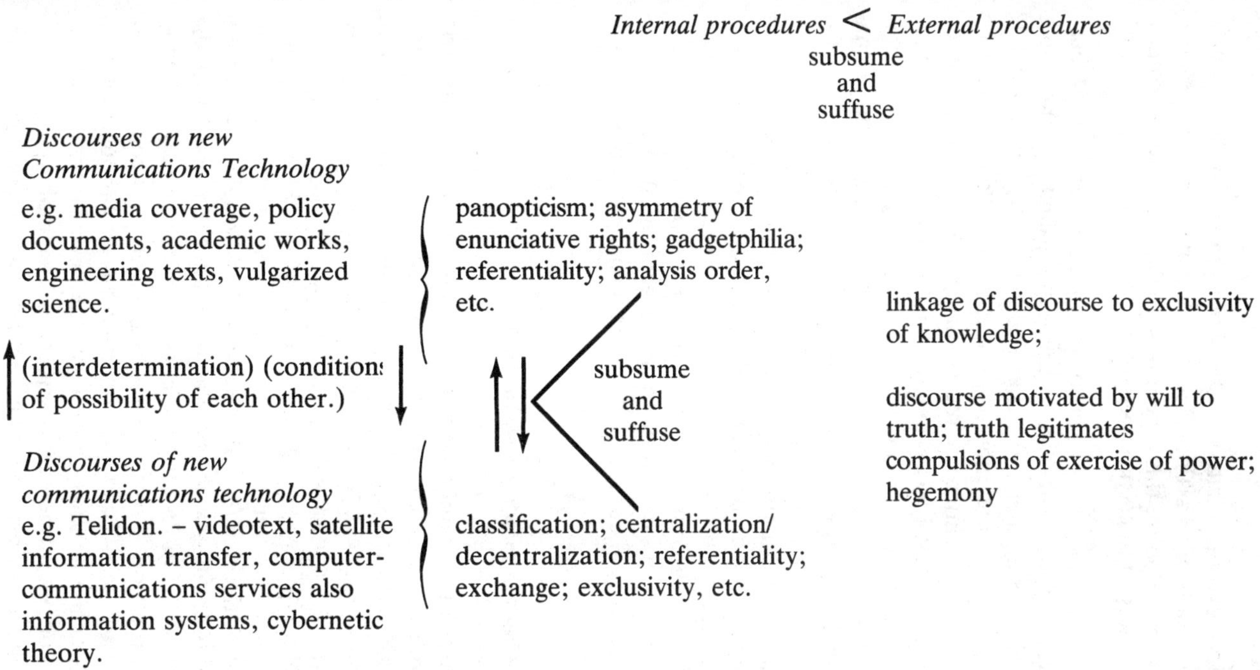

Internal procedures < *External procedures*
subsume
and
suffuse

Discourses on new
Communications Technology

e.g. media coverage, policy documents, academic works, engineering texts, vulgarized science.

panopticism; asymmetry of enunciative rights; gadgetphilia; referentiality; analysis order, etc.

(interdetermination) (conditions of possibility of each other.)

subsume
and
suffuse

Discourses of new
communications technology

e.g. Telidon. – videotext, satellite information transfer, computer-communications services also information systems, cybernetic theory.

classification; centralization/decentralization; referentiality; exchange; exclusivity, etc.

linkage of discourse to exclusivity of knowledge;

discourse motivated by will to truth; truth legitimates compulsions of exercise of power; hegemony

(The global procedure of discourse in classical science is that discourse/knowledge/power are exercised as a form of exclusive Domination and Social Control.)

key: x > y reads x subsumes and suffuses y.

Figure 3.1 The levels of discursive critique

> potentially extreme centralization of power to set the fact and value premises of all action. If a government or any other elite comes to monopolize this function, social control itself becomes monopolized. (Lowi/MIT 1981, 457)

Power to the people? Or power to the programmers? Or perhaps even power to those who dictate to the programmers, the upper echelons of the computer companies as described by Kidder (1981)?

However, this is not merely projectionism, which should be avoided if this text is not to be recast amongst those laid bare in the earlier chapters. Lowi has avoided projectionism while still being able to speak about the potential inheritors of power via new communications technology. He has avoided the amnesia that speculative futurology exhibited by recalling how, in the past, control of discourse has legitimated certain types of power groups. Thus, where a continuity of procedures is found, the existence of power lobbies might be suggested with more justification than prophetic speculation. On the basis of knowledge of historical relationships of discourse to power groups it should be possible to establish certain sets of criteria for the organization of communications procedures in order to fulfill certain configurations of power relations.

3.13 Revising the man/machine debate

Our regime of truth and power is centred in societies of discourse which produce and control the procedures of technological discourses. Another word for society of discourse is institution. The conflict is between these societies of discourse or institutions, not between man and machine.

In chapter one it was shown how the question of who has power in the computer-communications age was often reduced to variations on the man/machine debate. Localizing domination in such a confrontation only served to occult the question of the social locus of relations of domination. In other words, as long as we concentrate on a sort of wrestling match between man and machine, we lose sight of the contest between various groups in society to achieve social control in and through control of the discourses on and of new communications technology.

Marx, for one, instead of suggesting that the machine overpowered man, or vice versa, concentrated on the struggle between a) the society (class) that controls technology and b) the society which is subject to the control of the dominant group.

In volume one of *Capital* (342–581), Marx speaks of the process by which large-scale industry was developed on the basis of a division of labor according to the functions of the machine rather than to the function of human activity. This reduced the modern worker to an appendage of the machine. In the same section we also find a passage dealing with the 'inversion' of the relationship between dead and living labor and the various efforts by capitalists to increase productivity by placing an ever greater burden upon the worker in his relations to the machine.

In Marx's discussion of social exploitation, as it relates initially to the machine and then to technology in general, it might appear that he is describing the suppression of man by machine. A closer look reveals, however, that a certain 'class' – which we would prefer to designate more flexibly as a 'society of discourse' since this can also mean groups organized around a principle other than economics – manages to control the scientific discourse of technology and to impose it upon other societies of discourse which have no control. In other words, rather than insisting upon the man/machine struggle, Marx encourages us to view conflict and the potential for domination as a struggle for power between classes. We prefer to substitute the notion of various discursive societies from which the technological society is indissociable and in which it is perhaps even dominant. Obviously Marx meant certain things by class that 'society of discourse' effaces. To develop this would be to write a book on Marx and Foucault which is not our purpose here. Suffice it to say that we are trying to avoid certain *a priori* assumptions about the users of discourse as 'historical subjects' who are subject to dialectical laws.

3.14 Beyond the conspiracy theory of power grounded in an economic base structure

> The scope of determination of social relationships and cultural phenomena by authority and imperative co-ordination is considerably broader than appears at first sight. (Weber 1947–64, 327)

Power is not simply projected from some strategic group, capitalist or other. Power is a set of dispersed nodes of domination.

Although both Weber and Foucault insist that power and control are the central foci of discourses of knowledge, neither reduces power to the activity of a conspiring group. Power exists at many levels. In other words, though power is at the centre of knowledge it is not necessarily centralized. Foucault illustrates that for each epoch – episteme – there is not just one knowledge but many knowledges ('savoirs') organized on a variable hierarchical scale. This is one reason why Foucault concentrates his studies on 'knowledges located down low on the hierarchy' as opposed to 'high knowledges' (1980, 82). Thus, in an implicit reply to certain 'automatic' marxists,[5] he shows how power may be exerted as strongly in a discourse of and on sexuality as in a political text. The codes of conduct at a nudist beach are as much if not more a site of power over citizens (law and order) as is parliament (sovereignty). Rather than suggesting that the Rothchilds or the Rockefellers exercise power exclusively, Foucault shows the workings of power at 'much more minute and everyday levels.' Power occurs in the everyday discourses of society. For this reason we have sought discursive relations not only in policy documents on new communications technology but also in tv ads about this technology and in electronic video games. Consequently, what we have shown is not simply a change in a power elite, be it the government or a handful of capitalists, but an alliance of power relations and rules that exist as a whole discursive configuration, as an episteme that penetrates to very minute levels of speaking. Power in the fields of discourse around new communications technology has in this book been shown to be *complex, multi-faceted, dispersed, and localized:*

> Moreover, in speaking of domination, I do not have in mind that solid global kind of domination that one person exercises over others, or one group over another, but manifold forms of domination that can be exercised in society. (Foucault 1980, 96)

For example, in *Discipline and Punish,* Foucault suggests that power is exercised at very minute levels of bodily behavior:

> The object of the control (. . .) was no longer the signifying elements of behavior or the language of the body, but the economy, the efficiency of movements, their internal

> organization; constraint bears upon the forces rather than upon the signs; the truly important ceremony is that of exercise. Lastly, there is the modality; it implies an uninterrupted, constant coercion, supervising the processes of the activity rather than its result and it is exercised according to a codification that partitions as closely as possible, time, space, movement. (1979, 137)

One can very easily see the pertinence of this portrayal of power as dispersed, minute, and compulsive to the procedures of billing and cross-referencing in an information society:

> But in thinking of mechanisms of power, I am thinking rather of its capillary form of existence, the point where power reaches the very grain of individuals, touches their bodies and inserts itself into their actions and attitudes, their discourse, learning processes and everyday life. (Foucault 1980, 39)

Where a record is kept of one's every act of communication, where each piece of communication practice must be paid for, one can well imagine such penetrations of power at even the most intimate level of society. One need only think of the present day cable distribution and billing of pornographic films begun by Warner Corporation's QUBE systems.[6] What practices of sexuality are compelled? What representations are circulated? Who is allowed to speak, to initiate, and to submit?

Foucault insists that a theory of power as localized and compulsive is more akin to Nietzsche than to some marxists in that: 'power is not primarily the maintenance and reproduction of economic relations, but it is above all a relation of force.' (1980, 89–90)

Weber also analyzes power in a non-reductionist and non-deterministic way. He describes power and control as manifest within a vast network of various rationality-oriented and -motivated relationships, thus making power a function of knowledge and interaction rather than exclusively of economics:

> It is essential to include the criterion of power of control and disposal (Verfugungsgewalt) in the sociological concept of economic action if for no other reason than that an exchange economy involves a complete network of contractual

relationships each of which originates in a deliberately planned process of acquisition of powers of control and disposal. This, in such an economy, is the principal source of the relation of economic action to the law. But any other type of organization of economic activities would involve some kind of distribution of powers of control and disposal. (1947–64, 163)

Thus, Weber does not exclude economic exchange from power relationships. He simply does not reduce power to economics:

The concept of power of control and disposal will here be taken *to include* the possibility of control over the actor's own labour power, whether this is in some way enforced or merely exists in fact. (1947–64, 164, emphasis added)

For Weber, power through bureaucratic knowledge and organization is not reducible to political-economic domination:

Bureaucracy is superior in knowledge, including both technical knowledge and knowledge of the concrete fact within its own sphere of interest, which is usually confined to the interests of a private business – a capitalistic enterprise. The capitalist entrepreneur is, in our society, the only type who has been able to maintain at least relative immunity from subjection to the control of rational bureaucratic knowledge. (1947–64, 339)

The non-reducibility of power and domination to economic elites would suggest that the defence of new communications technology on the grounds of a decentralization of ownership and control is sufficient to guard against power and domination in the social sphere. Weber, like Foucault, insists that power is situated not only in the State and the economy but also in various specific institutions of discourse and knowledge, such as educational societies of discourse:

For instance, the authority exercised in the school has much to do with the determination of the forms of speech and of written language which are regarded as orthodox. The official languages of autonomous political units, hence their ruling groups, have often become in this sense orthodox forms of speech and writing and have even led to the formation of separate 'nations' (e.g., Holland and Germany). The authority of parents of the school, however, extends far beyond the determinations of such cultural

> patterns which are perhaps only apparently formal to the formation of the character of the young, and hence of human beings in general. (1947–64, 327)

One result of Weber's trans-economic analysis of power was that, unlike Marx, he predicted the massive bureaucracies of socialism. What this seems to illustrate is that social control has a telos above and beyond the economic. A mere change in economic conditions could not and will not correct some of the relations of domination that stem from new communications technology. New communications technology and discourses about it form a discursive configuration, an episteme, which circulates at all levels of society. These discourses are procedures of constraint that operate in relation to, but are not determined by, economic relations. For example, it is the imposition of certain discursive procedures in and through a standardized set of networking or storage patterns rather than purely profit-related concerns that dictates a technocratic, specialist-oriented relationship of domination, i.e., of division of labor. One proof of this argument is Foucault's demonstration in *The Order of Things* that the discursive procedures of exchange neither originated in nor were exclusive to the economic sphere. Rather, exchange was a discursive procedure of the episteme shared by various disciplines and social practices including grammar, painting, and biology. The relationship between a set of discursive procedures of knowledge and economic conditions of production becomes clearer in Weber's distinction between formal and substantive rationality.

3.15 Beyond economic determinism: bureaucratic organization and the formal rationality of economic action

Bureaucratic organizational tyranny is backed by the boundless technique of discursive procedures of rationality. Bureaucratic procedures belong to formal rationality. Particular concretizations of them are substantive rationality. Rather than making all action subservient to an economic base structure, Weber makes economic activity – substantive rationality – a particular case or realization of the formal procedures of rationality. By rationality, Weber means a set of knowledge procedures that are carried out within the language and interactive relationships of society. These procedures include qualification, systemization, referentiality, confidence in referentiality or factuality, and organized control, all of which we have seen to function in the discourses on and of new communications technology.[7]

Applied specifically to the economic sphere, these traits circumscribe what Weber described as the 'formal rationality of economic action.' This formal rationality becomes manifest or 'substantialized' as the following procedures of concrete practice:

> (1) The systematic distribution of utilities on which the actor can count;
> (2) The systematic distribution of available utilities and potential uses in the order of their estimated relative urgency;
> (3) The systematic production of utilities through 'manufacture';
> (4) Systematic acquisition. (Weber 1947–64, 166–67)

What is most important here however is not these precise manifestations of formal rationality in concrete economic practice but the fact that various forms of concrete practices in society, of which the economic is but one, are themselves all realizations – substantializations – of formal rationality. Jürgen Habermas, following up on Weber's distinction, suggests that many other types of activities besides the economic adhere to the discursive and knowledge procedures of instrumentalist rationality. Orthodox pedagogy, news broadcasting, traditional diplomacy, industrial management, and often interhuman relationships function according to this dominant form of rationality. The main characteristics of such a rationality involve a means-ends logic, a form of utilitarianism, and a judging of correctness based on true/false distinctions whereby all of these traits aim at one goal: *the effective control of reality* (Habermas 1968-70, 92).

To this list of activities governed by the procedures of formal rationality one can easily add practices of new communications technology. Earlier it was seen that the principle of instrumentality functioned constantly as a discursive procedure of both discourses of and on new communications technology. For example, efficiency-oriented systems organization is the rationale behind information or systems technology. Means/ends logic for effective control of reality cannot be denied in the numerous military sites which practise these procedures or in the business place where computers are used for greater control over the environment which in turn leads to greater productivity and efficiency. The quantitative bias of information processing systems is obvious given that the encoding is a variation of Boulean mathematics and that calculation is a principal function. One of the aims of computer-communications technology is clear in Weber's statement that 'the term "formal rationality of economic action" will be used to designate the extent of quantitative calculation or accounting which is technically possible and which is actually applied.' (1947–64, 184)

New communications technology, it would seem, qualifies as a form of substantive rationality in that it applies the technical procedures of calculation from formal rationality to the social sphere where certain ends, no matter what they be, are sought via formal, technological means:

> The 'substantive rationality,' on the other hand is the degree in which a given group of persons, no matter how it is delimited, is or could be accurately provided with goods by means of an economically oriented course of social action. This course of action will be interpreted in terms of a given set of ultimate values no matter what they may be. (Weber 1947–64, 184)

As regards efficiency, new communications technology is to communications practice what money is to economic practice: 'From a purely technical point of view, money is the most efficient and calculable means of accounting.' (Weber 1947–64, 184) Economics is not the dominant base structure of society. Rather, the procedures of knowledge constitute dominant configurations of reason which subsume economic practices.

While the dominant procedures of discourse can conceivably change from epoch to epoch, for Foucault the contemporary dominant form of discourse remains, despite a few 'modern' exceptions, the classical, rationalist one that he describes in all of his works. For Weber, the dominant form of rationality was formalist, bureaucratic, technological rationality based in an instrumentalist, means-end logic.

Thus it is the dominant form of rationality (or dominant episteme), and not merely the economic, that is responsible for the exercise of power:

> Bureaucratic control is inseparable from increasing industrialization; it extends the maximally intensified efficiency of industrial organization to society as a whole. It is the formally most rational form of control, thanks to its 'precision, steadfastness, discipline, rigor, and dependability, in short calculability for both the head (organization) and for those having to do with it (. . .)' and it is '*domination by virtue of knowledge,*' ascertainable, calculable, calculating knowledge, specialized knowledge. Properly speaking it is the *apparatus that dominates, for the control of this apparatus, based on specialized knowledge, is such only if it is fully adjusted to its technical demands and potentialities.* (Weber, in Marcuse 1968, 216, emphasis added)

3.16 Changing more than the machines or the economics

The implications of the distinctions between formal and substantive rationality for the debate on new communications technology and for policy-making are vast. Weber and Foucault both seem to be suggesting that in order to change certain social relationships one must change not only the substantive sphere of rationality but also and more importantly formal rationality itself. If one wishes to change the discursive procedures of new communications technology and of its social environment, one must change more than the substantive manifestations of that technology, be they 'machines' or their economic organization. The episteme itself, or what Weber would call 'instrumentalist rationality,' would have to be altered since it is the condition of possibility of the particular manifestations of technology. A major corroboration of this position does seem to lie in the remarkable similarity of the way in which both East and West view new technology: as a bureaucratic tool with which to gain more efficient production and greater control over the environment (Grant 1976, 124). Again the reader is referred to the Russian delegate Fedoseyev's treatment of technology at the UNESCO conference on the social implications of new technology. He does not talk specifically about economics or machines but rather about and according to a set of discursive procedures of instrumentalist rationality.

It is not our intention here to suggest that there is a reversal of the super-structure and base-structure allocation made by Marx whereby communication/discourse suddenly becomes the base-structure and economics a superstructural manifestation. All that can be said is that the procedures of rationalization in discursive interaction position or contextualize specific, manifest economic relations and actions. Communications relations do not change with a simple change of economic ideology. On the other hand, communications procedures of organizational knowledge are increasingly being recognized as the seminal procedures of economics/business, though those procedures are often erroneously reduced to information and hardware.

> All these developments result in the creation of new national and international *infrastructures* for the transmission and processing of digital data. They will soon have *truly global dimensions* when satellites will be increasingly used to provide access and link-up facilities from virtually all points of their range, without having to wait for special lines to be laid to remote locations by the various telecommunications carriers. (Gassman/OECD 1980, 12, emphasis added)

> Even where international trade involves non-information exchanges, it will nevertheless be totally dependent on information exchange systems. (Henry Geller, 'Introduction,' *OECD* No. 3 1980, 77)

At this point some explanation may be required as to why we have insisted so much on the non-reducibility of the sites of power to economic or class relations. Apart from trying to provide an adequate seat of power in technologized society, our motives are in the end interventionist. It would seem that over-insistence on the economic relations of technology and power has led to intervention purely within the economic sphere at the expense of intervention aimed at changing the very roots of control-oriented technological rationality. Not just economics but techne and episteme must change if relations of power are also to change.

3.17 Power: centralization vs decentralization

One of the main debates involving social relations and new communications technology has been over whether or not its implementation would lead to the centralization of power and social control in the workplace and in society at large. On the one hand, there are those, who like James Taylor (1981), insist that new communications technology will decentralize decision-making in the office, while others, such as Gilpin (1980), insist that the space- and time-binding power of new communications technology will lead to the expansionist centralization of both industrial and political decision-making.[8]

Does Foucault's theory of power as dispersed nodes of domination *preclude* any discussion of the powers of centralization by new communications technology? We think not. While Foucault does insist on the multi-faceted areas of power's impact he does not exclude the possibility of a centralization within certain societies of discourse, often referred to as institutions and disciplines. Indeed, a part of his project was to show 'the effect of the centralizing powers which are linked to the institution and functioning of an organized scientific discourse within a society such as ours.' (1980, 84)

In other words, the fact that Foucault sees power's effect as multiple and as occurring at many levels of social and individual existence does not mean that the practices of power are not centralizable. Control over discursive procedures of right and might may be concentrated in the hands of an exclusive community of the initiated, while the practice of the procedures may occur at various levels of social interaction. For example, where IBM designs, structures, deploys, and maintains certain types of communicational pathways and accessing patterns at the expense of others, there is no doubt that the effects of centralizing powers are in operation. The theories of knowledge and of social control of Foucault and Weber permit us to say something more significant about this question than a blanket, a-contextual declaration that new communications technology will or will not lead to centralization of control. The degree of centralization depends upon the degree to which inter-discursive relationships are hierarchically monopolized, i.e., the degree to which a particular society of discourse has managed to infiltrate the realms, spheres or fields of other discourses. Discursive hegemony is quite possible. Thus while in the communications society we might look for manifestations of power at the minute level of discourse, it may be the case that the technical instrumentalist discourse of systems analysis and calculation has in fact infiltrated these 'domestic' discourses. The degree of centralization of social control in the information society would depend upon the degree to which the discursive procedures raised in chapters one and two had infiltrated the discourses of other spheres of social discourse. One indicator that this infiltration may indeed be vast is the fact that we found the procedures of discourses on new communications technology to have been so regular for so many 'types' of discourse, including those of academia, journalism, commerce, and entertainment. We also uncovered *interdiscursive cartels,* the equivalent of nodes of concentrated power. Yet another indicator of the centralization of this set of epistemic procedures is the degree to which specific procedures of one form of new communications technology (e.g., IBM accessing, association, classification) manage to penetrate all other geographic and commercial spheres of new communications technology. The exportation of similar procedures of classification, association and accessing patterns throughout a world-wide network of computer-communications would be just such a hegemony.

For Foucault there have always been elite or privileged societies which dictate the procedures of the dominant episteme. Such societies may be likened to centralized nodes of power. Furthermore, Lowi suggests that given the ubiquity of new communications and computer technology the procedures of this discourse are indeed becoming the 'dictators' of the episteme, thereby creating a form of centralized control:

> The composition of the political elites, or top power groups, was altered still further with each successive influx of social or skill types, who were commanded. In recent generations we have witnessed an influx of such new elite members as the technical personnel (whose resources are specialized knowledge or expertise), the technological personnel, etc., and the technocratic or technetronic personnel (whose resources are information that passes through mechanisms and organization). (Lowi 1981, 455)

Harold Innis has shown how the control of a certain form of communications, such as the wire service, has led to centralization, to what he labels 'space-binding' and 'monopoly of knowledge.' His treatment of the relationship of political influence to the wire and newspapers systems in the US is one clearly documented case of centralization that stems from a monopoly of procedures of communication (Innis 1951–77, 156–89).

Gilpin, referring to Innis, Boulding, and Worsteller, suggests that computer-communications technology, by linking up the peripheries to the centre, has managed to reduce the significance of the distance/strength ratio, hence facilitating expansionist centralization or 'space-binding.'

> The impact of technological advance on political organization is largely owing to its effect on the ability of political centres to extend control over territory. As Kenneth Boulding has shown, the extent of control is a function of loss-of-strength gradient of a political centre, i.e., 'the degree to which its military and political power diminishes as we move a unit distance away from its home base.' (Gilpin 1980, 230)

> One school of thought holds that the computer and other modern technologies have undercut the significance of the loss of strength gradient thesis that power diminishes with distance. (Worstetter, in Gilpin 1980, 233)

One thing is certain, throughout history, regardless of new communications technology, where there is a monopoly over discursive/communicational procedures such epistemic centralization has been inflicted. The fact that we have so far found such regularities of procedures would indicate that new communications technology has done nothing to discourage a dominant episteme, based on classical, instrumentalist science which amounts to the centralization of a set of discursive procedures.

Weber also suggests that bureaucratic rationality has been sufficiently universalized to speak of 'monocracy' if not 'monopoly' or of the 'centralized' control of rules of reason:

> the monocratic variety of bureaucracy – is from a purely technical point of view, capable of attaining the highest degree of efficiency and in this sense formally the most rational known means of human being (. . .) The development of the modern

> form of organization of corporate groups in all fields is nothing less than identical with the development of and continual spread of bureaucratic administration. (Weber 1947–64, 331)

Lowi, quoting C. Wright Mills, insists that these modes of discursive control over information, these so-called 'management procedures,' are increasingly dominant and centralized within their particular sites: 'The regular *interchanges* among corporate, military and civil government elites proved the existence of a single national "power elite".' (Mills, in Lowi 1981, 456, emphasis added)

3.18 Decentralization: a contextually realizable possibility

Innis (1951–77), Carey (1969), Gilpin (1980), and Minc (1980) all suggest that new communications technology could potentially facilitate the unification of members of societies of discourse with common interests at the expense of hegemony.[9] They argue that:

> communications technology may be biased either way, in some cases encouraging political fragmentation and in others favouring centralization. (Gilpin 1980, 230)

> But where conflicts are many, varied and decentralized (. . .) the stakes are different (. . .) There is no one unifying core of conflict (. . .) clearly the modes of aggregation of various types of opposition which lie at the very root of political action, are bound to change (. . .) This will have marked consequences for the balance of socio-political forces. (Minc 1980, 158–59)

It is not the absence or presence of networking hook-ups alone that will determine whether centralization or decentralization will occur as a result of new communications technology. It is the extent to which particular groups may impose upon other societies of discourse an adherence to the rules and procedures of their own society of discourse. Thus, depending on the degree to which new communications technology encourages or discourages obedience to and internalization of a unified set of discursive procedures, this technology may be said to contribute towards centralization or decentralization.

Pleading the case of decentralization, Minc suggests that there will be both large, centralized corporate groups of communicational control and smaller and medium sized groups or discursive societies (Minc/OECD 1980, 158).

Pergler (1980) makes an even more interesting suggestion concerning the potential for the decentralization of communication and social control. He states that corporations will quite inadvertently impose new communications technology in domestic situations as two-way incasting and that, just as inadvertently, the discourses of these incasting technologies will follow corporate discursive procedures, unless the government intervenes to ensure otherwise. In other words, for Pergler, lack of government intervention will encourage centralization but its presence could diffuse the monopoly of 'corporate' discursive procedures over the domestic discursive sphere.

> Business interests, without caring much about social incasting, will probably soon introduce to Canada technologies and organizations that will make social incasting possible. Once introduced, social incasting will contribute to the pressure for improvement of over-government-citizen communication. If governments undertake a profound revision of this process of communication, incasting may marginally exacerbate the coming conflict. (Pergler 1980, xvii)

Lowi also deviates from the theory that centralization of corporate interests will lead to the discursive centralization of corporate control. He lists some traditional communicational procedures, which are roughly akin to the ones uncovered in the previous chapters, and then suggests alternative procedures. While we do not wish to indulge in futurist projections, it is worth noting that Lowi describes a possible change of power relations in society as dependent upon a possible change in communicational procedures of organization:

> External supervision through span of control, formal hierarchies of authority and narrow definition of jobs is being supplanted and may in the long run be replaced by other principles of organization, including professionalization, lower-level decision-making, networking patterns rather than simple vertical patterns of communication. (Lowi 1981, 464)

While the debate over centralization and decentralization rages on, we may at least make the following preliminary conclusions: centralization or decentralization occurs within a discursive community of knowledge seekers rather than within the physical force of an isolated individual. When we search for traces of centralization or its opposite we should look for them as relationships of discursive procedures and hierarchies. Here we would like to echo the disdain of Carey and Quirk (1970) for the great debate around the issue of centralization which treats the problem entirely out of context. In order to discover whether new communications technology leads to centralization or decentralization we can only look at concrete, discursive practices, and particularly at interdiscursive relationships within the context of specific social conditions of communicational praxis.

Future projections can only be vectors of observations concerning the types of inter-discursive relationships that have led to either centralization or decentralization in the past. Technological interdiscursive practices as have been described in Innis' analysis of the space-binding role of newsprint and wire services in American political alliances (1951–77, 156–89), in Foucault's examples of discursive hierarchies or in Weber's accounts of centralized managerialism in bureaucratic technology are quite revelatory and point to an increasing trend toward centralized hegemony.

3.19 Anonymity, paranoia or the sites of power

Many automatic marxists wish to make capital the source of social control. We wish to argue that, though capital can be active, it actually only associates knowledge/technology and surveillance in historically given sites or locations. It does not 'possess' power. If capitalism gave the discourses of technology a new twist, it was by letting them become dominant in specific loci which were different from those of pre-capitalist society. Capitalism is the site but not the source of power.

One may object that the preceding pages are defining some sort of metaphysical force: power which is never identified with actual entities in society. Is this critique of instrumentalist rationality nothing more than an escapist description of the social functioning of power, as paranoid 'theories of impersonal forces' manipulating society?

Lentricchia (1982) criticizes Foucault's theory of power from a marxist perspective because of its 'anonymity.' No single class is identified as the source and possessor of power. He calls this theory of a power unattached to identifiable actors a 'paranoid metaphysics.'

David Noble, in his book *America By Design,* also attacks any theory of power and technology which fails to identify specific human beings as the possessors of the power yielded by technology and as the source of that technology. Noble insists that people drive technology and that those people are engineers. In his work he describes the links between the educational institutions that train engineers and the managers of corporations at the service of capital. Noble's work is typical of a strain of traditional critique of the relationship of technology to social domination in that it attempts to identify certain subjects as the driving force behind the development of technology and the exercise of power, subjects who will the existence of certain social, scientific and discursive practices in order to substantiate their own power.

> And like any human enterprise it (technology) does not simply proceed automatically, but rather contains a subjective element which drives it, and assumes the particular forms given it by the most powerful and forceful people in society, in struggle with others (. . .) like all others, this historical enterprise always contains a range of possibilities as well as necessities, possibilities seized upon by particular people, for particular purposes, according to particular conceptions of social destiny. (Noble 1977, xxii)

We would argue that this thesis gives rise to a causalist version of the conspiracy theory whereby a certain faction in society requires a certain technology for its own power and causes it to be developed in order to possess it: 'the history of modern technology in America is of a piece with that of the rise of corporate capitalism. (Noble 1977, xxiii) For Noble, 'technology is people' (xxii). However, we have shown in our study that what particular institutions and individuals can say and discover about new communications technology is not just a question of people but of historically situated discursive procedures or patterns. People do not always drive and invent technology. Often technology is a case of certain already extant discursive procedures which are simply becoming manifest in other spheres.

Nevertheless, while we do not believe that this technology can be causally linked to any driving subject we do agree that it sounds equally paranoid to list discursive procedures of power without identifying some of the particular sites of their practice in our society. For Foucault, an author, a discipline, an institution are all sites in which discourses function and between which relationships of power exist, they are *not* the *sources* or the *possessors* of that power. Nevertheless relationships between these sites and the discourses that operate in them are relationships of power. What is more, power is a case of relationships that surpass while including the economic sphere. Although power relationships may be identified in certain sites of capitalist functioning, Foucault would avoid the foreclosure of stating that power and discursive procedures spring out of the head of capitalists. Foucault argues that power is not controlled by anyone or any group/class. Rather, Foucault allows that power can potentially exist in many sites as a product of relationships between many types of societies of discourse. However, this is not to say that for Foucault power can be just anywhere at any time. Power can and often does function as a network of relations from top to bottom, that is to say hegemonically. One goal of Foucault's research was to locate the specific sites and configurations of power in various epistemes. They were not always class alliances. In certain specific times and places, power relations exist in specific configurations, certain sites take precedence over others and certain societies of discourse, via their use of the procedures of discourse, appropriate more power over others.

Bearing in mind, then, that Foucault does allow for specific sites of power relationships, it seems that one way to avoid the seeming paranoia of describing an autonomous or disembodied form of power while also avoiding the subjective, causalist conspiracy theory, would be to treat the descriptions of the sources of power, as are found in Noble, Lenticchia, Galbraith, Lowi, Gilpin, and Braverman, not as identifications of *subject-sources* of power but rather as *consolidated sites* of power in the Western, bureaucratic, scientifist, capitalist episteme. Foucault's critique of the Enlightenment is indeed a critique of specific sites of power, i.e., science, and he explicitly stated his role as an intellectual as one of finding alternative, less hegemonic configurations of power. Our task, then, will be to see if, without reverting to a theory of the subject as the source of discourse and power, we can refer to these scholars' identifications of certain sites or societies of discourse that are today more hierarchically powerful than others.

Again these sites must not be identified as subjects of discourse or as sources and possessors of power but rather as *fields* or locations through which the discourses of power circulate. It may be argued that both Lentricchia and Noble betray, in their discussion of the sources of the exercise of power through knowledge, the possibility that these discursive procedures of science and technology existed before corporate capitalism and stem rather from a type of rationality than from a specific subject. Lentricchia's argument that capitalists are the driving subjects of technologies of discipline actually reinforces our own argument that capitalism is merely one of these technological procedures which inform instrumentalist classical, bureaucratic rationality. Lentricchia's own insistence that 'discipline promotes the efficiency that capitalism dreams of, and at the same time, dissolves the sort of consciousness that would turn workers into enemies' (1982, 45), effectively points to procedures which Weber and Foucault named in relation to the rise of bureaucratic and classical science, procedures that certainly are manifest in capitalist societies but that are neither exclusive nor generic to them. Lentricchia is admitting here that discipline serves instrumentalist reason whose procedures are the search for control and efficiency. Still, as Weber argued capitalism is but one site in which these procedures are substantiated.

Noble also demonstrates how engineers cum managers call for 'growth,' 'efficiency,' and 'control of nature and humans.' He argues that they do so *because* they serve capital. It may well be the case that engineers and managers 'in America by design' are the handmaidens of capital. However, our argument remains that these procedures of growth, efficiency, control, and rationalization have all been in operation since the pre-capitalist mid-seventeenth century and accompany a certain form of dominant reason rather than stemming exclusively from an economic relationship. Rather than making capital the source of the above mentioned discursive procedures, we would argue that management, corporatism, and engineering are all *interlocking sites* or media where the discourse of classical, scientific instrumentalism manifests itself.

Noble argues that technology does not have a life of its own, that people place that life into it, and that it is the product of some creator's will: 'modern technology in fact began to come under the conscious control of human authority in the specific form of private corporate capitalism' (xxvi). We would argue rather that technology as a set of discursive procedures and practices does indeed have a structure and life-process of its own. Social actors are not excluded from the practice of technology, they are merely the places not the sources of that practice. The procedures of technology lock actors into them, so to speak. Our analyses in chapters one and two have revealed that people are indeed spoken by technology. Technology speaks through them in the sense that so many have been caught mouthing discourses which do nothing more than repeat sets of procedures. Many more have been caught structuring electronic devices according to procedures as old as Bentham's panopticon. Yes, the discourse may serve these sites but they do not invent or control it. A discourse may serve a faction of society without having been invented or being controlled by it.

At this stage, bearing in mind the epistemological differences between our argument and those of Noble, Galbraith, and Braverman, it remains possible to refer to the studies of these scholars in order to identify sites of power relations upon discursive (scientific, technological) knowledge relations. Thus, we can refer to these extremely pertinent and historically informed identifications of the functioning of procedures of control and efficiency in managerialism throughout the history of the 1900s as a very illuminating identification of some of the social loci of the technological discourses of power.

3.20 Scientific managerialism: systems human engineering

Noble argues that technology has helped capitalist social relations to survive and has used the means of scientific engineering and management to make them fit corporate capitalism.[10] While Noble maintains that engineers are the source of the scientific procedures underpinning technology (growth, stability, and control), we would argue that the link between the discursive procedures of engineering and the corporate world is an identifiable site of the functioning of the procedures of instrumentalist reason, of technology, and of a relationship over those excluded from this site. We are not suggesting that 'technical reason' has strictly logical principles but that the principles of growth, stability, and control formed a part of rationality long before the rise of corporate engineering.

We could view Noble's enterprise as a sort of discursive genealogy in which he shows how the discourse of engineers linked the discourse of pure and applied science (control of nature) to the discourse of the behavioral social sciences (control of human nature) and forms the discourse of scientific managerialism.

> Thus, from the outset, they hardly proceed according to the dictates of some logically consistent 'technical reason,' blindly advancing the frontiers of human enterprise, but rather informed their work with the historical imperative of corporate growth, stability and control: as their technology progressed, so too did the science-based industrial corporations which they served (. . .) Forces of production and social relations, industry and business, engineering and the price system – the two poles in the dialectic of social production – collapsed together in the consciousness of corporate engineering, under the name of management. (1977, xxiv)

Herein lies what we referred to earlier as a discursive cartel of discursive societies. What is important for us, though, is that the fusion of the discourses of these various sites produces a discursive conglomerate akin to the discourse of classical instrumentalist rationality as described by Weber and Foucault.[11]

The specific sites that Noble identifies as those of the exercise of power through technology are the institutions (which we could call the discursive societies) of industry, education, as well as institutions such as licensing, patenting bodies, and research institutes at the service of industry:

> While some engineering educators continued to try to broaden the curriculum in order to enhance their academic and professional prestige, others, at the behest of the managers in the new science-based corporations, began to call for a 'liberal' education for other reasons. As one engineer phrased it, 'A liberal education gives power over men.' (Noble 1977, 32)

> As he strove to create a professional identity for himself, the engineer commonly tried to present himself to the public as 'technology' itself, the great motive force of modern civilization. In doing so however, he was forced to identify himself with corporate business, for that too was technology, as much a part of him as he was of it. (Noble 1977, 44)

Noble shows, with reference to various writings of the early American 'managers' how the coupling of engineering and corporatism created a location in which the principal discursive and scientific procedure was 'to systematize material and social production.' Here Noble, like Lentricchia, cites *'Taylorism'* as the typical site of the use of systematization and rationalization to control human activity. The procedures of managerialism are very similar to those mentioned earlier by Weber under the rubric of formalist, rational economic organization: (1) the accumulation of all pertinent information; (2) the systematization of that information according to comprehensible and applicable laws and formulae; (3) the scientific determination of optimum standards of performance for both machines and men; (4) the transfer of this optimum (substantial rationality) through the reorganization of the human and mechanical processes of production; (5) eliciting the cooperation of the workforce through the 'development of contented workers.' (Noble 1977, 264)

Lentricchia identifies Taylorism very clearly with the exercise of power through discipline as identified in Foucault's theory. Taylorism and the societies of discourse involved in its conception and implementation, then, may certainly qualify as a site of the technological politics of truth and science:

> Disciplinary theory and strategy appropriates asceticism and in turn will be appropriated by an emergent capitalism. The fruit of such appropriation, though Foucault does not mention it, is Taylorism, that fiendish mastery of detail provided by one of capitalism's darkest geniuses in the early years of the Twentieth Century. The phenomenon of Taylorism completes the portrait of discipline's long historical duration; its persistence grounded not in truth but in a power that produces truth as detailed knowledge of individuals in their everyday activities, a will to knowledge that endures from the medieval monastery to the factories of industrial capitalism because it is the chief weapon

> of domination and exploitation of class by class, group by group. (Lentricchia 1982, 44)

Interestingly enough, Lentricchia does show that the procedures of Taylorism go back to a pre-capitalist medieval world as do the Taylorist procedures of individuation and panoptic surveillance that arose out of classical liberalism and Sir Jeremy Bentham respectively.

Lentricchia presents Taylorism very lucidly, referrring to Braverman's *Labor and Monopoly Capital: The Degradation of Work in the Twentieth Century,* as a case of social control through organizational or systems science, in order to contain 'dangerous' workers and to ensure efficiency.

Noble also speaks at length about Taylorism saying that it began as a form of control over machines and materials but had to be revised to include social control or 'human engineering' since its very implementation in the machine shop and the assembly line had already caused a substantial worker revolt (Noble 1877, 274). Consideration of this human factor became a facet of corporate liberalism as defined by Wesley Mitchell: 'the twofold effort to understand and utilize the economic forces which control business activity and the psychological forms which control human behaviour' (in Noble 1977, 278). The revision of Taylorism entailed the systematization of the work force, the advent of 'human engineering,' 'corporate philanthropy,' 'behavioral psychology,' 'labor efficiency division,' and 'the philosophy of management' (Noble 1977, 286). Noble names persons and institutions as the sites of these procedures:

> Thus J.W. Dietz of Western Electric, Channing R. Dooley of Westinghouse, Albert Vinal of AT & T, Magnus Alexander of GE, and Mark M. Jones of Thomas A. Edison Inc. were all involved in the development of employment management in their respective companies. The corporation schools, which trained graduate engineers as well as apprentices were among the earliest departments of management to utilize procedures such as testing record-keeping, job specifications, etc., which ultimately came to be used for all employees. (1977, 295)

More recently Noble has written demonstrating that discursive cartels are still in full operation showing how certain corporations have bought controlling interests in technical educational institutions such as MIT in order to appropriate for themselves the power yielded up by science and technology (Noble, *The Nation,* Winter 1982).

3.21 Formalist rationality: its present-day concretizations

One can cite many of the discursive procedures of new communications technology as cases of scientific managerialism that substantively realize or concretize these formal rationalist traits. We will name but a few:

- Mass production and the continuous process of technology is manifest as the mass production of millions of identical silicon chips and of the constant thrust of research and development in the area;
- The rationalization of work organization and of 'scientific' management is evident both in the development of the technology whereby each engineer has a specialized and isolated task (Kidder 1982) as well as in the use of computer-communications technology for informatized decision-making models based on specialized information and systems programs;
- The common use of large limited liability companies as opposed to family units for the raising of capital is demonstrated in the tremendous advantage in economy of scale that IBM has over any surfacing small company (Gotlieb and Zeeman 1980, 3);
- Concentration of production and ownership is exemplified by the control that the transnational companies have over new communications technology and its markets (Mattelart 1979);
- The rise of managerialism is encouraged by computer-communications whereby technical specialization and technically specialized programs become the basis for problem-solving and legitimation of decision-making. DOC's seeding of the hardware industry indicates a marked increase in the degree of State regulation and control of the industry. The global expansion of the industrial or electronic economy is

both facilitated by the space-time-binding properties of new communications technology and manifest by the expressionism of the electronics industry;
- Robotization and office word processors are but two examples of a technology designed to replace men by machines, a trend which surfaces time and again in the debate on disemployment.

Some of these examples may be construed as purely economic in that they deal with the mode of production. However, though the sites of these procedures are economic, the procedures themselves are substantiations not of economic relations but of classical, instrumentalist rationality. These procedures conform very closely to the Weberian description of the procedures of bureaucratic reason:

> In the pure type of traditional authority the following features of a bureaucratic staff are absent: a) a clearly defined sphere of competence subject to impersonal rules, b) a rational ordering of relations of superiority and inferiority, c) a regular system of appointment and promotion on the basis of free contract, d) technical training as a regular requirement, e) fixed salaries (. . .) paid in money. (Weber 1947–64, 343)

Habermas (1968–70) also refers to Weber's description of rationality in identifying social relations of power in modern society. In an attempt to explain the student riots of the 1960s, Habermas suggests that the revolt was caused neither by class difference nor by economic deprivation but rather by the consciousness of yet another mechanism of social control, the technologized work-production ethic subsumed by the above-listed traits. In other words, regardless of real economic conditions, formal rationality had continued to dominate the social sphere. Students were objecting to being educated as technical experts with no reflective or critical criteria of evaluation or program formation. They resented being told 'Learn a trade, learn a skill.' so that they would fit nicely into the workings of bureaucratic and technocratic rationalized society. Indeed, since the sixties, universities have been swinging more towards the applied sciences: tradesmen, calculators, technical specialists in a particular field are all churned out, regardless of other facets of knowledge. Despite the obviousness of the 'affluent society,' students were still told that they had to work very hard to maintain the structure of society, that they had to be ever-more productive and profitable for the sake of progress, production, efficiency, and profit. The students, according to Habermas, were revolting against the universalization of formal rationality and instrumentalization of their educational systems, labor possibilities, and social and family spheres at a time when substantive economic conditions no longer required it. Lowi also points to those phenomena which Habermas claims provoked the riots of 1968:

> Institutions of higher education are already becoming technocratized. More and more frequently do we find university curricula being judged by criteria of 'relevance.' (1981, 469)

A glance at the job ads in today's newspapers and at the priorities of funding in university programs in the eighties would suggest that the revolts of the sixties accomplished nothing. Systems analysts, informatics experts, applied engineers and MBAs are in increasing demand whereas social scientists find either their budgets or their contracts cut. Four-year-olds are having their computer skills honed in school and at summer camp since their country will require this 'work force.' They must learn to think in computer language, and quickly.

The computer-communications society is substantially reinforcing the discursive procedures of formal, instrumental rationality: means-ends oriented work, technical specialization, utilitarianism, and the supreme valorization of productivity. And, should academia still provide any vestiges of immunity to these technocratic tendencies of science in fields of both education and employment, it is certainly not in a position to be complacent about it.

Furthermore, even if the professional in rationalized society is not simply a calculator (or a redundant, unemployed person), he remains a socially-controlled victim of the work ethic whereby his every process is directed towards production, productivity, and profit. The new communications technology user is defined in relation to work, even if his leisure and athleticism are primary concerns:

> As the workplace is decentralized, work itself will become

> harder to define (. . .) there will be more leisure for everyone but those at the top. Some will not be able to find a place in the new system and social unrest will increase. The most crushing demands will fall on senior executives who will have to become all-round intellectual athletes. (Russel 1978, 41–2)

Not only does the work ethic prevail but leisure and work become confused with one another in the descriptions of the 'communications society.' Doing accounts with a computer-communications system, playing video games on your new home computer or sending electronic mail all become part of the same work/leisure electronico-technocratic complex in which the common standards are efficiency- and goal-oriented.

3.22 Reason's exclusivity: the technostructure: 'discursive hegemony'

It is feasible to demonstrate through discursive analysis that these technological discourses are today under the control of a few centralized societies: universities, private and crown corporations, government and media. If the question of centralization and/or decentralization in relation to new communications technology is a question of which societies of discourse dictate the dominant episteme, then it becomes possible on the basis of the results of a discursive analysis to argue the existence or non-existence of what many have called the 'technostructure.' Galbraith defines the technostructure as follows:

> Eventually not an individual but a complex of scientist engineers and technicians; of sales, advertising and marketing men; of public relations experts, lobbyists, lawyers and men with a specialized knowledge of the Washington bureaucracy and its manipulations; and of coordinators, managers and executives becomes the guiding intelligence of the business firm. This is the technostructure. Not any single individual but the technostructure becomes the commanding power. (1973, page xx)

Here Galbraith is also identifying some of the sites of the use of technology to yield power and hegemony. He suggests that technology is not used for power by any single man but rather by institutions which he groups under the heading, 'technostructure.'

> the firm is required, with increasingly technical products and processes, increasing capital, a lengthened gestation period and an increasingly large and complex organization, to control or seek to control the social environment in which it functions. (Galbraith 1973, 39)

Galbraith points to the instrumentalist procedure of growth (for growth's sake) and shows how with this growth 'in larger and highly organized form authority passes to the organization itself – to the technostructure of the corporations' rather than to any individuals or holders of capital (1973, 39). Galbraith provides an answer to Lentricchia when he states that this power is no longer exclusively at the service of capital but of a whole rationale known as the principles and procedures of the technostructure. While Lentricchia argues that what is missing from Foucault's analysis of power is the argument that the 'worker is stripped of power by management,' Foucault refuses 'to totalize the economic relationship' (Lentricchia 1982, 49) and Galbraith brings to the surface some information which seems to reinforce Foucault's stance:

> At the highest level of development – that exemplified by General Motors Corporation, General Electric, Shell, Unilever, IBM – the power of the technostructure, so long as the firm is making money, is plenary. That of the owner of capital, i.e., the stockholders, is nil (. . .) In this process the power of the capitalist is, if anything, diminished. It is that of the technostructure that is enhanced. (Galbraith 1973, 40)

Galbraith refers specifically to institutions that make up the technostructure, insisting upon the multinationals which operate according to a planning system which seeks control of all environmental and human variables. The planning systems include communications media and extend beyond national boundaries:

> needs of the planning system are the norm to which discussion eventually returns (. . .) And this is the norm to which editors, publishers and broadcasters in the absence of thought to the contrary also repair. (Galbraith 1973, 159)

> But the planning system also transcends the national state to create an international planning community. (Galbraith 1973, 164)

While Galbraith bases his argument in part on economic figures, we can see here that he certainly recognizes the role of knowledge in the constitution of this hegemonic power base. Rather than calling this configuration of power the 'technostructure,' however, we would prefer to call it a *'discursive hegemony'* based on a monopoly of the discursive procedures of rationalism and classical positivism.

Whereas Schiller, Mattelart, and other proponents of theories of media imperialism claim that the technostructure exists because of economic links or military exploits, we would suggest that it has grown out of the inter-discursive relations outlined in the preceding chapters. The technostructure exists where classical, rationalist, technological discourses dominate the episteme. In that these procedures occasionally belong to the parties which are commonly declared members of that economically circumscribed society of discourse known as the military-industrial complex, while simultaneously being imposed on the rest of society as the procedures of right legitimating power, it is perhaps fair to say that these parties centralized power in the technostructure.

3.23 The military industrial complex

Macoby is another who identifies control-seeking cartels of societies of discourse: business and the military which provided the initial motivating drive (not the existing discursive procedures) behind the research and development of new communications technology:

> It is complex, but it comes out that this technology is developed as a response to two basic factors. One is a market mechanism that wants things that increase predictability and control. New electronic technology is sold to other companies or government bureaucracies, which buy it because they believe it will increase predictability and control. The other type of technology development is determined by governments, in terms of national security. When you look at the variables that determine national security, they have to do with issues such as glory, power and protection, all of which are limitless. (Macoby/*Daedalus* 1980, 11)

Examples abound of the fact that government, military and industry all thrive on the same discourses of technology. However, we might cite just one which makes government and industrial bureaucratic discursive links quite explicit:

> A major objective of the Telidon program has been to encourage the transfer of the technology to private industry as quickly as possible so that it can develop its own systems in response to market. (DOC/Fox, *Blue Paper* 1979–80, 13)

It is a matter of fact that by far the largest portion of the communications research budget of developed countries goes to the military.

Mass production and distribution of micro-chips based on a number of hooked-up prototypes (in the name of efficiency) dictates that the whole of society must adopt the procedures of the chip: 'The cost of tooling up to produce the first chip is very high indeed. The more chips he sells the lower the price per unit of production.' (Madden 1980, 45)

In other words, to abide by the rationalist discursive procedure of the quest for productivity, efficiency, and profit, the whole society must, we are told, adopt the discursive procedures programmed into a few varieties of mass-produced chips. The centralization of power in the hands of the technostructure amounts to a centralist-oriented expansion of the procedures of instrumentalist reason, throughout the episteme and throughout all rationality. In order to study whether or not the technostructure exists one has also to discuss whether or not *interdiscursive relationships* are allowed along corporate and technical interests. Kari Levit, for one, in his article 'The Hinterland Economy' (1970), argues that lines of delimitation are no longer national but corporate and multi- or trans-national:

> The international corporations have entirely declared war on the 'anticipated' nation-states (. . .) The charge that materialism, modernization and internationalism is the new liberal creed of corporate capitalism is a valid one. The implication is clear: the nation-state as a political unit of democratic decision-making must, in the interests of progress, yield control to the new mercantile mini-power. (1970, 163)

Lowi expresses his awareness of technostructural domination in and through instrumentalist communication even more precisely: 'Management is a political phenomenon at the macro-level; it is merely a name for *the effective use of information as a power source.'* (1981, 465, emphasis added)

The discursive procedures referred to earlier as part of an inter-discursive cartel, i.e., enunciative a-symmetry, universalization of rationalist instrumentalism, and exclusivity, are indeed concentrated nodes of power. Furthermore, both the centralization of management as well as expansionist growth are incontestably linked to its facilitation by the exclusive centralized networking of computer-communications; you cannot control an arena if you cannot communicate with or about it from a distance: 'Only with the computer can occur what Hymer calls the "law of tendency since the industrial revolution for firms to increase in size."' (Gilpin 1981, 237)

The technostructure is a case of societies of discourse such as private and public corporations turning new communications technology into a discourse of reason and an apparatus of power to serve them while compelling the rest of society to adhere only to their procedures.

3.24 Extra-discursive relations

While we would not base the existence of the information society solely on the basis of GNP, it consolidates our argument to note that, despite the exaggerations of Machlup's *The Production and Distribution of Knowledge in the United States* (1962) and of Porat's statement that the service and information sectors account for the largest part of GNP, industry is spending more and more energy and money in developing communications and knowledge control capacity. Industry has become quite aware that discourse and knowledge both procure and legitimate power. GNP is one extra-discursive indicator of this. Thus, military and corporate societies have sought to monopolize communications as they seek to expand (Finlay-Pelinski 1982b).

As Gotlieb and Zeeman's study of the major companies involved in new communications technology indicates, IBM has a ten to thirty per cent advantage in economies of scale over its rivals, requiring any would-be competitors to put one billion dollars up front (1980, 3). The technocratic oligopoly of discursive procedures of rationality, instrumentalism and control are also related to extra-discursive economic factors.

Gilpin reconfirms the input of economies to the monopoly of control:

> Through its effects on economies of scale, comparative advantage and economic efficiency, technological advance can recast the national and international division of labor. (1980, 231)

Such an advantage also enables companies to develop technological procedures, i.e., soft and hardware, and to impose them on the market/society. Funding patterns for R&D linking the Pentagon and informatics research institutes such as MIT constitute further extra-discursive evidence of hegemonic discursive control (Licklider 1980; Noble, *The Nation,* Winter 1982). Thus, extra-discursive relations attest and contribute to the constitution of the discursive hegemony known as the technostructure.

3.25 The dinosauric strain: power to the status quo

Allowing that there is some discursive and extra-discursive evidence for the existence of discursive hegemony in the sphere of new communications technology, one must now question whether new communications technology in any way affects which societies of discourse enjoy the dominance.

Lowi projects what Foucault has described and demonstrated as a reinforcement of the currently dominant episteme. He projects that the control of the groups already in power will be augmented since it is they who possess the discursive-technical means to design and access the new information resources. We must not forget that in seventeenth century France and England everyone could potentially 'benefit from' the judicial, criminal, and medical institutions but these institutions did nevertheless reinforce the status of the 'scientific institutions rather than of the average citizen.' Thus, Lowi seems to concur with Weizenbaum who suggests that computer-communications have served primarily to preserve dinosauric institutions such as the military and the banks.

Without new communications technology, these institutions would have strangled themselves in their own bureaucratic complexity and size. Both Weizenbaum (1976) and Lowi question whether this conservational function of new communications technology is not more regressive than evolutionary. It would seem to make sense that if new communications technology follows basically the same procedures as existed in industrialism, then the same societies of discourse will continue to maintain power.

To summarize, then, we can say that while Galbraith's notion of the 'technostructure,' Macoby's demonstration of the 'military-industrial complex,' and the insistence of both Noble and Lentricchia upon the discipline and social control of human subjects, does show that concrete institutional sites of power relations do connect closely with capitalist institutions, they also point to 'something else' which transcends capitalism in the strict sense.

That 'something else' is simply the set of procedures of an episteme/rationality that are not necessarily within the control of individuals but which nevertheless are quite concretely manifest. Hence the analysis of their power is not metaphysical paranoia. As the preceding chapters have shown, these procedures supervise and speak through managers, capitalists, media people, teachers, and even the machines themselves. They speak in certain sites at certain times; they serve certain societies of discourse in specific places, including capitalist sites, though they do not belong exclusively or generically to capitalism. Simply stated, the words issuing from the mouths of capitalists and new communications technologists have already been spoken by bureaucrats, pure scientists, and the military long before the existence of corporate capitalism. And they continue to emanate from the mouths of Soviet technocrats. Perhaps intervention should begin not with the mouth but with the discourse.

Thus the drives of new communications technology as discourse and of discourses on new communications technology are not those of a specific individual or corporation but of a whole discursive and scientific community which legitimates its power and control relations by means of constraining procedures that permit certain discursive or technological practices and prohibit others. New communications technology is epistemically driven. But this is not to deny that the 'technological imperatives' are themselves a part of the instrumentalist, scientific, classical episteme.

3.26 Grounds for a critique of power and domination

What is wrong with power and social control? Why condemn them at all? Did we not say earlier that no society could exist without order, without constraints on the practices of its citizens? So how can we now set out to criticize those constraints that favor the exercise of power for social control?

The first answer provided by the Frankfurt School is that there is an excess of power relative to what is required for the maintenance of social order. But then what is wrong with an excess of power? The Frankfurt School's answer to this question is that such an excess is contradictory.

3.26.1 The role of contradiction in critical theory

Being content with exposing the inherent contradictions of classical discourse is not at odds with the tenets of critical theory or 'critique' as elaborated by the Frankfurt School. Max Horkheimer, in his essay 'Traditional and Critical Theory' (1937–1972), constantly reiterates that the role of critical theory is to expose contradictions, to become aware of the tensions of theory and society: 'The two-sided character of the social totality in its present form becomes, for men who adopt the critical attitude, a conscious opposition.' (1937–72, 201) Furthermore, Horkheimer concurs with our argument that the fundamental contradiction at the heart of classical, liberalist reason lies in the tension between an ideal of human emancipation and equality, on the one hand, and the exercise of power, on the other hand:

> the fundamental difference between a fragmented society in which material and ideological power operates to maintain privileges and an association of free men in which each has the same possibility of self-development. (1937–72, 219)

However, for Horkheimer critical theory also served to point out and transcend the contradiction between the subject's reason and his social relationships of production:

> Critical thinking, on the contrary, is motivated today by the effort really to transcend the tension and to abolish the opposition between the individual's purposefulness, spontaneity and rationality, and those work-process relationships on which society is built. (1937–72, 210)

While we disagree with the necessity of maintaining the individual subject at the centre of a social critique, which is to remain within the humanist paradigm of which liberalism itself took advantage, we do find the Frankfurt School's critical project of exposing contradiction to be one of the best starting points for a critique of the discursive procedures of rationalism, and more specifically of new communications technology.

The judgment that critical theory passes on society, then, is one which demonstrates how certain social relationships, and, in Horkheimer's case, certain economic relationships, ceaselessly generate contradiction and tension:

> But critical theory of society is in its totality, the unfolding of a single existential judgment. To put it in broad terms, the theory says that the basic form of the historically given commodity economy on which modern history rests contains in itself the internal and external tensions of the modern era; it generates these tensions over and over again in an increasingly heightened form. (Horkheimer 1937–72, 227)

But as we have argued, it is not only economic commodity relations which ceaselessly generate this tension. It is all relations of classical knowledge and social practice, particularly bureaucratic knowledge or the organizational sciences. Furthermore, critical theory exposes tension and contradiction with the ultimate goal of surpassing them and of generating an alternative knowledge and social practice. Horkheimer explains the generative capacity of the critical project as follows:

> (the) presentation of social contradictions is not merely an expression of the concrete historical situation but also a force within it to stimulate change, then its real function emerges (. . .) a process of interaction in which awareness comes to flower along with its liberating but also its aggressive forces (Horkheimer 1937–72, 215).

Before one can say what alternative episteme might be possible, one must open a way for it by showing that the old one is in a state of crisis because of the tensions with which it is wrought. This is the negative dialectical project of Adorno, to make way for 'utopia' by exposing the unresolved antitheses of non-utopian, extant discourse and practice. It is, however, insufficient to remain at the level of negative dialectics. An alternative, which does not necessarily just evolve out of the contradictions exposed in the negative stage, must surpass the negative dialectic.

3.26.2 It does not work anymore

The most that we can do in our search for a ground upon which to criticize discourses of social control is to show how these discourses simply do not work anymore. They could provide social direction so long as their inherent contradictions remained unrevealed. The doctrines of Adam Smith and John Locke could suffice as social regulators, so long as society remained unaware that classical rationalist liberalism was at one and the same time calling for individual freedom and equality yet practising the exclusive will to truth and power in and through exclusive procedures of discourse. So long as people believed in Hobbes' notion of natural law with no awareness of the machinations of manipulative subjects of science, then it too could serve to guide social law. So long as no one realizes that new communications technology is calling 'revolutionary' precisely those procedures that industrialism called revolutionary, then the progressivist, evolutionary, historicist view of social progress can explain and legitimate social policy and technological 'progress.' But the liberalist claim to equality and individual freedom no longer functions once it is seen that the scientists making the claims themselves enjoy power over others.

To summarize, the very foundation of our critique of the discursive procedures of rationalism is based on an awareness of their intra- and extra-discursive contradictions. Here we must stress, however, that a discourse may continue to function in a society though it be full of contradictions. It will certainly not serve society well if it is contradictory but it is the awareness of the contradiction that should, hypothetically if not predictively, compel its discarding and the search for an alternative. First, one exposes the contradictions in the external and internal procedures of discourses on and of new communications technology. Then, one shows how the system can no longer contain them. These contradictions have been made apparent by our discourse analysis which has shown up the paradox between apparent content and actual practice especially with regard to claims of equality and freedom. Exposing these contradictions in discursive procedures is already in itself a beginning of critique. Furthermore, a critique must reveal the contradictions between the claims or promises of discourses of rationalism and the extra-discursive performance of those discourses. We will see that despite a claim to individual autonomy and equality there exists discursive hegemony and that socio-economic conditions betray the absence of equality, the absence of natural laws of progress, and the presence of tremendous inequality.

In addition, it must be borne in mind that we are not criticizing any and all forms of social order but rather a specific form of social control, namely classical, liberalist, rationalist social order. The test is not to set out to establish anarchy but to suggest alternative forms of discursive, epistemological and social order.

Discursively and historically, the old episteme does not work any more. But we cannot go back to the moralist position of liberalism in order to criticize it. We can show its contradictions, we can show its failure, both discursive and extra-discursive, to live up to some of its own claims, but we cannot posit some absolute position such as confidence in individualist freedom by which to judge it.

Rather than seeking such a moralist position from which to criticize liberalism, a position which risks falling back into classical liberalist confidence in rationalism, we can simply expose the crisis of the episteme out of which may arise a different kind of social order with different constraints or discursive procedures, one which, it is to be hoped, could avoid the internal and external paradoxes that today face its predecessor.

3.27 Weber's and Husserl's critique of rationality – the non-pertinence of technology to the social life-world

The discursive procedures of classical rationality and the values they imply exclude all other procedures and values. The politics of administration reduces reason to technique – episteme to techne – thus ending any discussion of the legitimating knowledges that might exist.

> The reason why I argue on every occasion so sharply and even, perhaps, pedantically against the fusion of Is and Ought is because I cannot stand it when problems of world-moving importance, of the greatest intellectual and spiritual bearing, in a certain sense the highest problems that can move a human breast *are transformed here into questions of technical-economic 'productivity' and are made into the topic of discussion of a technical discipline, such as economics.* (Weber 1924, 202)

Very often, and thanks to Talcott Parsons, Weber is understood in Western societies as the unambivalent defender of rationality. Weber argued, however, that the procedures of instrumentalist rationality allowed only means to be discussed. Many have interpreted this to mean that there could at any rate be no rational discussion of ends. If we recall that Weber argued for the existence of many kinds of rationality, including 'Wertrationalitaet' – goal or value rationality – then we can see that this conclusion is not quite faithful to his critique. If one thinks of present-day expressions such as the 'rationalization of decision-making or production' and 'rationalization of communications,' expressions found in business faculties as well as in the press, one can easily understand how the notions of rationalization and rationality are unthinkingly associated with positive values.[12] Marx did not help to clear up this confused worship of rationality. He, too, praised progressivism and rationalism. After all, the revolution depended on these two factors (Kumar 1978, 109). Rationality and technology (a form of rationality) were not the problem for Marx. The problem was, rather, irrationality in the form of 'bourgeois contradictions' that made society senseless and necessitated revolution. 'For Marx, the rational elements of society were the means which served, yet which increasingly contradicted, unmastered and irrational elements.' (C. Wright Mills, ed., in Weber 1946–78, 68)

3.28 The irrationality of rationality

For Weber, however, rationality was not such an unequivocally positive set of procedures of knowledge and political organization. For example, while Weber viewed rationality as quite necessary for survival and for defence, including military instrumental bureaucratization, he also strongly criticized the link between rational bureaucracy and domination and legitimation, as for example, in Bismarck's Germany.

For Weber, then, there really was nothing blanketly wrong with rationality but there was something wrong with rationality becoming the blanket form of reason. What he objected to was the extrapolation of bureaucratic, instrumentalist rationality to all spheres of human existence at the expense of traditional, religious, affective, and value-oriented rationalities. To put it mundanely, while instrumentalism may be highly commendable in the face of the beef steak on one's plate, it is not always so commendable in the face of one's fellow man. For Weber, there are other ways of reasoning with the world than this means-ends, efficiency-oriented, control-aimed form of rationality. Current talk of 'informatizing society, the home, school, etc.,' in other words, of rendering ubiquitous new communications technology, only reconfirms the pertinence of Weber's critique of the instrumentalist rationality over all spheres of existence.

Weber also criticized the increased rationalization of society because it led to mechanization which resulted in depersonalization and oppressive routines. Rationality, for Weber, in this context is seen to be the adverse of personal freedom. He describes the man created by the bureaucratic work ethic of protestantism as: 'The Puritan willed to be the vocational man that we have to be.' (1947–64, 50) For this reason C. Wright Mills interprets Weber as deploring the type of man that bureaucratic reason selects and forms, i.e., 'a cog in the wheel' of efficient productivity.

Weber enumerates at least four social consequences of substantive reason as bureaucratic control: levelling, plutocracy, lengthy technical training, impersonality and emotional monopoly:

> 1) The tendency to 'levelling' in the interest of the broadest possible basis of recruitment in terms of technical competence;

> 2) The tendency to plutocracy growing out of the interest in the greatest possible length of technical training;
> 3) The dominance of spirit of formalistic impersonality without hatred or passion and therefore without affection or enthusiasm. (1947–64, 340)

As a further criticism of rationalism, Weber makes it quite explicit that the 'alienation' of man, the divorcing of the worker's means of production for himself, is provoked more by technical reason than simply by capitalist class structure:

> The expropriation of the individual worker from ownership of the means of production is in part determined by the following purely technical factors: a) The fact that sometimes the means of production require the services of many workers, at the same time or successively; b) the fact that sometimes sources of power can only be rationally exploited by using them simultaneously for many similar types of work under unified control; c) the fact that often a technically rational organization of the work process is possible only by combining many complementary processes under continuous common supervision; d) the fact that sometimes special training is needed for the management of co-ordinated processes of labour which, in turn, can only be exploited rationally on a large scale; e) the fact that, if the means of production and raw materials are under unified control, there is the possibility of subjecting labor to a stringent discipline and thereby controlling both the speed of work and standardization and quality of products. (1947–64, 248)

Weber suggests that technical reason is a true form of human bondage whereby rationality becomes what Innis would call a 'monopoly of knowledge' and Foucault the 'dominant episteme':

> Joined to the dead machine (bureaucratic organization) is at work to erect the shell of that future bondage to which one day men will perhaps be forced to submit in impotence, as once the fellahs in the ancient Egyptian State – is purely, technically good, that is, rational bureaucratic administration and maintenance is the last and only value which is due to decide on the manner in which their affairs are directed. (Weber 1920, in Marcuse 1968, 151)

In the above passage, Weber criticizes technical reason as an extremely efficient form of social control and domination. Thus, the Frankfurt School's continuation of Weber's critique of bureaucratic reason is far more faithful than Talcott Parsons' version which would see in Weber the champion of 'rationalization.'

Marcuse, in his essay on Weber, 'Industrialization and Capitalism in the Work of Max Weber' (1968, 223ff), states that where Weber refers to 'rationality,' he means a 'specific form of unacknowledged political domination,' embodied by 'capitalist economy, bourgeois private laws, and bureaucratic industry.' (Marcuse, in Habermas 1968–70, 81–2)

Habermas is even more critical than Marcuse of instrumentalist rationality when he suggests that this form of rationality excludes ethical reflection, concerns of public interest, and above all any intersubjective interaction. Instrumentalist rationality 'removes the total social framework of interests in which strategies are chosen, technologies applied, and systems established from the scope of reflection and rational reflection.' (Habermas 1968–70, 82) And, speaking of rationality, though without actual reference to either Weber or the Frankfurt School, Weizenbaum echoes this rejection of a reduction of all forms of human activity and knowledge to instrumentalist reason:

> When instrumental reason is the sole guide to action, the acts it justifies are robbed of their inherent meaning and this exists in an ethical vacuum (. . .) it is the act itself that matters.
> (Weizenbaum 1976, 276)

Furthermore, in the following quotation, Marcuse explicitly draws the link between the procedures of rationality and those of new communications technology. This linkage provides further evidence and substantiation of our constant insistence that the focus for new communications technology as it is so far both preached and practised, is that of power, social control, and domination:

> The formal rationality of capitalism celebrates its triumph in electronic computers, which calculate everything, no matter what the purpose, and which are put to use as mighty instruments of political manipulation, reliably calculating the chances of profit and loss, including the chance of the

> annihilation of the whole, with the consent of the likewise calculated and obedient population. (Marcuse 1968, 234)

The computer 'expert,' Weizenbaum, tends to repeat Marcuse's analysis of the relationship of this computer-communications technology to instrumentalist reason as a mechanism of social control:

> Of course, the introduction of computers into our already highly technological society has, as I will try to show, merely reinforced and amplified those antecedant pressures that have driven man to an ever more highly rationalistic view of his society and an ever more mechanistic image of his society. (Weizenbaum 1976, 11) [13]

3.29 Rationality's self-legitimation

The highest political good is technological reason and its progressive application in society; so at least bureaucratic reason would have society believe, and with substantial success.

> the particular claim to legitimacy is to a significant degree and according to its type treated as 'valid'; that this fact confirms the position of the persons claiming authority and that it helps to determine the choice of means of its exercise. (Weber 1947–64, 325)

Rationality justifies the exercise of power; technology is rationality. A perfect example of the self-legitimation of technology may be found in *Time* magazine's (27 April 1981, 14–15) coverage of the space shuttle program, 'Touch-down, Columbia. The Shuttle is "right on the money" and gives the US a mighty lift.' In the spring of 1981, the US was in trouble. President Carter had made shameful moves in foreign policy. The economy was sluggish and President Reagan had come to power on a 'laissez-faire' platform. The technological feats of Columbia had become the new slogan for legitimating the system. This can only be the sequence of technocratic discourse. Technocracy is a form of society whose rule is legitimated by technology: progress and science exist for their own sake. Management legitimates its own power on the basis of 'good procedures of management.' Such is the tautology of rationality's self-legitimation. The tautology is exclusive, it declares decision-making rights for bureaucratic reason and institutions at the expense of power to the public sphere:

> the notion of a public could be blurred by the simple continuity of management once it is established as a separate function with its own highly legitimate access to all decision centres (. . .) government would also be subordinate to management (. . .) But the emerging power holders and the emerging structures of authority depend little for their claim to control on traditional roots that are grounded in sovereignty, citizenship and loyalty. This transformation could easily reduce government to a position of mere power based on superior resources and more elaborate management techniques. (Lowi 1981, 466)

Although at times technocratic discourse attempts to win over public legitimation by saying that what is good for the corporation is good for the public in general, as is borne out by our discursive analysis, the public is omitted altogether except as the one who should kneel in awe before technology and worship it for its own sake. 'Look at the wonder-technology we are. We are the most progressive and advanced in the world, the most scientific and rational and therefore we have the well-deserved power.' Such is new technology's slogan of self-legitimation whether it be found in a Seiko watch ad, in top-of-the-line computer communications systems, in electronic arms guidance systems, or in the discourse of the two super-powers locked in cold war.

3.30 A good society for science

Weber described how, since feudal times, political power has relied increasingly on experts who enjoyed special knowledge about information leading to and legitimating decision-making. These experts were the beginnings of the bureaucratic structure. Weber then showed how these experts gradually ceased to serve the political decision-makers. The decision-makers became so dependent upon the technical specialists that social decision-making was rendered subservient to the science that had originally been set up to serve it. Politics served science. Society had no choice but to legitimate science because science had legitimated and declared itself to be the sole legitimation of society. Political power in society was reduced to a choice between management science and political economics. Science became a tool of rulers but rulers were also the handmaidens of science. A blanket acceptance of political economy and of technical experts as the major directing forces behind current social decision-making illustrates perfectly the reversed roles of science and society.

Where 'power (. . .) is presumed to accompany such a science' (Foucault 1980, 83–4), society must conform to the rules of that science. Taken as a discourse of science, technology on the one hand legitimates itself while on the other hand demanding that society must conform to its rules. The legitimating relationship between scientific discourse and social control is clearly illustrated in the cliché 'a good society for science.'

The tautological self-legitimation of science in society is one of the dominant principles of the classical liberalist discourse of science. Foucault refers to this tautology when he states that in classical science to say the truth one must already be in the field of truth. Discourses rarely legitimate technology by first addressing the question of social needs and then tailoring technology to fit those needs. The use of knowledge to legitimate and exercise domination would not only be a change of focus but would also be a deviation from rationality.

3.31 Discourses of rationalist classical science are undemocratic

The relationship of the public sphere to freedom and control is of course the question of democracy. Any contemporary theory of communications must address the relationship between bureaucratic, rationalized, scientific discursive societies and the public.

Consequently, in trying to interpret the social effects of new communications technology, one must address the problem of the relationship between the discourses of classical scientificity, rationality, and democracy. In order to respond to this question one must first look at the position of the public sphere in relation to discursive control.

The will to truth and power of certain exclusive societies of discourse seems very remote from a common sense understanding of democracy. In fact, the word democracy is conspicuously absent from Foucault's vocabulary as he describes the functioning of classical discourse. Weber argues that rationality is far removed from democracy even though it may describe itself as democratic. Weber saw the surfacing of a form of democratization in the still pre-bourgeois State. He also showed how this democracy changed into plebiscitary dictatorship with the irrationalization of rationality manifest as charisma. When democracy becomes charismatic, society ceases to be 'democratic' in that it no longer has a public-oriented and -controlled government. (Weber 1947-64, 286) According to Weber, a rationalized society which uses charisma to legitimate its power as a 'plebiscitary democracy' is nothing more than a monopolistic form of social control.

To ascertain new communications technology's role in democracy, one must look at what rationalized discourse and science have done in the past. Rationality's past is linked with charisma. Plebiscite and social procedures of control appear to belie rationality's claims to further democracy in the future. Weber reportedly wrote to Ludendorf that in rationalistic societies the people 'democratically' choose a leader who then looks after the society while at the same time muzzling them (C. Wright Mills in Weber 1946–78, 42).[14] In a rationalist society the leaders do not represent the will of the electorate. Both electorate and leader are subjugated to the will of bureaucratic and technical experts. Contrary to the steering function that Weber believed experts should play, they became the power base of social organization. They became the agents of 'techno-cracy' and 'bureau-cracy.' (C. Wright Mills in Weber 1946–78, 43) Weber clearly established the connection between bureacratic or technocratic rule and the undemocratic centralization of power which legitimates itself by claiming that its discourse is that of 'reason.'

> The primary fact underlying representation is that the action of certain members of a group, the 'representative,' is binding on the others or is looked upon as legitimate so that its results must be accepted by them. (Weber 1947–64, 417)

Weber places the term 'representative' in quotation marks because he argues that bureaucratic society functions neither with 'instructed representation' nor with 'full representation' but with 'appropriated representation.' (Weber 1947–64, 417)

Since industrialism society has claimed to be a democracy of 'instructed representation.' Technical experts and bureaucratic managers, however, have not been appointed on the basis of any public accountability. They have been appointed, rather, on the basis of the legitimation claims of knowledge, managerialism, efficiency, and technical expertise. Given this, the 'instructors' of society are chosen based on a system of 'appropriated representation.'

Weber has criticized and demystified bureaucratic and industrial discourses' claims of 'democratizing' society by demonstrating that 'democracy' for rationality is never ruled by the 'undistinguished mass.' Nor is it an increase in societal participation in decision-making. Rather, it is a system of representation that legitimates itself by controlling public opinion in its favor. This control is a form of interiorized consensus engineered with the aid of discourses which convince of their right to be right and to exercise might.

> the term 'democratization' can be misleading. The *demos* itself, in the sense of an inarticulate mass never 'governs' larger associations; rather, it is governed, and its existence only changes the way in which the executive leaders are selected and the measure of influence which the *demos,* or better, which social circles from its midst are able to exert upon the content and the direction of administrative activities by supplementing what is called 'public opinion.' 'Democratization,' in the sense here intended, does not necessarily mean an increasingly active share of the governed in the authority of the social structure. This may be a result of democratization, but it is not necessarily the case. (Weber 1946–78, 235)

Much later, Kumar demystified the cliché that industrialism and new communications technology would democratize society.

> democracy inevitably comes into conflict with the bureaucratic tendencies which, by its fight against rule, democracy has produced (. . .) levelling of governed (not governing and governed). (Kumar 1978, 74, note 64)

Wherever power and social control is the ability to manipulate the discursive and scientific procedures of decision-making (i.e., efficiency, reduction of risk, control and productivity, and exclusivity), democracy in the sense of accountable representation of the

polis and to the polis, no matter how regionalized and distinguished the polis may be, is ever more remote. Such is the case because
(a) technology specializes knowledge and leads to specialized groups of discourse users;
(b) democracy seeks equality across all groups of discourse users;
(c) users of a dominant discourse resist democratization;
(d) their resistance takes the form of specific types of democratic representation; and
(e) they legitimate their resistance by pointing tautologically to their own discourse.

3.32 Remedies defunct, tried and failed

Liberalism failed to solve the problems of the industrial revolution. Why should we have confidence in it to solve the problems of the information age?

Minc suggests that one of the main changes necessitated by new communications technology will be a shifting of the location of conflict. Consequently, the liberalist model of centralized control for problem-solving will no longer suffice.

> We are coming to (an era) where the numerical quantitative aspects – hence the conflictual ideological consequences in the world of industrial production will change. It is clear that in the long term the core of social conflict will not lie in industrial production. (Minc/OECD 1980, 158)

It is not sure that the real locus of interdiscursive relations and conflict, let alone power, will/have change(d) with the advent of new communications technology. It can be said, however, that past economic and liberalist social theoretical models no longer suffice given their failures. Liberalism's impotence vis-á-vis the past problems of industrial society is apparent.

The UNESCO hearings on communication and the new world information order – The MacBride Report (1980) – as well as the recent Canadian hearings on monopoly in the press – The Kent Commission (1981) – are two of many studies which point out vast inadequacies in past and present records of liberalist policies of 'free-flow,' 'laissez-faire,' and 'private enterprise' in the sphere of communications. These reports are critical because they demonstrate that the public interest, that is equality for all individuals was sacrificed to the power of exclusive societies of 'scientists' who espoused and generated the liberalist doctrine. Smith (1980) points out that no other country has been so committed to, yet suffered so much from, the liberalist doctrine of free-flow as Canada. Third World countries, however, have been pressing UNESCO to revise the 'free-flow' doctrine.

In terms of communications theories there has been a shift away from the liberalist, rationalist approach to communications of Schramm and de Sola Pool – the modernization theorists – towards a sort of neo-liberalist cynicism. Schiller, Mattelart, Hamelink, and before them all, Adorno question the ability of liberalism's evolutionary self-balancing principles to serve the rights of society in a non-hegemonic fashion.

3.33 Neo-liberalist cynicism

Neo-liberalist cynics have criticized the liberalist faith in naturalist balance, progress, and the self-correcting capacity of society and nature in the face of disruption. Their cynicism was largely directed towards the sphere of mass communications which they saw as being more and more unilaterally manipulative.

> Neo-liberalism, born in a time when the manipulation of symbols is highly organized and often monopolistically controlled, puts far less faith than classical liberalism in the efficacy of the self-righting process. (Jensen 1976, 184)

> The Weltanschauung of neo-liberalism emerges as a denial of the historic liberal tradition, the lifestyle of Mass Democracy as a denial of the historic idea of democracy. The two stars of the liberal crown – 'respect for reason and respect for the individual' – (Gotshalk) have fallen into ignominy, their eminence and repute usurped by the irrationalism, relativism and collectivism of neo-liberal thought. (Jensen 1976, 200)

Habermas suggests that a contradiction has arisen between certain disenchanted worldviews and the continuation of liberalist, rationalist claims to legitimacy based on technological progress.

> To the extent that technology and science permeate social

> institutions and thus transform them, old legitimations are destroyed. The secularization and 'disenchantment' of action-oriented world views, of cultural tradition as a whole, is the obverse of the growing 'rationality' of social action. (Habermas 1968–70, 81)

Liberalist classical science as the father of technology enters into a state of crisis when one begins to question whether technical progress and the rationalization/instrumentalization of all processes and power are sufficient to legitimate a régime, to encourage its citizens to work and respect the State and the law. Although Soviet and American régimes continue to legitimate themselves by touting their wondrous technical feats, both have difficulty with legitimation or with what Habermas calls a 'motivational crisis' (Habermas 1975). Lack of voter participation in America, low worker efficiency and productivity in Russia, and passive and active industrial sabotage in both East and West provide yet further evidence of delegitimation. Lowi describes the delegitimation withdrawal aspect of the anti-liberalist movement in strict connection with his remarks concerning the social implications of new communications technology and managerialism.

> A third side development is also likely. This is the likely increase of withdrawal movements. These arise out of a felt inability to pierce established institutions or a genuine distrust of the available information flow. (. . .) They become politically significant mainly when withdrawals are so numerous that they raise serious questions about the legitimacy of society and government. (Lowi 1981, 641)

While neo-liberalism[15] is fathomable as a revolt against liberalism and as a placing into crisis of the ability of this social theory to cope with present-day society, it is still nothing more than a negative determination (Adorno 1966–73) of the discursive procedures of liberalism with all of the social impotence that this implies. In that neo-liberalism is really just a negatively rendered prisoner of the liberalist episteme, it does not avoid the procedures of classical science. It merely negates their content. It does not provide any alternative procedures.

If the inability of liberalism to cope with 'post-industrial' society is to be overcome, an alternative discursive practice must be found. The onus would be placed upon such a discursive practice to do more than just criticize the classical liberalist episteme. It would have to practise other procedures that would be genuine transformations of the classical liberalist episteme.

Liberalist social theory is no longer pertinent to the political, social, economic, and discursive order. Furthermore, it can no longer explain, let alone remedy, anything. This does not, however, preclude the fact that many still use the discourses of liberalism in order to attempt to explain present-day social procedures. Indeed, as we have seen, the discourses on new communications technology do little more than reiterate the basic premises of liberalist theory. Therefore, the question is not whether liberalist and neo-liberalist discursive procedures are still in circulation today, but whether they are relevant to contemporary discursive and non-discursive problems. Moreover, liberalism enters into a state of crisis equal in proportion and intensity to the degree that society recognizes the inherent contradictions of its discourse and its failure to resolve social problems. Once this crisis is fully exposed to society by theoretical and other forms of intervention, it will demand an alternative form of reason and discourse. Change can only begin with critique.

3.34 Conclusions

By combining Foucault's attribution of relationships of domination and the control of the discourses of knowledge with Weber's theories of domination (as grounded in the control of technically specialized knowledge and communication forms), an interpretative framework has been formed for the discourses on and of new communications technology. This framework demonstrates how technically specialized knowledge and discourse yields relations of social control and domination. The benefits of new communications technology have clearly been shown as have its deprivations, inadequacies and contradictions. To change these types of social relationships in order to resolve the crisis, one would have to change more than a machine or a set of economic conditions. One would have to change the procedures of discourse, the modes of communication, the dominant episteme, the instrumentality of rationality. It is this particular form of discourse linked with knowledge that has given both technology and society their particular forms of practices. The search for alternative discursive procedures of reason and for an alternative communications practice, technological or other, is the challenge which we now face.

CHAPTER 4

Alternative procedures of discourse: epistemologies of technology

> The design of a public language, then, is a serious task, one pregnant with consequences and thus laden with extraordinary responsibility. (Weizenbaum 1976, 102)

> True participatory democracy in which the town meeting concept is extended via the ubiquitous computer networks to embrace whole nations, and even perhaps the world (Godfrey & Parkhill 1980, 70–71)

Is an alternative possible?

How can alternative procedures of communication be suggested? By whom? Where and how could transformations of the classical, liberalist episteme arise? First of all we have attempted to show that the ways in which technology and society have been spoken and conceptualized are highly determinate of the types of social and technological intervention that are possible and/or implemented.

As alternative discursive procedures are suggested, their capacity for practical realization must be borne in mind. How can we insist upon this practical aspect? If indicating discursive procedures is the theoretical component of communications work, then policy advocacy and policy implementation may be considered the practical components of such work. The move between theories for alternative procedures and the practice of these procedures can be achieved, at least in part, via policy or regulatory advocacy. We must state immediately, however, that the nature of this policy formulation and implementation must be very different from traditional advocacy if we are to avoid falling back into the trap of scientific experts and bureaucratic organizations dictating relations of social conduct to the public sphere. Thus, in what follows, the reader will find not only an attempt to indicate alternative discursive procedures but also recommendations for the implementation or practice of these procedures; in short, some specific policy and meta-policy recommendations.

Lest the reader wonder how the shift from a discursive critique towards policy study is justified, one might characterize communications policy study as the critique of communications practices in the form of discursive procedures. In this regard, the preceding chapters could also be called policy study. However, policy advocacy and policy-making would be the direct suggestion and implementation of discursive procedures felt to meet the aims of 'public interest' and the discouraging of those which do not:

> By policy we mean both the development of rules, regulations and guidelines that direct human communication behaviour and the directives that determine the development and distribution of communication technology. (Harms et al. 1977, 25)

Policy advocates on the basis of which it legitimates or delegitimates extant discursive practices but it may also suggest or uncover alternative practices and generalize advocated procedures therefrom.

Theory so far has taught us that mechanical changes alone will not alter the communications horizon but that an alternative practice of discourse based upon the recognition of the failings of the dominant practice will. Even Pool concedes the first half of this argument: 'no mechanical device will by its mere existence solve the problem of making bureaucrats responsive.' (1973, 241) In other words, theory and praxis meet as the practices emerging out of very profound changes in the intellectual and discursive underpinnings of communication in general, including new communications technology. It is with the aim of 'closing the gap between intellectual speculation on this right (to communicate) and concrete realities of communication in the world today' (UNESCO/'Interim Report on Communications Problems in Modern Society', September 1978, 76), that we hazard the leap from theoretical discursive critique to practical policy suggestions for alternative discursive practice.

The main dangers which we hope to avoid are those of technological determinism and of the technological imperative. We have seen that the traditional classical discourses employed to talk about new communications technology typically led to a belief that machines would change society or, alternatively, that society must at all costs adopt machines. These two positions in turn left very little leeway for intervention in the interface of society and technology. Machines were introduced without the desired changes or with changes only for those whose interests they served, and various groups (parents of school-aged children, libraries, television viewers, etc.) were blackmailed into adopting new technologies for fear of being left behind. Neither of these constitutes a sound or satisfactory basis for reflection or policy advocacy.

4.0 From critique to policy

In the preceding pages we have seen that the discourses of and on new communications technology operate according to certain procedures which delimit specific forms of knowledge. These specific forms of knowledge are in turn accompanied by particular types of social relations and hierarchies of control and domination. In other words, we have seen that the discursive and intellectual practices of new communications technology are part of an episteme, a particular form of rationality, which conditions certain relations of power in society. Furthermore, we found these discursive procedures and power relations to be inherently contradictory and therefore in a state of crisis. The main contradiction lay in the opposition between, on the one hand, the overt semantic claim that new communications technologies would promote a liberalist doctrine of free and equal individualism, and the actual practice, on the other hand, of procedures of social exclusivity and will to power in both discourses on and of new communications technology.

The epistemic crisis calls for the elaboration of new procedures to provide social direction and to legitimate an alternative politics of truth. These procedures must, however, avoid reproducing the classical liberalist contradiction between claims for freedom and the unequal and exclusive practice of power. On what grounds, then, could such an alternative discourse reside?

The call for such an alternative is not entirely foreign to communications circles. Jean d'Arcy, who coined the term 'the right to communicate,' calls for an alternative conception of communication which would lead to an alternative practice:

> New thinking is now due. A new philosophy and a new approach to communications issues would lead to studies for the reshaping of both national and international communications structures. (in Harms et al. 1977, 52)[1]

In *Power/Knowledge* (1980), Foucault suggests precisely this. Once the contradictions of the classical episteme are pointed out, 'the role of the intellectual is to seek to construct a new politics of truth' that avoids the hegemonic order of its predecessor. For Foucault, mere criticism of the existing discursive and social order is insufficient. One must go a step further and search for or lay the groundwork for an alternative discursive order in which alternative practices of knowledge and power would be possible. The challenge remains in finding a new communications relationship which does not reproduce earlier contradictions.

One alternative is nothing more than a regression. While opponents of technology such as the Luddites have convincingly criticized the futurological bent of post-industrialist literature, their remedy is no less historically naive: return to the cottage industry organization of pre-capitalist society, learn to live without electricity, to repair your own shoes, and to grow your own food in order to become independent of centralized nodes of domination. These are the kinds of suggestions put forward by Kumar in the last chapter of his book, *Prophesy and Progress* (1978), and whose naiveté is so thoroughly exposed by Slack (1984). The suggestion that man could suddenly and willfully reverse the procedures of an episteme and thereby return to an earlier one is merely the measure of our ignorance of social anthropology (Gehlen 1980) and also leaves little opening for technological intervention. Once discursive procedures have entered a sphere of discourse, it is not possible simply to reverse them. They can be changed only by much more complex transformations which are not necessarily the initiative of an individual subject. Kumar's theory reverts to the notion that a few elite subjects can radically alter, or indeed reverse, the episteme for everyone. It is a utopian dream that betrays a will to power by the few. The possibility of a complete reversal of discursive procedures is as naive a historical view as the evolutionist line of perfectibility that holds progress to be always good.

If we are not to contradict the alternative theory of power which we wish to elaborate, we must reject such simplistic solutions and ask the following questions:

- How can we avoid positing freedom as a semantic principle and then impose functional constraints?
- How can we ourselves resist succumbing to the temptation of playing the role of the elite scientific subject who exclusively imposes a discursive and political order upon society from above via policy or policy advocacy, while declaring everyone to be equal?
- How can a new ground for an alternative discourse be proposed without falling back into the moralist position of a scientific subject dictating the 'good' moral order for society, and imposing values and practices upon others?
- How can we avoid once again linking scientific truth to political power where the truth of an elite is the legitimation of the imposition of rules upon the citizenry?

Finally, although we have seen that classical discourse is claimed to have arisen from a single scientific subject, its practice in effect grows out of transformations and regularities that constitute the episteme and whose source is social practice rather than some subject of enunciation. How can an alternative discursive practice be posed while avoiding the claim that it arises from the individual subject, from the 'author' of this book? As potential alternative discursive practices are suggested, we must also look to ways in which these alternatives and their very postulation avoid the pitfalls of the classical episteme. Moreover, the potential for the regress of transformational discourse to the contradictions of the classical episteme must constantly be guarded against.[2]

4.1 From legislation of the uses and abuses of technology to the legislation of technology itself

The gist of this whole study has been to argue against this kind of statement:

> There is no distinction between the electronics that make demagogy easier and those that make responsible politics easier. It is the men who use them that make the difference. (Pool 1973, 242)

We have shown that technology is not neutral and that it does not suffice simply to regulate its uses and abuses.[3] We showed this by demonstrating how new communications technology – the hardware, the software, and the services – is itself a set of discursive procedures which are an integral part of the discursive episteme that situates it. In order for these procedures to change substantially those of the whole discursive-epistemological environment would also have to change. Our position is not one of technological or social determinism but rather one of epistemic-technological interdetermination. The discursive procedures are a part of the inherent design and structure of new communications technology, and these specific or localized procedures are conditioned by the procedures that dominate the episteme. If we wish to change some of these procedures, it does not suffice to legislate merely the social uses of technology while not also legislating the design of technology. Both the technological procedures per se and the episteme or the accepted and socially generalized discursive procedures would have to be changed. But, just how one would go about changing these procedures is not so easy. Certain discursive practices may arise in a specific field which are transformations of the generalized dominant procedures and which may eventually extend their sphere of influence to change and even substitute themselves for the traditional procedures.

For example, were the dominant episteme to be constituted of instrumentalist discursive procedures, it is highly likely that so would new communications technology. It would be unlikely for most new communications technology procedures to be non-instrumentalist. However, specific practices of new communications technology which did not reconfirm instrumentalism but which practised interactive communication could, conceivably, be designed and deployed. These alternative practices would not be mere uses of the technology. They would be procedures designed and deployed in the inherent structure of the technology. They would be irregularities or transformations of the procedures of the dominant episteme and they could possibly contribute towards changing more dominant procedures and eventually altering the dominant episteme. Thus, as Table No. 1 illustrates, there is a mutual interdetermination amongst the procedures of the dominant episteme and procedures of new communications technology as well as amongst the irregularities in either the limited sphere of new communications technology or the dominant episteme. A transformation in one sphere depends in part on the other, while a transformation in a limited sphere may eventually become dominant in the whole episteme.

In what follows, we will suggest how communications policy might encourage or reinforce certain transformations of the inherent discursive procedures in both spheres, as opposed to restricting itself to a merely functionalist intervention.

For example, it does not suffice to legislate the uses of new communications technology in order to protect oneself against the tyrannical ubiquity of surveillance. One must first recognize the fact that the discursive procedure of surveillance is reinforced by a society in which the onus lies on the individual subject's control of private property as well as on bureaucratic control, hence making surveillance a socially desirable and necessary procedure which would probably thrive even without technological means. Thus, as long as surveillance is a procedure of the dominant episteme, the very design and structure of new communications technology will include and reinforce procedures of surveillance, as was shown in chapter two. Surveillance is reinforced by the inherent procedures of new communications technology and by the social context or 'technocracy' that situates this technology. There could conceivably, however, be a transformation of these procedures in either sphere which could discourage surveillance procedures on new communications technology and eventually within the whole episteme.

This chapter will concern itself precisely with the aperception or suggestion of such transformations as would alter the classical episteme and classical practices of new communications technology. Our policy suggestions are aimed at pointing to ways in which

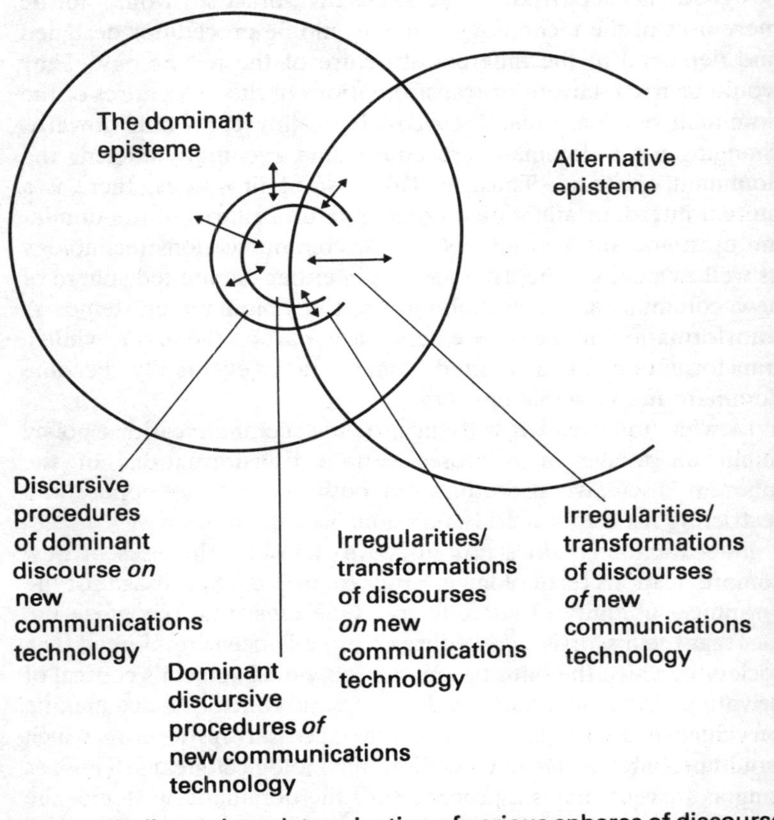

Figure 4.1

such transformations may be reinforced or generalized so that they may become the dominant discursive procedures thereby replacing classical, instrumentalist, rationalist discourse.

4.2 Institutions

Some argue that neither technology *per se* nor its uses and abuses should be legislated but rather that our regulatory efforts should be addressed to social institutions. 'The key solution of social problems is not change in technology per se, but changes in the institutional structure.' (Naselund, in Harms et al. 1977)

While we hold that the procedures of technology cannot simply be put to one side, we find no objection to the statement that institutions also must change. In that we have defined institutions as a rather rigid configuration or society of discourse, they may be seen as particular fields within the dominant episteme and hence as the regulatory nodes of discursive procedures which are subject to transformations. Indeed, it may be the case that the current social formation is composed almost exclusively of subfields of discourse which are institutional. Indeed, institutionalized discursive procedures would most certainly have to change, and changing the episteme implies just that: changing the discursive practices of the dominant institutions of society. One might even go so far as to declare the field in which new communications technology is practised an institution as well. It is certainly the case that any policy legislation must recognize the role of institutions in the hegemonic social order. For example, Slack's suggestion (1984) for the abolition of 'patent law' protection in the case of communications software attacks the institution as a part of the larger social order of capitalism. Society is a very complex network of interdiscursive relations where one practice cannot be expected to change without being a condition of possibility and a determinant of other procedures. Moreover, for any great change to occur, such as the much touted democratization of society via new communications technology, the whole episteme must change.

In a very insightful remark, the Canadian Telecommission Study, *Instant World,* also recognizes this 'va-et-vient' between alternative technological procedures and alternative procedures (habits) of general social interaction when it states that: 'More is needed than just harnessing new services – new and more satisfying ways of life and habits of mind (are required).' (*Instant World,* in Fisher 1982, 12) UNESCO and Harms et al also recognize this contextual interaction of technology and society as a whole when they state that the 'communications environment,' and the 'socio-cultural context' come to 'form part of the communication issues under discussion at any particular time' (in Fisher 1982).

4.3 To regulate or not to regulate? Presence or absence of discursive constraints in an emancipatory society

In the last few decades an international debate has been raging between those countries which hold on to the liberalist concept of freedom as the absence of all constraints – 'laissez-faire' – and therefore favor a reduction in policy control over new communications technology, and those countries which claim that more regulation is necessary in order to ensure their freedom and autonomy. It was partly this debate that gave rise to the MacBride Commission.

Neither side would ever suggest that the desired end of their cause is not human emancipation which has been put forward in Article 19 of the United Nations declaration of human rights as the fundamental human right. The crux of the argument resides in whether or not emancipation is defined as the absence of any and all constraints, communicational or other, on the individual, or whether it is precisely the presence of constraints which creates the condition for human emancipation. Again, it must be pointed out that emancipation is not the creation of the liberalist doctrine of political philosophy though it was primarily liberalism that defined it in terms of the absence of constraints upon the individual.

Emancipation in the Greek and Judeo-Christian tradition, Habermas would argue, has often been defined as rule in, by and for the public, the polis. This notion of freedom invokes the necessity of rules and constraints on behavior precisely in order to ensure participatory rule. Marcuse also agrees that freedom involves the imposition of certain constraints along with the elimination of others. Emancipation and the democratic interest did not imply absolute 'laissez-faire' for Marcuse but rather that which he called 'discriminating tolerance':

> However, the alternative to the established semi-democratic process is not dictatorship or elite, no matter how intellectual and intelligent, but the struggle for a real democracy. Part of this struggle is the fight against an ideology of tolerance which, in reality, favors and fortifies the conservation of the status quo of inequality and discrimination. For this struggle, I proposed the practice of discriminating tolerance. To be sure, this practice already presupposes the radical goal which it seeks to achieve. I

> committed this petitio principii in order to combat the pernicious ideology that tolerance is already institutionalized in society. The tolerance which is the life element, the token of a free society, will never be the gift of the powers that be. (Marcuse 1965–68, 122–23)

There is a political and a sociological argument to be made in favor of a form of emancipation which does not exclude all notions of constraint. Foucault has also made an historical, epistemological argument against the very possibility of the total absence of constraint. While Foucault criticized the constraints of the classical episteme as internally contradictory and as highly exclusive and hierarchical in their distribution of power, he also argued that throughout history there has never been and never can be a discourse without procedures or constraints. There have been different kinds of constraints leading to different kinds of social order, but a discourse without constraints is impossible. The liberalist doctrine is an historical and epistemological impossibility. Thus, rather than trying to abolish constraints altogether and to create anarchy, Foucault suggests that the intellectual should contribute towards the formation of alternative discursive procedures which would render a less contradictory, less hegemonic and more localized politics of truth.

However, just how these transformations appear or occur is a major question. They may appear in a disorderly manner as chaos or they may appear purely by chance, neither of which would make them any less valid as alternatives. Indeed, one of the greatest challenges of the scientist and the social scientist is perhaps to construct an alternative (social) order out of disorder (Dupuy 1982). Furthermore, we cannot overemphasize that these transformations do not spring out of the head of the singular author of this book. They are communications practices which already exist or which have been pointed to as contrafactually possible.

4.4 Discursive transformations

It is to these transformations that we now turn as alternative discursive procedures of knowledge and technology and as alternative rules for the practice of communication.

4.4.1 Ontological shift of the definition of communication: from object to practice

We have seen that the debates on access, free flow, content-carrier, and billing, all stem, at least in part, from a reduction of communication to the status of an object.

The first alteration of discursive procedures that could be suggested would be to consider communication as an activity or a practice rather than as an object. This change of procedures would result in the shifting of issue formation from one of trying to ensure equitable distribution of commodities to one of trying to guarantee an equitable order of communicational practices and processes.

Discourses which have already pointed out this ontological shift include not only Foucault's, which suggests that the only thing we can say about discourse is that it 'is practised,' but also policy advocates, such as the Harms group on the 'right to communicate' whose inspiration is derived from Bert Cowlan's statement that communication 'should be treated only as a verb' (in Harms et al. 1977). In these discourses we find a shift from an emphasis on the object nature to an emphasis on the actional nature of communication.

As regards specific legislation which this ontological shift would seem to require, we would pay attention to the regulation of practices rather than of objects. Instead of permitting or forbidding satellites, for example, certain satellite transmission practices (which could be built into satellites) would have to be legislated. Human communications rights would shift from the right of access to certain hardware to the right to carry out certain communications practices for which hardware and software may or may not be necessary.

We have already seen in chapter one that the content debate and the separation of content from carrier, the access debate and the hardware/software distinction, the regulation of transborder data flow as commodity, and the free flow debate all arose, in very large part, from a reduction of the ontological status of communication to that of object. Even Herb Schiller recognizes the disadvantages of reducing communication to an object when he states as a policy objective the need to resist the penetration of commodity values into this sphere (Schiller 1974, in Fisher 1982). Legislation should deal with what happens in the hardware processes rather than with the equitable distribution of that hardware.

Communication rights must shift away from the question of access to objects and towards the right to practise certain discursive procedures. This involves rights to communicate in certain ways, with certain stylistic, lexical, and even political competences. As we shall see later in this chapter, the right to access would have to come to mean, within the perspective of communication as practice, a certain communicational competence, i.e., a right to communicate in certain ways. A very practical policy suggestion which would shift the perspective from communication as object to communication as practice, would be a review of the recently passed 'freedom of information' bill (an act not yet in effect). The revised bill would insist not only on rights of access and of protection but also on rights to communicate.[4]

We would suggest that the control goes much deeper than the State's capacity to divulge or withhold information. Rather than a simple information ombudsman, we require a 'communications ombudsman' to decide about the justice not only of withholding or releasing information but also of the ways in which information is processed, of the ways in which interaction takes place, of degrees and areas of control conquered through allowed and disallowed constraints on communication. And, of course, the ombudsman is supposed to serve the public interest. At the end of this chapter, we will discuss further the types of communication rights, as opposed to information rights, that such an agency would be expected to ensure.

Finally, as we mentioned earlier, the free flow debate is also a determination of an objectal view of communication. We had not noted that since the 1970s several discourses were beginning to suggest transformations of the free flow debate. Most of the members of this commission, with the exception of the Canadian representative whose objections are copiously documented, agreed that democratic communication was not a mere case of free flow of information objects but rather of certain rights of communicational exchange and process:

> Access is only a part of the democratic process. The democratization of communication means more vast possibilities for nations, for political forces, for cultural

> communities, for economic entities and for social groups to exchange information on a more equal footing, above and beyond the domination of the weaker partners, and without discriminating between any of them. (MacBride 1980, 173)

Here we see two discursive formations. Firstly, there is a move away from treating communication as the transfer of an object and a move towards viewing it as interaction and exchange. Secondly, there is a move away from the liberalist doctrine of the free and unconstrained individual and towards a recognition of social groups as the pertinent participants of communication and the interests to be reckoned with.[5]

4.4.2 Beyond control by the individualist subject

In liberal democracies such as Canada and the US, there are already several cases in which free flow is limited by law. For example, legislation has enshrined restriction on free flow when nation-building and Canadian autonomy were at stake. The free flow information at the whim of individuals was compromised in Canada and the United States as soon as the airwaves were declared public property falling under the jurisdiction of the federal government. In Canada, the rights of individuals were sacrificed to those of groups when both the Broadcasting Act (1968) and the constitution established the rights and tastes of small groups as the legitimate province of broadcasting rather than the rights and tastes of just any individual in the mainstream of a 'mass' Canadian population.[6]

What is extremely important in the Broadcasting Act is that while private interests are acknowledged, the possibility of a conflict in the communicational sphere, arising from the encounter of private and public interests, is foreseen. Such conflicts are to be resolved in the public interest. The public interest is to prevail over all others under the Broadcasting Act:

> i) where any conflict arises between the objectives of the national broadcasting service and the interests of the private element of the Canadian broadcasting system, it shall be resolved in the public interest but paramount consideration shall be given to the objectives of the national broadcasting service.

Thus, Canadian communications and cultural policy has for some time legitimated a departure from an individualist 'laissez-faire' posture. Rather than conceiving of the communications horizon in Canada as an amorphous mass of atomized individuals, Canadian policy recognizes that communicators have a responsibility (not merely an absence of constraint) to meet the public interest consists of various sub-groups with distinct identities, interests and needs.

4.4.3 The privacy debate: an epistemic impossibility

However, despite the general departures from the liberalist doctrine of the individual on the part of cultural and communications policy, there remain many vestiges of it. The most striking is perhaps the continuing obsession with privacy throughout the Western developed world. A recent public survey undertaken for the Canadian Department of Communication showed that the new communications technology issue which most concerns citizens is privacy. The privacy debate flourishes as a carry-over of the belief that communication (literary, political, technical) originates with the individual subject/machine of enunciation. However, scholars of communication such as Foucault, Weber, Peirce, and others, have shown that discourses of knowledge are always the conditions of possibility of certain collective agreements (conventions) to what constitutes saying the truth in any particular time and place. As we have argued throughout this book, discourses, including new communications technology, reflect the constraints, laws, rules and transformations permitted by a group of speakers within a given time and place.

Foucault discusses this most explicitly in his text which relativizes the principle of the individualist author (1977, 113–38). Weber shows how what the individual can say and do is conditioned by a form of rationality. Peirce suggests that the truth is what the community of investigators considers it to be within any given epoch.[7]

In other words, the procedures governing the circulation of information and other communicational practices in any society, have never been controlled by an individual source, despite the fact that liberalism claimed them to emanate from a genius subject. The procedures derive from a collective agreement and the collectivity which agrees can be a quite small and limited elitist grouping, such as scientists. Hence the crux of the privacy debate, i.e., control of the procedures of communication by the individual, is an epistemic impossibility. The very hypothesis of individual control is itself a condition of the liberalist episteme, and not of any individual. In other words, cold warriors would do well to recognize that the paradox of individual vs collective communication rights does not really exist. The possibilities of individuals are circumscribed by collectively accepted discursive procedures.

Basically, what is being argued here is that there exist many threats to personal and social freedom other than the threat to privacy. The potential for social management, i.e., for surveillance and control of social groups by means of new communications technology, the new form of bureaucratic capital, not only invades an individual's privacy but also compels or represses the behavior and conceptualizations of entire societies. It would be to misconstrue this discussion to understand that we are suggesting the absence or elimination of privacy. We are objecting to the exclusive centrality that the issue occupies and to the liberalist, individualist way in which privacy is understood.

4.5 Must we abandon the democratic emancipatory principle once and for all?

If, as seems to be the case, we are evolving towards a totalitarian society in which the State apparatus is omnipotent, the ethics necessary in order to ensure the survival of the true, free, and human individual can be named as follows: lie, cheat, wangle, trick, hide, counterfeit documents, and build your own new improved electronic gadgets that surpass those authorized by the authorities. If your tele-screen surveys you, take advantage of the one night when you are allowed to turn it off in order to rewire it so that the police monitor which watches over you will end up seeing a picture of his own living room.

The preceding paraphrase of science fiction writer P.K. Dick's political philosophy of technology suggests that the only alternative to social manipulation is to manipulate the manipulators to use the surveillance capacity of new communications technology against itself, to watch the watchers. Dick suggests a totally immoral position in face of increasing social control. Fight fire with fire, so to speak. Citizens should be as ruthless and as cynical as the designers and implementors of technologies of social domination. Is this the only alternative to liberalist, classical discourse? This is probably the most difficult and most important question faced by any proposals for social and technological intervention.

If we reject the liberalist conceptualizations which underlie the issues of free flow and privacy, are we also abandoning the democratic principle or the struggle for control of enunciation? Hardly, but the foundations of democratic principles need to be reevaluated.

As we saw in the previous chapter, the Frankfurt School's critique of domination relied heavily upon the ideal of the free, individual subject. Indeed, Marcuse (1968) even states explicitly that the ideal of the freely developing individual must be salvaged from liberalism and that it is certainly worth more than any notion of the 'Volksgeist.' The Frankfurt position is adopted also by Bacon, Comte, Hobbes, and Descartes, and is perhaps nothing short of a revival of nominalism.

In partial keeping with the Frankfurt School, however, Jürgen Habermas has suggested that the ground principle of freedom is not particular to liberalism or to classical science, but has always been implicitly affirmed throughout the Greek and Judeo-Christian traditions of civilization. Critical theory cannot do without the presupposition of freedom: 'the concept of necessity in critical theory is itself a critical concept; it presupposes freedom, even if a not yet existent freedom.' (Horkheimer 1937–72, 230–31)[8]

The dilemma we are now faced with is the following: a) must we abandon the almost universal principle of emancipation as the ground for discursive and social practice in order to avoid the moralizing contradictions of classical discourse and liberalism or b) is there any way to retain this historically persevering principle without becoming embroiled in the types of internal and external discursive principles, contradictions, and hegemonies we criticized earlier?

In our criticism of liberalist, classical discourse we did not attack the emancipatory principle *per se,* but rather the actual inequalities of subjects who had been declared free and equal. We criticized the hegemonic exercise of exclusive discursive constraints by knowing elites. It was not emancipation but contradiction and a particular hegemony that were the brunt of our attack. Therefore, it is contradiction and particular hegemonic practices, not the principles of democratic freedom itself, that must be abandoned, unless, of course, it can be shown that such a principle cannot be maintained without either theoretical or practical contradiction.

Democracy and liberalism

We cannot overemphasize that a reconfirmation of the democratic and emancipatory ideal is not tantamount to a reconfirmation of liberalist political philosophy. C.B. Macpherson in *The Real World of Democracy* (1965) demonstrated that the concept of democracy was born far before liberalism and that liberalism itself arose independently of democracy. Liberalism merely borrowed from democratic theory at a later date when it had difficulty in assuring mass legitimation of the free enterprise system. Moreover, Macpherson shows how many present-day non-liberalist régimes lay claim to being democratic. Liberalism has neither a claim to primacy nor to exclusivity of democratic political philosophy:

> The idea of democracy goes a long way farther back than the period of liberal democracy and the modern non-liberal notions of democracy are plainly drawn from that original notion. (Macpherson 1965, 36)

What Macpherson does find to be constant throughout the historical and ideological variations of democratic political theory is the wish 'to provide the conditions for the full and free development of the essential human capacities of all members of society.' (1965, 37)

But why struggle to maintain the emancipatory principle in today's social discourse on new communications technology and in discourse in general? Habermas and Horkheimer both suggest that the notion of a 'better way of life,' of a rational society (not based on instrumentalist rationality) depends upon the principle and practice of emancipation:

> Unless there is continued theoretical effort, in the interest of a rationally organized future society to shed critical light on present day society and to interpret it in special sciences, the

> ground is taken from under the hope of radically improving human existence (. . .) The issue, however, is not simply the theory of emancipation; it is the practice of it as well. (Horkheimer 1937–72, 233)

Indeed, for Habermas, hypothetically if not prophetically, the emancipatory interest itself grows out of the contradictions and tensions in social theory and practice as exposed by critical theory. Constraint gives birth to the desire for freedom in its prisoner. For Habermas, the emancipatory principle is the one universal and non-negotiable principle in a rational society, though within the emancipatory framework all other discursive practices are negotiable.

There most certainly still are difficulties and paradoxes involved in trying to ground philosophically any principle of democratic emancipation. There are, however, strong political grounds for maintaining such a principle; they may serve until all of the philosophical problems are resolved, if ever. Few Western societies would overtly reject the democratic, emancipatory ideal. We may simply take these political declarations at their word and then attempt to hold society to them. Thus our task is not to find a substitute for the basic emancipatory tenets of democratic theory but rather to find an alternative to the liberalist version of democratic theory. It is on such a foundation that we could subsequently identify and encourage communicational procedures or rules that would realize emancipatory democratic aims.

We may return yet again to the political reason (though this is not a 'Grund') for upholding the democratic principle of emancipation. It is really quite futile to profer policy-oriented discourse which has no chance of any social reception. In this society, to elaborate a rhetoric which abandons the aim of improving human life by increasing the degree of democratic emancipation in social interaction would condemn one to a dialogue of the deaf. As long as society, including its regulatory agencies, public spheres, and most of its science, continues to cling to this principle, perhaps the best first step towards changing the dominant form of discourse is to maintain this principle while attempting to suggest or to reinforce particular discursive practices which do not contradict it. Furthermore, in that the emancipatory, democratic ground has been shown by Habermas and others to go back further than the origins of liberalism to the underlying concept of the Greek *polis,* its retention cannot be dismissed as the maintenance of the essence of the classical, instrumentalist episteme. Indeed, we might understand the traditional liberalist episteme as setting into circulation procedures designed to counter the overt claims of democratic emancipation. Some of these counterproductive procedures stemming from classical, liberalist discourse will have to be evacuated from the discourse based on principles of democratic emancipation: individualism, imposition of the ideal by a select few, definition of freedom as the absence of constraints, etc. It is still possible, one hopes, to suggest or point to alternative discursive procedures consistent with the democratic, emancipatory principle.

Finally, we cannot but agree with William Leiss' criticism of the emancipatory principle as elaborated by the Frankfurt School. Leiss (1974) argues that for the past fifty years history has not given us a glimpse of an emancipatory subject arising dialectically from the contradictions of capitalism. However, even Leiss does not abandon the quest for a ground on which to justify the emancipatory principle. He closes his criticism of the Frankfurt School with the hope that some other route to emancipation and autonomy may be found than through the machinations and purgatory of capitalism.

4.5.1 Autonomy and responsibility

What alternative discursive procedures could replace those of liberalist individualism and the absence of constraints for an identification of the participation roles in communication? Researchers of a recently inaugurated group known as the 'Centre pour la recherche sur l'épistémologie et l'autonomie' (CREA) have been struggling with precisely such questions, especially as regards the role of science and technology in society. While one may often detect in their writings a regression to the liberalist notion of the self-determining individual, they do seem at times to be tending towards a replacement of the notion of free unconstrained individual with that of a self-determining social sphere. (By social sphere, we are not necessarily referring to a collective subject as a class of individuals but to sets of discursive practices, social habits, rather than groups of individuals). A self-determining social sphere is an autonomous one which sees the capacity to act not in terms of the absence of constraint but in terms of social practices and habits subject to self-determination. While we are not certain that even a collective subject has such a great degree of self-determination, CREA's position is a step towards retaining simultaneously a concept of emancipatory social order, collective discursive constraints, and the responsibility of individuals to the collective order:

> 'Epistémologie et autonomie,' le mot d'ordre que nous nous sommes donné signifie pour nous, une connaissance véritable de l'homme et du social qui se doit de déboucher sur un accroissement de la capacité d'agir (. . .) En fait, les sciences sociales en leur état actuel entretiennent avec l'autonomie sous tous les aspects, autonomie de la personne, autonomie du social et autonomie politique, des rapports étranges, ambigus, contradictoires même, et pourtant fondamentaux.
>
> Ainsi l'autonomie de la personne est liée aux sciences (. . .) La sociologie ne trouve partout que contraintes sociales, influences et déterminismes, s'il le faut même pulsions psychologiques. Depuis deux siècles l'histoire des sciences de l'homme est l'histoire de l'extension impitoyable du déterminisme à l'ensemble des conduites humaines. Ces sciences vivent, semble-t-il, de la négation de l'autonomie de la personne qui fut la condition sociale de leur naissance. Ce n'est là pourtant qu'un aspect du problème. Si on s'intéresse maintenant à l'autonomie du social, on verra que la façon dont l'aborde la réflexion moderne est tout aussi étrange et contradictoire. Les sciences de la société telles que nous les concevons aujourd'hui commencent lorsque les hommes renoncent à poser l'origine et le sens social hors d'eux-mêmes dans une divinité transcendante qui le fonde et le défend. C'est-à-dire lorsqu'ils commencent à concevoir que l'organisation sociale est une auto-organisation, une auto-institution que les hommes se donnet à eux-mêmes.
>
> On saisait aisément les liens qui rattachent cette idée de l'auto-organisation du social à l'autonomie de la personne. Dês qu'il devient possible d'initier de nouvelles pratiques sociales et donc de remettre en cause l'organisation de la société, par le simple fait des transformations qu'introduisent en elles les pratiques nouvelles, cette organisation ne peut apparaître

> bientôt que comme le pur résultat de l'activité humaine. Mais l'origine purement humaine de l'organisation sociale sans poser de sérieux problèmes, rien ne garantit plus contre l'arbitraire des lois ou le règne du plus fort. (Domenach, Dumouchel, Dupuy 1981, 'Projet de Création d'un Centre d'Epistémologie des Sciences de l'Homme á l'Ecole Polytecnique.' CREA: Centre de reserche sur l'epistémologie et l'autonomie, ms., 3–4)

An *autonomous* society with *responsible* individuals, such is the new order proposed for social and scientific practices by the CREA. The implications of their key words are that somehow the social must organize itself and transform itself in ways that also respect personal freedom. 'Autonomie' would be neither anarchy nor chaos but rather systems of constraints designed to give the social sphere a less deterministic hierarchical organization and more possibilities for human action. Autonomy is not the absence of constraints upon individuals. It is, rather, a socially-selected and socially-serving constraint/organization.

CREA is referring here to the notion of autonomy as employed by Castoriadis. An autonomous society is one whose order would be based on an immanent rather than an external, transcendent instance, one in which no external tyranny would dictate order. Autonomy would be the exemplary democracy in the sense of rule by and for the polis. Like Habermas, Castoriadis refers us to the ancient Greek ideal of democracy and philosophy whereby society would have the privilege and the responsibility of making and reexamining its own laws. Social order would be based on openness and propensity towards self-generated change rather than on self-reiterating closure. Such a society would not refer to a higher level for its laws and norms but would make its own and accept responsibility for them. We will see later in this chapter that Habermas has provided a communicational framework for autonomy whereby the polis must form a consensus concerning the rules of communication but must also follow certain rules of communication in forming that consensus. These rules are variable, ever subject to reexamination by the polis. The rules of communication of their validity claims would not be imposed from on high.

What we must bear in mind concerning the notion of autonomy, however, is that neither Castoriadis nor Habermas maintains that such a social order has yet been realized. We are still engaged in thinking this relationship and its realization would not occur in a single stroke. It would be a constant process and struggle. Furthermore, both realize that an autonomous or immanent consensually determined social order is not only a new social order but also a new discourse, a new knowledge, a new politics of knowledge, and a new relation to nature.

Social autonomy and communications policy

What are the implications of the departure from the individualist procedure of discourse for policy-making? The concept of autonomy implies certain policy structures organizing communication in society. These structures are ordered and hierarchized in such a way as to favor social participation and freedom rather than individualist 'laissez-faire.' Furthermore, the notion of autonomy implies that no individual can impose an order, even the autonomous order, on society. Rather, structures or discursive procedures would have to arise in society as socially encouraged transformations:

> If we add to it the notion of 'appropriate spheres of autonomy,' such structures can have both the management power of hierarchy, while at the same time preserving high levels of participation and freedom. However, such models cannot be forced upon society. They are unfamiliar images with no cultural context to encourage them. But let us hope that they will be given a chance to evolve naturally, not stamped out by the prevailing ethic of hierarchical control and bureaucratic nationalism. (Seeley 1980, 47)

We prefer, however, to speak of transformations rather than of 'evolution' where discourses are concerned.

4.5.2 Universality of the right to communicate

One case of policy advocacy that takes its distance from free flow individualist policy is the UNESCO and Harms declaration that the right to communicate is a *universal* right which can only be assured by regulation. Harms (et al. 1977) states that the first concern for policy-makers should be to ensure universality, i.e., the right of everyone to communicate, including specific controls or rights to equalize the rights of groups of unequal strength. Very specific collective communication rights are called for by Harms who suggests that Article 19 of the United Nations Charter of Human Rights is simply too broad and fails to acknowledge the procedures required to ensure more than individualist free-wheeling.

Again, instead of defining communications freedom in terms of the absence of any constraints, alternative specific constraints are called for. Schiller speaks of the need 'to make explicit the rules and assumptions by which message creation and message flows are undertaken in the nation.' (in Fisher 1982) And, Canadian policy-makers have called for the universalization of the right to communicate:

> that communications are of the people, by the people and for the people. If it be accepted that there is a right to communicate, all Canadians are entitled to it. (*Instant World* 1971, 12)

4.5.3 Towards a contextual consideration of differing communication procedures for differing regional or cultural groups

The principle of universality, however, does not mean massification or majoritarianism. Harold Innis' call for a respect of the peripheries as opposed to a centralization of power is one example of the call for an alternative communicational order. Innis made his mark in international communications scholarship by suggesting that rules and modes of communication should not be within the control of a centre but should vary with the divergent interests of peripheral groups. Regional, ethnic and cultural groups must be given autonomy to set some of their own discursive rules and constraints in order to have some say in the determination of their own social order. This, of course, is in direct opposition to the hegemonic set of discursive organizational procedures referred to earlier as instrumentalism.

Redefinition of the individual: a legal revision

When is an individual not an individual?

Often in discussions of new communications technology we find the recognition of two types of individuals: flesh and blood natural individuals and juridical 'corporate' individuals, which include individuals but not necessarily human groups. The latter are dubbed individuals so that they may benefit from the doctrine of the absence of constraint upon the individual. When these two types of individuals come into conflict in legal or policy situations, the former is no match for the latter. Yet very often when non-corporate groups wish to act as a unit, as in the case of class action suits against corporate individuals, the case is thrown out of court, as when in 1982 an Ontario-based group of car owners attempted to take General Motors to court. The confusion and sham of the individualist doctrine is flagrantly apparent in this legal distinction between individuals:

> Communication between natural and juridical persons, i.e., persons, organizations, institutions of mass communications or groups within a given society, lies in the competence of a given State (Klienwaecher 1978, in Fisher 1982, 24).

Rather than artificially considering corporations to be individuals, we would suggest that society be considered as made up of sets of discursive corporations with differing interests which would be allowed to interact and to dispute as groups rather than as omnipotent corporate individual vs flesh and blood individual. In this case, policy would also be responsible for ensuring some degree of symmetry of communicational strength between these groups by placing limits on resources usable by each party in a debate.[9]

Such redefinitions of individualism and of group symmetry are often attacked, however, on the grounds that they are simply 'inefficient,' depriving corporate individuals of the control they need to meet their production goals, etc. It is to this instrumentalist assumption that we now turn in a search for an alternative set of epistemic procedures.

4.6 From instrumentalism to interactionism

The most fundamental discursive-epistemological transformation that we will suggest is the move away from what was earlier called instrumentalist rationalist procedures and towards interactionist discursive procedures of knowledge.

4.6.1 Theoretical background of the shift to interactionism

Karl-Otto Appel[10] has suggested that we have recently seen a shift in first philosophies, away from a theory of consciousness or a theory of syntactico-semantic referentiality, towards one of pragmatic interaction. He refers to such thinkers as C.S. Peirce, the Wittgenstein of the *Investigations,* and, of course, Jürgen Habermas, as proponents of this shift. Several thinkers whom he does not mention, but who also reflect such a discursive and epistemological shift of perspective, include George Herbert Mead and the Chicago School of symbolic interactionism to which Habermas referred in his doctoral dissertation, as well as Gregory Bateson and other members of the Palo Alto School. Out of these schools grew an ethics and an ecology of interactionism. All of these thinkers claim that we can only know by virtue of interacting with what we wish to know and that in so doing we relinquish control over the object of our knowledge, allowing it to play an active, participatory role. For them, an active subject, referring adequately to a passive object has become an epistemological impossibility, and for some an ecological disaster.

Jürgen Habermas in one of his more recent works, *Communication and the Evolution of Society* (1979), argues that the move from an instrumentalist practice of knowledge and communication toward an interactionist one means that we no longer seek to act upon an object ourselves but rather to communicate with that object, realizing that it too can act upon us, enunciate itself as an 'I' addressing us as a 'you':

> Only if men could communicate without compulsion and each could recognize himself in the other, could mankind possibly recognize nature as another subject: not, as idealism would have it, as its other, but as a subject of which mankind itself is the other. (1979, 88)

Habermas' attempt to provide not only an epistemology but also a theory of society based on interactional relationships must also be seen as an attempt to complement a materialist theory of society based on work relationships with a theory of society that examines all forms of communicational interaction, as well as relationships of production. Even before Habermas, the Russian theorist of language Mikhail Bakhtin, a member of the Tartu

School, suggested much the same communicational complement to a materialist theory of society. Bakhtin (1977), who dared not sign his works, advocated dialogistic, polyphonic forms of communication allowing for the participation of many unreduced voices rather than monophonic or subsumed, reported discourse (Bakhtin 1977, *Marxism and the Philosophy of Language*).

4.6.2 Dismissal of the man vs machine debate: interaction replaces control

Instead of repeating *ad infinitum* the age-old metaphysical debate between man and machine, in which the factions for and against machines are locked into an instrumentalist, control-oriented discourse, one might take a more interactional perspective and ask how nodes of power and domination may be diffused in the ways in which man and machine interact in society. The primordial ceases to be 'who will dominate whom?' and becomes 'what procedures must be encouraged in order for democratic, interactional and social participation to be fostered?' The aim is not for man or for machine to control the other but for machines to be programmed and designed in such a way as to encourage social, human interaction. Thus, for example, the protocols of closed user group networking must definitely be seen as anti-interactive and pro-exclusivity and should, consequently, be discouraged. Furthermore, interactionism should be insisted upon at all costs as far as machines are concerned. It is in this light that we agree entirely with Weizenbaum's opposition to the coupling of the electronic and the biological in matters of intelligence precisely because it militates against democratic social interaction.

4.6.3 Why interaction?

The interactive model for discourse and knowledge is not equivalent to an abandonment of all discursive constraints nor of social order. It is an attempt to dismantle the existing hegemonic hierarchy of social domination in favor of a set of rules which would give more control to the public sphere and more equity of control to participants in social interaction. Salter suggests that Innis saw such discursive power-sharing as necessary because the absence of interaction, taken as a separation of production and reception in communication, encourages monopolies of knowledge whereby 'monopolistic knowledge robbed the public of legitimacy in the interpretation of its own experience.' (Salter, in Salter, Melody, et al. 1981, 206).

The right to communicate

Nasselund argues in a similar vein when he justifies raising the issue of the 'right to communicate.' This 'right' was a draft proposal made by the Harms group at the request of UNESCO. It sprang from a recognition that communication, and, consequently, communication policy formulation were/are highly constitutive of social culture in the past, present, and future (Nasselund, in Harm et al. 1977, 65).

The 'right to communicate' was first written about, although in a very embryonic form, by Jean d'Arcy in an article entitled 'Direct broadcast satellites and the right to communicate' (*EBU Review,* No. 118, 1969, 14–18). Concern for this new policy issue grew out of a growing dissatisfaction on the part of non-aligned nations with extant orders of communication which seemed to be reinforcing domestic and international inequalities rather than living up to liberalist claims of freedom and equality for all. The formulation of a specific right to communicate also grew out of a felt need to specify which discursive procedures would be involved in a more participatory, interactive, and democratic society.

The right to communicate was a case of policy advocacy designed to protect specific, heretofore underprivileged groups in society and to encourage an interactive, participatory form of social interaction:

> Everyone has the right to communicate. It is a basic human right and is the foundation of all social organization. It belongs to individuals and communities, between and among each other. This right has been long recognized internationally and the exercise of it needs constantly to evolve and expand. Taking account of changes in society and developments in technology, adequate resources – human, economic, and technological – should be made *available for all mankind for fulfillment of the need for interactive participatory communication* and implementation of that right. (Harms and Richstad 1977, 89–90, emphasis added)

The call for the right to communicate is a call for sets of

communicational rules which would be an alternative to those flourishing under the liberalist doctrine while still maintaining the democratic and participatory aim:

> The aim is to create more democratic relationships by integrating the citizen into the decision-making process of public affairs. (MacBride 1980, 200)

> The right to communicate (is) seen as a capacity for interaction and dialogue, a facility for access and participation and *involving obligations and responsibilities.* (Fisher 1982, 19, emphasis added)

For the advocates of the right to communicate, this right is not simply a vague, blanket absence of constraint. The right to communicate is built up out of specific communicational rights and freedoms:

> the right to communicate implies a two-way communication give-and-take, an interrelationship. It involves several fundamental freedoms affecting not only individuals but also groups and nations. (MacBride 1980, 16)

These specific freedoms may be qualified in terms of specific interactive rules of discourse favoring participation, in terms of a dialogical practice of discourse, but first it is essential to state clearly what they are not.

4.6.4 Beware of false interactive participation

'Please, no cameras in the courtroom.'

One of the most important tasks of communications scholars and regulators alike is to beware of and to expose hierarchical domination that is touted as participation. Even administrative researchers like Pool admit that there is a great likelihood of manipulation under the guise of 'furthering democracy.' 'Electronic national town meeting(s) may work far better for the leader than for the led.' (Pool 1973, 239)

Just as a medium does not guarantee transparent mediation between representation and its correlatum, so the simple addition of a distributional response element to cable television systems does not ensure dialogue or participation in the social sphere. Habermas suggests that we must first expose the distortions of participatory communication before beginning to realize participatory communication *per se*. Such distortions would be, for example, the a-symmetry of minority rights to participate or administratively regulated censorship.

> but this (information) explosion constitutes no guarantees that the hitherto neglected majority, ethnic minorities or any particular group of people be reached by the flow of information from outside – or may take part in dialogue. On the contrary, opinions and dissenting voices are in fact often silenced by means of administrative rulings or commercial interests. (Zachrisson, in Fisher 1982, 54)

Obviously, for genuine participation, two-way technology must be made available but the technology itself is insufficient to guarantee participation. It is only a tool. We must constantly look towards other constraints and procedures which exist outside the technology but which can all too easily be built into it.

In systems such as Telidon, then, what degree of public participation is actually permitted? An added channel with a keypad provides for a non-visual, printed, and limitedly coded (i.e., yes/no, 1 to 10) enriched feedback loop. This videotex system assigns us to the very passive role of reactive consumers. We can only consume already furnished information with little say, other than in market surveys, as to what type of information should be made available. Indeed, low public demand for the extant Telidon protocol in comparison to the home computer indicates the limited nature of videotex choice. The technical specifications of the feedback channel demonstrate that a far more complex and nuanced reply is feasible. Participation is unequal.

The *trivialization of feedback* is yet another danger of false participation. One of the major difficulties facing any two-way system will be how to avoid the 'radio talk show' phenomenon. The banalities set forth as discussion topics and cliché level of discussion in most AM radio talk shows are a case in point of the communicational depoliticization of public participation. This trivialization springs mainly from the prevalent view of the public as an indistinct mass. Little attempt is made to solicit opinion on specific issues from various social groupings implicated in the public sphere. So far, some of the intended uses of videotex services for participation reflect a reinforcement of this trivialization, especially the reduction of feedback to consumerism – electronic banking, pornography, video games – as much of the Telidon publicity amply demonstrates. One exception, where a special faction of the public sphere is addressed and a response solicited in relation to interests and needs may be the Ida/Telidon Grassroots project in Manitoba where agricultural information is shared, provided, and solicited. Even this information, however, is devoted almost exclusively to instrumentalist, economic concerns.

False participation will always be touted as real participation as long as the sole criterion by which to judge participation is the presence or absence of a distributional feedback channel. The hypocrisy of moralist philanthropists is also a case of false participation. Tate (in English 1973) argues that telematized community services such as education, health care, legal and consumer advice, and safety surveillance, give power to the community. Again, this is a form of false participation, indeed of social control exercised by a bureaucratic elite. Foucault has been cited time and again in this book as showing how the State, instead of physically repressing and punishing its citizens for fear that they would revolt, prefers to watch over and control their cognitive frameworks by means of so-called philanthropical social services which simultaneously served to classify and hierarchize (i.e. control) them.

Electronic polling and 'think tank' electronic interaction are two other forms of computer-communications practices which cannot be accepted as contributing to participatory democracy. We would express strong reservations over the Institute for Research in Public Policy's optimistic confidence in the possibility that democratic participation will be encouraged in and through new communications technology:

> However, as the present means of obtaining citizen input on social and political questions is widely regarded as inadequate, it seems equally likely that the systems will be developed to allow social and political polling as well. Current and extensive calls for direct democracy could well be fulfilled through incasting television transceivers.
>
> Members of political assemblies could receive constantly updated polls of their constituents' attitudes. The public service,

> at present widely thought to be out of touch with the public's needs and general attitudes could profit from incasting information. The possibility of citizen feedback reaching directly the political representative or public servant would well spark greater citizen interest in social issues. (Pergler 1980, xvi)

Marketing/polling surveys have usually served to give more control to the producers over the consumers rather than to represent consumer interests. Also, electronic polling does not give the citizen an equal role in defining the issues to be discussed in the first place, it merely allows him or her to react to pre-enunciated issues. A typical case of this *a priori* definition of the issues by hegemonic discursive societies is that of new communications technology itself.

As long as the public is not allowed to formulate the issues to be discussed, one cannot speak of democratic participation. Moreover, electronic polling would not really cause any great change in the discussion of issues in society today. Edelman points out in *The Symbolic Uses of Politics* (1964) that, from a host of possible issues, only a few are chosen for discussion in the public sphere and this for very strategic reasons. Selective agenda-setting is a far cry from democratic participation.

Another example of communication falsely considered to be democratic can be found in Sheridan's description of the harnessing of electronic communication to carry out Delphi-type surveys. Sheridan describes this survey as one where:

> 1) Leader defines a question;
> 2) the participants reply;
> 3) the participants hear each others' replies; and
> 4) modify their own position respectively. (Sheridan, in Pool 1973, 238)

However, he who begins the act of enunciation, he who initiates the questions dealt with in the dialogue, has an immediate advantage over the respondent since the initiator makes the respondent follow the rules of his communication game. A reasonable degree of equality can only be achieved when the respondent has the communicational competence to switch roles, in which case the vested power swings back and forth between participants (Barthes 1978, 12–13). The electronic Delphi survey is but a repeat of communicative a-symmetry.

4.6.5 Beyond a mere simulacrum of dialogical participation

Sometimes, in the mail, we receive flyers which announce that some event is going to take place. We may read them with interest and then never hear of the event again. This phenomenon has been dubbed the 'flyer feature,'[11] and has been likened to what often happens in debates over democratic, participatory communication. Throughout history we have had various technological 'advances' sold to us with flyers announcing an increase in democratic participation. And then, the event never seems to enter our sphere of experience.

A simulacrum is the imitation of representation of something which never existed. Despite his 'believers,' Weizenbaum claims Eliza to be a simulacrum of psychiatric dialogue. We would presume to say that all of this talk about new technology, communicational or otherwise, is a simulacrum of democratic dialogue. Democratic dialogue does not exist substantially in this society yet new communications technology purports to represent it. Raymond Williams discovers much the same simulacrum when speaking of modern broadcasting systems. Television produces in represented debates only simulacra of debates which never really occurred:

> telecommunication is an apparently public form, in which there is reactive and speculative discussion of a decision-making process which is in real terms displaced or even absent (. . .) a public process at the level of response and interrogation is represented for us by the television intermediaries. (1974–78, 52)

More socially manipulative than tyrannical is a form of rule which uses technology to produce a simulacrum of participation where none exists. Therefore, the onus is on communications scholars and regulators to expose such simulacra both in the communication practices which they regulate and in the very practice of regulation itself, i.e., not to put forward as social decision-making those decisions actually taken only by a scientific elite. Government agencies for promoting research should give strategic grants to study new technology in society and should give special priority to exposing simulacra of participation rather than to propagating them. Steps towards instigating procedures of democratic, participatory communication can not begin until complacent acceptance of simulacra is renounced. Cameras in court rooms do not make for participatory justice nor do two-way television channels make for public involvement in decision-making.

4.7 Degree of interactivity: towards a full-dimensional dialogue

Interactivity may be defined in various ways which range in degree of complexity from very limited reactive feedback (electronic voting), to selective feedback (passive consumption of material stored in data banks), to user information provision, and to forms of dialogue unbound by preprogrammed sequences.

In currently available videotex publicity (such as that for Telidon), only the first two types of feedback, reactive and selective, are advertised as feasible. One may use a votaphone to punch in on a scale of one to ten or on a yes/no keypad whether or not one likes one's MP. Or, the keyboard can be used to order a certain type of information, provided it is already contained in the data bank. It is quite obvious that the degree of public feedback is extremely limited.

And yet, the biggest promise by the proponents of videotex services is that we will all be information creators, that the public will be an information provider. As it now stands in Canada, however, VISPAC, whose two major contributors are the press monopolies Torstar and Southam, is the principal information provider. Furthermore, as we saw earlier, the prohibitively high cost of electronic page creation discourages individual or private citizen information provision. Public information provision would have to be subsidized and the subsidies would have to be directed towards the provision of types of information other than those which are most profitable to corporations, i.e., information which would not simply be usable as more market profiles.

Furthermore, if the much publicized diversity and creativity of communications technology is actually to occur, localized information provision must be encouraged not only through anti-trust laws and subsidies but also by ensuring the public's control over its own discursive procedures precisely so that public creation and dissemination of information can occur. By control over procedures, we mean:

(1) the availability of a reachable and pertinent audience/receiver;
(2) the capacity to program the message in a sufficiently nuanced way;
(3) the possibility of self-expression for the sender;
(4) the capacity for dialogical symmetry in terms of mastery of style and rhetorical expertise; and
(5) any other forms of communicative and knowledge competence deemed necessary by participants to ensure equitable participation.

To enjoy equitable status as an information provider requires a discursive capacity that includes but is not limited to a capacity to use the technology. Although such a discursive capacity is acquired partly through some form of programmer training, a far broader form of discursive education is also required in order to acquire stylistic and rhetorical skills equivalent to those of potential dialogue partners such as social parties or corporate information agents. Regardless of the availability of a channel, it still holds that in order to say the truth one must be in the field of truth, i.e., one must have mastered the discursive procedures that are considered to be those of truth in any given community.

Apart from these initial three degrees or dimensions of interactivity, there are two vaster aspects of feedback which are extremely pertinent to new communications technology's claims to favor democratic participatory feedback.

Fully interactive communication would allow for participation in the determination of the rules of communication programmed within the hardware and software structure of the system. In that we have demonstrated that the structure, design or discursive procedures of the technology itself are highly determinate of the potential practices of communicative and social interaction, it follows that the public, to be equitably participative, must have some input into the decisions concerning aspects of technical design. For example, the public should have a say in the decisions concerning networking patterns, types of nationally and domestically pertinent software, available and desirable programs, etc. Later on we will suggest a form of administrative and discursive mediation which could ensure a meeting point between public expression on these questions and scientific engineering in order to realize public demands.

A fifth level where interactivity must also be manifest is that of *functional freedom.* Functional freedom would permit users actually to expand the system's repertoire of operations, by constructing whole new programs. For example, the public would have the right to reject certain functions, such as surveillance of the coupling of electronic brain with biological organisms. The public would be in a position to request that optimal support be allocated for the furtherance of certain socially pertinent functions such as services for ethnic groups or for the handicapped.[12]

Again, such a weighting of preferences would imply dictates imposed upon the very structure and design of the technology. Seeley gives an example of functional freedom. He suggests that what is called a 'knowledge dialogue' should occur where the system would attempt to model how the user sees the world, i.e., to represent the user's contextual worldview, as opposed to forcing the user to adapt to the machine's contextual worldview. Unfortunately, we can only refer to Winograd's own admission that such a degree of contextualization is still very far from being realized in the sphere of artificial intelligence.

These last two dimensions of interactivity involve a degree of participation in communication practice that goes far beyond reaction, consumption or provision of content. However, the degree to which the technology, and the socio-economic institutions of that technology, provide an equivalent opportunity is still very limited. Hence, the degree of participation is still very limited.

The potentiality for such fully interactive participation has been suggested elsewhere. In 'The neglected majority: mass communication and the working person,' H. Mendelsohn and Hans Knecht suggest that social groups, in this case the working class, should actually participate in the production of the communication tools which they use to cope with their environment (in Pool 1973). Above and beyond public participation in the franchising regulation of new communications technology, which is Tate's meagre suggestion for community participation (in Pool 1973), these other structural and functional dimensions of public participation would be required if 'public dialogue' is to be a meaningful concept. This functional and structural participation would also imply lessening the gap between techne and episteme. Another example of the attempt to have public input into the very design of new communications technology is the contributions to *Reset,* a newsletter sharing and encouraging alternative programming.

It is only in this fully dimensional set of feedback procedures that actual dialogue can occur and that the very classical goal of economics and politics, that of social choice, can even begin to be realized. Social choice arises only when the public has a mastery of the communicative procedures and competences on an equal footing with that of those who hold power in society. We believe that it is this degree of functional and structural competence that Goldhaber is referring to when he states that communications technology must be at the service of the needs and practices of the discursive society so far excluded from participation in society:

> If one has access to job-related information, why not try to get politically useful information? If one can communicate with co-workers on some issues, why not on others? (. . .) Why not integrate life with work and consumption in concord with the larger community? The computer network would create possibilities of interconnection and co-ordination of action, and could heighten conscious integration of the altered class. Achieving this is a political matter. (Goldhaber 1980, 24)

Of course, it is not only a 'political matter,' but also a matter of changing technologies, of feedback at the level of functions and structures, since, in its present form of design and deployment, videotex is not capable of encouraging horizontal links of participation and collaboration.

4.7.1 The procedures of dialogism

> Since interactivity is a key watchword with the new computer-mediated services, we should also take a careful look at the role of dialogue in interactive computing. (Seeley 1980, 48)

For Bakhtin, an emancipatory society exists only where different communicators allow their discourses to be open and receptive to the discourses of the other, rather than trying to separate off, to subsume, and to control the other. Bakhtin refers to such a receptive form of discourse as quasi-direct discourse, in opposition to reported discourse. He also suggests that quasi-direct or mutually receptive discourse is the form of communication that encourages dialogism or polyphony, a procedure allowing many voices and perspectives to speak without reducing them to one another. Full dialogue consists of a genuine receptivity of one to another. Furthermore, Bakhtin suggests that while dialogism and polyphony characterize certain types of literature and drama, namely the Socratic dialogue, the Menippean satire, the carnival, Rabelais and Dostoievsky, they also characterize a relativistic epistemology and a tolerant social order.

Bakhtin apprehends these interdiscursive relationships in punctuation, vocabulary, verbal tense, and pronominal and adjectival indicators. In computer-communications we can perceive interdiscursive relations not only by means of linguistic indicators but also in terms of the relationship of computer jargon to everyday language, the networking relations between closed or open user groups, and the receptivity of programs to alterations and innovations by users.[13]

Receptivity of computer-communications discourse to non-technologized discourse would be essential in participatory communication. When one participant has the right to initiate communication (technological or other) while the other has not; where one participant is always in the position of answering questions in terms posed by the other; where one party is in control of the context that situates another's information; where some are excluded by encryption or closed user groups from certain fields of discourse; there we are very far indeed from democratic, participatory, dialogical communication.

> The high volume one-to-many systems, such as broadcast systems, are hardly free at all, since only a privileged few can provide information. Bureaucracies that have much flow of data but restrict access of citizens to it, are not very free and are engaging in power accumulation. (Seeley 1980, 47)

If we had to give a general definition of dialogue pertinent to new communications technology, then we might say that dialogue is the capacity for response unbound by pre-programmed sequence. Dialogue exists when software and hardware are user-oriented and -designed as a function of the user's interests. It exists when 'command languages' are written and designed to adapt to the public. It exists when source-oriented and -controlled restrictions are not placed on the functions and structures of new technology.

4.7.2 Communicational competence: in command of certain discursive procedures

Dialogism, the 'right to communicate,' democratic participation, all these concepts, though they stem from various theoretical and policy sources, share something in common: democratic communication is understood in terms of a certain equality of rights of controls over rules of social interaction or over discursive procedures.

For equitable or democratic communication to occur, the participants must share a certain parity of discursive conditions of possibility, i.e., of 'communicational competence' (Habermas 1979).[14] Where these procedures are directly lodged in the technology, then technology must be designed which makes it possible for users to participate on an equitable basis. The public must not only be a user but also a programmer. But for this to occur substantial technological redesign is necessary. What is more, since the set of communicational rules or discursive procedures is not restricted to the site of technology, parity of discursive procedures in the environment of technology, i.e., in society at large, must also exist. These communication rules or discursive procedures which include, but go far beyond, the specific procedures of new communications technology may be referred to as pragmatic rules of the greater social language game or as rules of communication competence. For example, one such rule might be that of enunciative symmetry whereby each participant has equal right to initiate discourse on a level and about a subject he or she sees fit. Hence the user of a computer-communications terminal would have the right to ask questions on his or her own terms, and in his or her own language, as opposed to being under the obligation to adopt the lingo and the protocol of the machine.[15] Furthermore, communicational competence would also involve possession of stylistic, rhetorical, and even political communicational capacities. It is not simply a question of altering or fulfilling the communicative competence of an individual or of a restricted group. In order for interactive communicational competence to reinforce communicational democracy, the fundamental discursive procedures of society, i.e., the episteme, would have to be radically transformed.

Harms and Richstad, in their work on the 'right to communicate' (1977, 101–2) have also recognized the discursive aspect of the right to communicate. They include not only 'communications resources' but also 'communications skills and style development,' 'participation,' and 'control of communication' among the constitutive elements of this right.

Jürgen Habermas, under the influence of American pragmatics and Anglo-American speech act theory, suggests that the validity of a communicational practice depends upon its having fulfilled the contextually specified rules of communicational competence, or the validity claims of the speech act. He is careful to say that these claims are not absolute: they are defined and refined by each new communicational community. To a certain extent in this book we have looked at the validity claims of the speech community dealing with new communications technology.[16]

Habermas goes on to say that certain other rules of communicational competence would have to exist for a democratic form of communication to be realized. These, he says, do not really exist now nor have they ever existed. They are what he calls 'quasi-universal' in that they have no ontological status and are posed 'contrafactually,' i.e., as rules that should characterize particular acts of communications by virtue not of what is, but of what could be. The set of rules of communicational competence that contrafactually would be the conditions of possibility of a democratic, emancipatory practice of communications are elaborated by Habermas under the rubric of 'the *ideal speech situation.'*

We will not enter into the more philosophical aspects of the ideal speech situation, since this has been done at length elsewhere (McCarthy 1978; Finlay-Pelinski 1982a). The ideal speech situation is a sort of ideal standard by which to measure actual communication practices for their degree of democratic participation. If emancipatory, democratic society is claimed to be desirable then it is also the standard towards which communication must tend: to be truthful, sincere, to have a propositional content, and so on. It must also offer the following rights to all participants of the speech community:

(1) equal rights to appropriate discourse;
(2) equal rights to change the propositional content of discourse;
(3) equal rights to change the level of discourse, i.e., from object-level to meta-level where the rules of communicating are re-evaluated;

(4) equal rights to place into question and to reformulate the laws of communication followed at any particular time by a particular society of communicators.[17]

Habermas (1968–70) insists that in 'democratic decision-making we must distinguish, at least for analytical purposes, between a) the discussion of proposals and justification and b) the demonstration of a decision with reference to the preceding arguments.' Furthermore, in a democratic society, according to Habermas, each claim to validity, each justification of a decision taken would have to be established by a consensual norm on the basis of free and open discussion (not to be equated with 'free flow') following the rules of the ideal situation. It would not be democratic for an elite, with the types of exclusive rights over discourse that we saw in discourse on new communications technology, to make decisions that implicate society. This set of rules for the ideal speech situation does not mean that all of them must always be followed in every communicative act. For example, not everyone will be expected to be sincere all the time. Habermas is not suggesting such a boring and moralistic conception of communication and he does have an appreciation of certain contextualized distortions such as irony. Indeed, the rules of actual communication practices may change from society to society. The sole invariable would be the set of rules for deciding upon the rules of certain social games, especially decision-making games. Those meta-communicational decisions must be arrived at on the basis of the rules of the ideal speech situation. For example, if it is decided that a single voice will represent the group, thereby suspending the right of all others to speak on all subjects at all times, this decision must be arrived at through a consensus based on a discussion in which all participate on an equal basis. Hence, before bestowing upon scientists, bureaucrats or politicians the right to decide about new communications technology and society, this right must be legitimated by consensus based on free and open public discussion.

4.7.3 'The ideal speech situation' and 'the right to communicate'

The Harms group insisted that the right to communicate was not simply an unqualified blanket right. It consisted rather of specific sub-rights pertinent to the practice of communication in particular contexts. We would now like to show how several of the above-mentioned rules of communicational competence find a parallel in specific aspects of the 'right to communicate' as elaborated by the Harms group and UNESCO's MacBride Commission. These texts evidence a transformational consistency in the development of a counter episteme in suggesting that communications rights become the broad guidelines for further policy-making in the field of new communications technology. The similarity between Habermas and MacBride would seem to illustrate that such a theoretical shift in the conception of the social role of communications finds its counterpart in policy advocacy and eventually it is to be hoped in policy implementation.

For example, the right for participants to discuss and revise the specific rules of communication practice finds expression in Hammarskjold's report to the UN:

> the social dynamics set in motion will automatically invalidate it (e.g. poverty line) and thereby help to raise it further, especially if a society can summon up the will and the means to institutionalize a *political procedure based on dialogue* between the different levels of the community – local, regional, national and no doubt international – (. . .) each level can assume responsibility for those decisions which concern it (Hammarskjold in Harms et al. 1977, 72).

The first of the rights, to enter into communication on an equal footing, is also stated by the 'MacBride Commission':

> The right to impart; to give others the truth as he sees it about his living conditions, his aspirations, his needs and grievances. Whenever he is silenced by intimidation or punishment, or denied access to the channels of communication, his right is infringed. (MacBride 1980, 113)

Other rules of the ideal speech situation reconfirmed by the 'MacBride Report' include the right to enunciate any particular propositional content or to change the content or the level of the content as well as the obligation to veridiction. The following statement is remarkable for the awareness which it evinces of the relationship of power to discourse and knowledge, and hence of the recognition that for social relations to change so too must social discursive relations:

> 2) The right to know; to be given, and to seek out in sure ways as he may choose, the information that he desires, especially when it affects his life and work and the decisions he may have to take, on his own account or as a member of a community. Whenever information is deliberately withheld, or when false or distorted information is spread, this right is infringed. (MacBride 1980, 113)

Above all, there is a coincidence between Habermas and MacBride in that both insist that although particular rules of communication may vary from community to community, and from epoch to epoch, these rules must be agreed upon by consensus arrived at in and through free and open discussion. Both agree that democratic communication and legitimation of power can be arrived at only through free and open discussion of the rules of communicational practices to be followed:

> c) The right to discuss; communication should be an open-ended process of response, reflection and debate. This right secures genuine agreement on collective action and enables the individual to influence decisions made by those in authority. (MacBride 1980, 113)

Both UNESCO and Habermas recognize that the 'right to communicate' is a set of validity claims by means of which decisions arrived at in and through communication are legitimated by communities, that is to say the means by which they are considered to be 'rational.' However, this is a meta-communicational right: it is a recognition that specific communication practices may, and indeed, must change. The rules of communication must be sufficiently open-ended for participants to discuss them at a meta-level and change them later on in their communicational practices. Participants must decide which rules should hold in which situations and how they may be revised.

Yet another indication of UNESCO's recognition of the need for a relativization of the rules of communicational practices above and beyond the general rights of the ideal speech situation is its insistence upon the multi-cultural nature of the right to communicate and the need to respect divergences (cf also IBT Working Committee, Cologne 1975, in Fisher 1982). In other words,

UNESCO suggests that above and beyond the general human right to communicate, other rules of communicating, ranging from grammar through politeness to the relationship between communication and science may be determined differently by different cultures, provided they are agreed upon consensually.

A statement by the Canadian UNESCO working group also insists upon the necessity of recognizing the relativity of validity claims within different cultural contexts provided, once again, that they are arrived at following the rules of the ideal speech situation based on free and open discussion:

> Freedom is not a context-free notion and is not a fixed state, but exists only in action, that is, in the constant exercise of free choice between alternatives. The availability of alternative structures applies as much to the protection of cultural groups as it applies to the protection of the rights of individuals. (in Harms et al. 1977, 43)

For both Habermas and the advocates of the right to communicate, its ultimate pragmatic rule is *the maintenance of discussion* rather than of exclusivity as regards the right to intervene.

> It is the duty of communication to raise the level of social conflict 'from the plane of violence to the plane of discussion.' (Jensen 1976, 191)

> But in participant society all the modes of decision are under the pressure of full and free comment by those who would like to affect the process. (Pool 1973, 244)

Habermas, with the democratic ideal still in mind, suggests that rules of communication are seminal to a democratic legitimation of power and decision-making. Only decisions arising out of participative discussion involving all those effectively implicated by the decision are democratically legitimated:

> A political battle cannot be waged legitimately unless the fundamental decisions depend on a discussion between all of the participants – here what is more, and especially here, there is no special or privileged point of access to truth. (in Preface 1971–73)

What is more, Habermas is fully aware that distorted communication which does not correspond to the ideal speech situation is often claimed to be democratic. This serves to give the impression of freedom while containing any possible revolt that might arise from its absence. The key task for the communications researcher and policy-maker is to recognize distorted communication, to name it as such and to unmask the claims to democracy proferred by the practitioners of distorted communication. On the basis of such a critique one may *tend* towards the ideal speech situation.

For both Habermas and Dag Hammarskjold, the keystone of democracy resides in equitable communicational competence:

> A democratization of power is only possible through a thoroughgoing decentralization aiming at allowing all those concerned, at every level of society, to exercise all the power of which they are capable. In other words, each local community should be able on the basis of self-reliance and ecodevelopment to manage its own affairs and to enter into relations on an equal footing with others, in order to solve their common problems. (Dag Hammerskjold Report, 'What now' 1975, 7th Special Session, UN General Session, in Fisher 1982, 39)

4.8 Some specific preliminary and very inconclusive policy suggestions for participatory, dialogical communication in the realm of new communications technology

On the basis of the preceding description of democratic, emancipatory, dialogical communication, we may now suggest complementary concrete discursive practices that should be encouraged by specific policy recommendations.

However, before even beginning to make such suggestions, one caveat should be borne in mind. The following are merely suggestions for policy transformations. They are not dictates issued from a scientific subject. Furthermore, these suggestions, seen as transformations of dominant practices of communications, must not be seen as arising solely from the author of this book. They arise from various practices, policies, and theories of communications that have already been put forward or hinted at as deviating from classical, liberalist communication. Also, these policy recommendations are not being suggested as practices or guidelines that should immediately be made 'law.' On the contrary, the decision to adopt them as regulations would have to be arrived at through consensus based on free and open discussion of the various social groups involved. Thus, the following suggestions should not be construed as further moralizing of a scientist seeking to consolidate power through knowledge.

4.9 Strong anti-trust legislation: necessary but insufficient

4.9.1.

The alternative order for which we have been arguing disperses discursive hegemony and guarantees public interests of the right to communicate. It does not treat corporations, the government, the State, private institutions, and the public all as individuals on an equal footing. We have seen the link between the monopolization of the procedures of the discourses of new communications technology and corporate hegemony. If the institutions that develop, design, and deploy computer-communications technology were to monopolize the accessing patterns through hardware and software design, then policy-makers would be left with no choice but to legislate in a way that meets the demands of certain social and public interests, the development not only of hardware but also, and perhaps more crucially, of software. Anti-trust legislation such as the separation of content from carrier as well as the 'arm's length' subsidiary policy go at least part way in assuring that private interests do not monopolize the procedures of communication. Still, it is a fact that Canada has had anti-trust legislation in the field of communication for quite some time and yet the problems of discursive hegemony remain unsolved. One may, as does Kent in regard to the press, call for the extension of anti-trust legislation. However, though necessary for a dispersion of discursive power, anti-trust legislation is certainly not sufficient. It is perhaps as regards the sufficiency of such specific legislation that we differ with the conclusions of Slack's excellent study. She argues that patent law is on a structural causal relationship to capitalist society and that to disallow patent law in the field of new communications technology is to begin to change capitalist convergences of technology and society (Slack 1984).

4.9.2 Refocus away from seeding the hardware industry towards encouragement of pertinent public interest software development

As we have seen, the discursive procedures of new communications technology are highly reliant on software development. Recently, there has been much criticism of DOC's policy which consisted in seeding primarily hardware development at the expense of pertinent software. Gotlieb and Zeeman (1980) have shown how unlikely it is that less economically powerful countries like Canada will be able to compete with the industrialized giants such as Japan, the USA, and West Germany in the development and marketing of hardware. Furthermore, they have shown how Sweden faced up to such a disadvantage and thus cut its losses by concentrating on the development of pertinent software.

More recently, videotex systems have not 'taken off.' Customer demand in France, England, and Canada is low. The Canadian Department of Communication is tacitly admitting that Telidon is not likely to become a commercial success as a home videotex service because its protocol is too limited when compared to those of home computers and because it has not been specifically designed to meet diverse public needs. Several DOC subsidized researchers are now looking for ways to make Telidon compatible with foreign home computer software. What this amounts to is the public subsidization of a highway for the invasion of discursive procedures which are foreign, corporate, and not necessarily in the Canadian public interest.

If this situation is to be rectified, then not only must the government seed hardware which could favor public interests, i.e., horizontal networking patterns, but it must also seed research and development of software embodying discursive procedures that favor national economic and cultural goals, public participation, and regional respect. Just as legislation prevents Canadians from deducting foreign broadcast advertising time as a tax expense, so could legislation favor the development and use of nationally designed software. Also, Gotlieb and Zeeman recommend the use of the famous Japanese trade weapon, non-tariff trade barriers (NTTBS), as yet another way of protecting national computer-communications development. We favor these policies as means of giving less powerful countries the breathing space they need to reflect upon, research, and develop communication systems pertinent to their needs. But they are only means providing leeway for a far more radical reevaluation of the ends of technology design in society. We must now look more specifically at precisely what types of software procedures should be seeded and developed in Canada.

4.9.3 Monoperspectivist vs multiperspectivist

In that we have recommended that the discursive procedures of dialogism, polyphony, and interaction replace those of monophony, monologue, control, and instrumentalism, it follows that software must permit the self-expression of a multiplicity of voices from their own perspectives. The software must not reduce these voices and perspectives to some singular, quantificational language. The networking and programs must permit a multiplicity of inputs and must avoid reducing biases or perspectives to each other.

On the one hand, legislation should ensure that software packages and networking hook-ups do not permit the decontextualization of information, such as the shifting of medically perspectivized information to job-focused or taxation-oriented information. On the other hand, work must be done to discover if it is possible to develop software that allows the user perspective to remain whole rather than being reduced to some control-oriented command language. That is what Seeley has in mind when he speaks of 'command languages' being written to reflect the user's worldview and of developing software for 'knowledge dialogue' where the system would attempt to model the user's view of the world. The likelihood of such development is still dubious given the embryonic stage of attempts on the part of AI researchers to make computers context-sensitive.

4.9.4 Centralization to give way to contextualization

In the preceding chapters, we showed that at present new communications technology contributes to the hegemony of instrumentalist, bureaucratic reason. In a participatory, democratic society this hegemony must be broken and replaced by a publically interested hegemony. New communications technology which reinforces instrumentalist hegemony should be legislatively discouraged while forms of technology that would encourage contextually relative discursive procedures should be encouraged both legislatively and fiscally.

Innis made a plea for a balance of the time and space organization of the peripheries and of the centre. Communications technology must be organized and developed in such a way as to respect peripheral contexts, i.e., the right to spatial and temporal diversity in communication practices and in social interaction. Contextuality is a call for a recognition of and respect for regional disparities, cultural diversity and the integrity of communities.

4.9.5 Respect for regional and cultural diversity

As mentioned earlier, the notion of ideal rules of communication does not mean that everyone must communicate according to the same discursive procedures but rather that the procedures of the ideal speech situation should be followed in order to decide which procedures will be used to communicate in specific situations. Indeed, the ideal speech situation is built upon respect for cultural and regional diversity.

Canadian regulation of telecommunications companies in order to ensure the provision of service at fair rates to all parts of Canada goes some way towards recognizing the need to respect regional diversity and equity. Canada also demonstrates an attempt to decentralize technological production: 'To ensure that rural areas become attractive sites for development of new technology' (CRAB 1979, 15–16).

Policies concerning the extension of services throughout Canada, catalyzing and catalyzed by the Anik satellite and pilot projects, also show a concern with provision of services to all regions of the country. Still, many are worried that the same limited services will be provided to all Canadian groups regardless of their diverse needs.

The recent testing of the satellite communications in the North, however, suggests this may not be so. The Inuit used these systems to produce some of their own programming and hunters have used them to keep in touch with their home base (Inuktituk 1980 and 'Broadcasting in the North', CRTC Hearings, Fall 1980). This is an example of actually altering technological design and deployment to meet regional and cultural requirements.

An alternative to using a system designed for a one-way flow of content, such as the Annenberg project for the 'University of the Air' in Third World Countries, would be the design of two-way systems to serve for such innovative projects as 'Challenge for Change.' This internationally renowned Canadian experiment provided underdeveloped and remote marginal constituencies with communication aimed at equipping the participants to seek change rather than to submit to the status quo. Available alternative channelling was used for alternative education. This can only be called a 'good' Canadian communication policy and practice. It involved equipment networking and use-plans explicitly designed with the aims of participation, public equity, and social change in mind. 'Challenge for Change' is the model of communicational design and deployment that future policy-makers should look to notwithstanding the project's various difficulties.

Furthermore, Canada has constantly iterated in its advocacy and in some of its legislation a multi-cultural policy to respect regional, cultural, and ethnic diversity. In Computer/Communications Policy (1973) it is advocated that communications be developed to 'contribute to the flow and exchange of regional and cultural information' and 'reflect Canadian identity and the diversity of Canadian cultural and social values.' Nevertheless, some would see rather cynically, a form of apartheid inherent in Canada's multi-cultural society. Despite the semantics of the good intentions there are still many inadequacies and operational weaknesses with this policy on both technical and political levels. We must not elide the fact that Canadian policy on regional/cultural diversity is intended specifically to breed allegiance to the central government – not to the regions – and that only those most subservient groups and activities, i.e., ethnic dance troupes, etc., ever get much funding. Altogether, the threat of international hegemony serves to justify internal monopolies.

One of the main objections to a recognition of the need for different communication systems and services for different regional, cultural, professional, and ethnic groups is the communication suppliers' fear of 'audience fragmentation'. However, fragmentation is feared primarily because it reduces the hegemonic discursive societies' respective share of the target audience or market. Were one to deviate from the instrumentalist episteme then this concern would no longer be a viable objection to the furnishing of diverse services. It is precisely against this instrumentalist, control-oriented worldview that the implementation of the provisions of the Canadian Broadcasting Act must militate. The Broadcasting Act committed itself to providing Canadians with a wide variety of programming to meet the varied needs and tastes of diversified groups. The Act recognizes that society is a mosaic of groups with different interests and that each should be respected provided it is not exploiting the other. This praiseworthy guideline could give rise to two specific policy recommendations in the area of new communications technology. First of all, decisions such as the CRTC's refusal to license CBC2 should be reversed given that adequate channel capacity exists for alternative programming, that there still exist marginal tastes not being served, and that the CBC has a specific mandate to meet them. Instead of abandoning programming to foreign sources, as is the case for all intents and purposes the case with pay tv, pertinent national programming aimed at a variety of population segments must be supported by policy and subsidization. Secondly, respect for the diversity of community needs must filter through into the computer-communcations industry. Instead of selling a few, generalized programs and vertical incasting to customers, what is required is specialized networking to unite interest groups and specialized programs designed at the request of specific groups in view of their interests and needs. Foreign computer education programs are not the answer to national educational needs nor is a blanket, uniform national educational packet necessarily the answer for all kinds of Canadian citizens. Networking and software design must be sufficiently diversified as to respect these differences. Pergler (1980, xvii) also suggests a diversification from American standardization by virtue of an insistence on a privatique type scenario with specialized software developed for each system: 'The overwhelming cultural and economic influence of the United States in Canada might be lessened by incasting systems.' We see in the Appelbaum-Hébert Report's (1983) advice to decentralize production and provide the resources for regionally-oriented production the beginning of a communication policy that recognizes communication not as object but as practice whose role in participatory democracy would involve networking patterns linking interest groups rather than geographically circumscribed entities such as nations.[18]

4.9.6 Avoid funnel effect in networking

Types of networking play a very large role in any attempt to avoid monopoly and centralization. To avoid the vertical, funnelling or gatekeeper effect, networking patterns would have to be far more horizontal, less linked to central data banks, and less circumscribed by incasted frontiers than the telematics scenario allows. The 'privatique' scenario would be a partial step towards decentralization. As it now stands, the networking patterns suggested for the Telidon videotex system would have to be radically decentralized and made far less dependent upon the centralized banks of institutions such as Sears, Time/Life, and VISPAC/INFORMART than is currently projected. Subsidization of smaller information providers linked to a decentralized 'réseau' would have to counterbalance the current centralized information provision which enjoys such a tremendous advantage in economies of scale.

4.9.7 Diversified software

Networking diversification alone does not suffice. Simple control of networking patterns does not guarantee against the emergence of centralized nodes of control over knowledge and communication. Even the privatized networking system of the home or self-contained office computer-communications hook-ups could lead to nodes of hegemonic communication and knowledge procedures. In that it has already been demonstrated that software largely dictates the discursive procedures to be followed by its users, any standardized, mass-produced software packages developed outside the region concerned risk imposing upon user groups discursive procedures which homogenize their own procedures and which impose a discourse from without.

One absolute priority of any national legislation of new communications technology should be to ensure national and regional software production for national use as well as to ensure a diversity of software production, even if this implies protectionism and subsidization to protect industry against competition which enjoys huge economies of scale. The Brazilian micro-computer industry can serve as an example on this point. There must also be regional, public input into software design and development or what we referred to earlier as 'functional and structural' dimensions of feedback. It would be inconsistent to suggest which precise structures and functions should be singled out since the important thing here is that the public participate in the task.

4.9.8 Identification of the agents involved in communicational interaction: re-evaluation of sites of domination

As we have stated over and over again, it is impossible to do away with all forms of discursive control. However, as we have also seen, it is a violation of democratic principles to identify individual citizens with large corporations and to suggest that both stand on an equal footing and enjoy equal rights. The a-symmetry between the individuals, corporate and human, is evident in the results of most struggles between them. The same applies to countries which cannot all be considered to be participants of equal standing. Policy-makers must take a close look at the present hierarchy of the sites of power and decide whether or not it actually stands for democratic equality leading to emancipation and self-development for all. Once the participants and their power status are identified, policy must be invoked to reestablish the balance. What is more, the *conditio sine qua non* of a democratic society is that the balance be tipped in favor of the citizenry, no matter how many groups make it up. The liberalist doctrine, that if left to their own devices things will reach such a balance, has been disproven historically. Policy-makers have an obligation to give more discursive power to some than to others in order to right the balance. One way of doing this is to include those (i.e., public interest lobbies) who are now excluded from direct involvement in decision-making concerning the design and deployment of new communications technology. Later, we will suggest the public inquiry as a means of so doing.

4.9.9 The right and socio-technical feasibility of unplugging

The ideal speech situation suggests that we should have equal rights to initiate communication but perhaps a corollary should be added declaring that we have equal rights to desist from a particular communicational game until the rules are rediscussed and altered. This right would involve the freedom to exempt oneself from the exercise of power via certain discursive procedures. The lesson that we hope to have illustrated in preceding chapters is that the obligation to participate in certain discursive practices, to follow certain procedures, may submit one, however unwillingly, to the relations of domination that those procedures entail. Where it becomes obvious that certain procedures lead to certain relations of domination which do not favor key sectors of the public sphere, those participants must have the right to opt out of participation at that particular level until such time as the rules have been discussed and revamped at the meta-levels. For example, as long as one is not assured that information about oneself or one's group will not be decontextualized, one should have the right to acquire public social services such as medical care or banking services without plugging in, i.e., without having information about oneself electronically processed and telecommunicationally transmitted.

The right not to participate is suggested in the historical awareness that in some discursive practices the right to speak is nothing more than the right to give away hostages, to submit to the rules of another's language game, to play the game with no control over the rules or the relations of domination that ensue.

Thus the 'right to communicate' must paradoxically include a right *not* to communicate in certain cases unless one has an equal say in the determination of the rules to be followed in that practice of communication. Habermas suggests that we have an obligation to participate in the discussion to decide upon those rules. Diplomats constitute one society of discourse which fully realizes the need for a preliminary discussion in order to determine the rules of communication to be followed in a political negotiation. Most sectors of the public sphere are far from enjoying this privilege.

Third World countries have increasingly been demanding the right not to communicate. They claim this right in order not to have to submit to the discursive procedures of the developed nations. Many of the contributions to the Brazilian INTERCOM conference on new communications technology (Summer 1983, Bertioga) demonstrate convincingly the types of communicational and epistemological, not to mention economic and political, constraints imposed upon Third World countries by dominant countries which export communications technology. Those who claim the right not to communicate, the right to unplug, be it at the domestic, national or international level point to articles twelve and twenty of the Universal Declaration of Human Rights as having already suggested this right in that they provide for the protection of minority cultural and religious beliefs from foreign inundation.

The right to unplug might also resolve the billing issue discussed in chapter one. Unless it can be *technically* rather than just legally guaranteed that no traces of information consumption will be available for surveillance purposes then no billing can occur and information must be conceived of on a different plane from that of exchange value. Only technical design and structural change can ensure the uncoupling of billing and surveillance.

We have mentioned above several verbalized and even concretized/practised transformations of the instrumentalist classical episteme to be found on the communications horizon. It appears then that some of the flyer advertisements on the right to communication have indeed come to fruition. However, in Canada, despite its favorable international reputation, studies such as the Clyne, Davey and Kent Commissions each begin by lamenting the fact that their predecessors were not not acted upon in the form of concrete policy-making. In other words, it could be said that in many cases the discursive procedures have been enunciated but the communicational practices have not been concretized. Policy advocacy may only be a transitional stage between old and new epistemes directive of communication practices. However, for these transformations actually to occur certain profound political changes would be required so that the concretization of policy advocacy in the public interest need not depend once again upon the bureaucratic organization of society but rather upon the public sphere itself. And yet, for such political change to occur alternative communication decision-making is essential. A vicious circle? In other words, the dialogical and *accountability* structures of consensus formation must already be practised in the communication policy-making sphere for the policy and practices themselves to begin to follow the alternative, democratic procedures. Briefly we will set up a dare which might approximate this prerequisite.

We will now conclude our suggestions for alternative discursive procedures with a brief discussion of the role of the ideal speech situation in decision-making as well as of the role of decision-making in tending towards the ideal speech situation.

4.10 An alternative politics of communication

In chapter one it was suggested that the compulsions of 'gadget-philia,' among other procedures, serve to depoliticize the debate surrounding new communications technology. Issues are trivialized, suggested uses are banalized and priorities are not realistically hierarchized to reflect the primacy of military and corporate sites in the development of this technology. Placing smart toasters and surveillance satellites on the same footing in regard to social implications has the impact not of depoliticizing new communications technologies *per se,* but of depoliticizing the public debate concerning this technology by making it seem politically insignificant. There is no reason to depoliticize either technology or the debate. It has become quite evident that our form of political organization is closely linked to our form of communication, technological or other, and that in order to change this organization, discourse must also change.

Jean d'Arcy, when he suggested the 'right to communicate,' also stipulated that such a change would simultaneously imply an alternative form of power, more specifically, an alternative form of government: 'A new form of government must be found, a government which no longer derives its power from the refusal or the control of communications.' (in Harms et al. 1977)

This is not a chicken and egg question of which comes first, the alternative government or the alternative form of communication, since both are in a relationship of 'entangled hierarchy'. In this same vein, Innis suggested that new communications technology could conceivably change power relations in favor of less centralized control were the actual communicational order to be altered.

The reciprocal nature of that relationship of political and discursive control is also pointed to by Foucault when he suggests that the duty of the intellectual today is to seek a new discourse and a new politics of truth, one which, in both cases, would disperse hegemony and replace instrumentalist, bureaucratic reason with an alternative episteme, and corporatism with an alternative politico-economics: 'To the vast new techniques of power correlated with multinational economics and bureaucratic states, one must oppose a politicization which will take new forms.' (1980, 190)

Thus the whole attempt to suggest a more interactive, participatory and localized control of discursive practices as an alternative to the classical, instrumentalist hegemony of domination exercised through discourse is not only a communicational but also a political change. We are suggesting that the pluralist model of Western theories of democracy has failed to produce the type of communication or social order that it promised and hence discursive procedures of 'laissez-faire' and individualism should be abandoned in both the discursive and political spheres. This *pluralist* model of the relationship of State to society has become what the political scientist Alfred refers to as *'pathological'* in that, despite its claims to pluralism and equality, one group or society of discourse, namely the scientific, bureaucratic, military complex, enjoys exclusive control. This hegemony permitted little or no autonomy in social choices.[19] 'Managerialism,' 'Taylorism,' 'corporatism,' 'technostructure,' 'technocracy,' all these terms reflect the awareness that Western theories of democracy had led away from the pluralism they promised and towards pathology or hegemony. The suggested alternative, that of interactional, participatory communication based on consensus formed in the public interest through free and open discussion, would be a transformation of this classical liberalist model of communication and power-sharing.

We must again beware, however, of those who take even this transformation, the dialogical alternative, and depoliticize it, returning it to a reinforcement of the same old discursive hegemony. Pool did just this in his anthology *Talking Back* (1973). He began by stating the desirability of participatory communications and pointed to its roots in democratic political philosophy. But he then reverted to a full-fledged criticism of politicization through participation of the polis in decision-making. He states that there is a 'need for authoritarian citizen organization' (244), and that if every issue were to be politicized chaos would result. Pool had no intention of decentralizing decision-making either in the sphere of communication policy-making or in other political spheres.

He was not suggesting that participatory communication decentralize decision-making but rather that the degree of participation be sufficiently contained so as to maintain the participation of the polis in politics at a very 'low level,' while allowing 'representative' decision-making to continue at the national level. He criticized 'devolution of focus of politics to local and interest group level,' saying that it would lead to a 'lack of social cohesion at the national level.' And, in another article, co-authored with Murray and Dobbs, Pool reduced participation to the dimensions of reaction whereby participants are little more than sophisticated information consumers: 'audiences (which can control) timing and content of messages received.' (Pool 1973, 266)

Pool's argument would hold, of course, if, when speaking of participatory media in the political sphere, we were to imagine millions of individuals rushing to their consoles in order to vote as an amorphous mass on each and every parliamentary issue with no background information on which to base a judgment. Of course, we can only agree that this scenario would indeed cause chaos. However, such a vision is once more a prisoner of the vision of society as an amorphous mass of atoms.

By localized participatory communication and decision-making, Foucault, Gramsci, and Habermas all recognize that the agents participating are the respective social groupings that constitute society. For these theorists, it is at this localized level that meaningful decisions about social order may be made with the participation of those involved by them. They suggest that communication practices must be opened on these levels in order to facilitate significant citizen participation and intervention in social decision-making. Were participatory media designed for and geared to community, educational, and constituency-oriented political exchange, then new communications technology might possibly be said to contribute to the further democratization of society. But this would involve a very different structure of technology.

Innis (in Salter, Melody, et al. 1981) and Lowi (in Forrester 1981, 453–472) argue that new communications technology designed to favor these community exchanges could indeed contribute to a mobilization of localized and special interest groups as power lobbies that could influence even the national course of events. But the fulfillment of Lowi's global predictions would depend entirely on the participatory episteme as a whole overruling instrumentalist, discursive hegemony:

> with access to political institutions still limited, however, ease of communication and heightening of expectation through social

> movements and special interest groups (. . .) (t)hese tendencies will manifest themselves in changing power relationships, institutional patterns and government policies. (Lowi 1981, 460)

The discursive transformations we suggest aim at nothing short of an alternative form of communication that could lead to a social order where public power-sharing would replace bureaucratic, scientifist and corporatist hegemony.

Moreover, we are not suggesting the abandonment of all communicational constraints but rather the consensual acceptance of different discursive constraints.

Rather than usurping the public's voice in the political sphere, one of the procedures of discourses on and of new communications technology, indeed of the alternative episteme, would be the responsibility of and accountability to public participation in decision-making on all political matters. The real test of the democratic claims made for the new communications society will be to see whether or not its procedures make possible practices that provide a forum for public decision-making.

4.11 Public decision-making concerning questions of new communications technology

Given that democratic participation is one of the promises made by so many of the discourses on new communications technology, our major policy recommendation is to set up trial communicational situations which would actually 'call the bluff' of these political stakes, i.e., make their advertisers accountable.

4.11.1 Public inquiry as an exercise in participatory democracy

In Western democracies, one of the main exercises in participatory democracy is considered to be the public inquiry and its contribution to decision-making.

Liora Salter and Debra Slaco have recently published an excellent investigation into the degree of participation involved in *Public Inquiries in Canada* (1981). They show quite convincingly how public inquiries arose in great part due to the aim of participatory democracy but also how complex factors often mitigated against full and nuanced public participation in the inquiries and subsequently in decision-making. Salter and Slaco ask 'How in the hearing or development process (can) people share power? (1981, 169)[20]

While this is not the place to review extensively the factors which play a decisive role in the degree of public participation in decision-making, we may note that many more conditions need to be met than the mere provision of a forum for public input. The degree of public access to legal knowledge, technical information, capacity for 'official' expression, and other resources equivalent to those of governments and corporations determine the degree of public input into the planning issue-raising stages as well as into the testimony and political impact of the inquiry. The techniques used by public inquiries are, to varying degrees of success, largely dependent upon the mandate of the inquiry which sets the expectations and limitations as well as the possibilities of the inquiry.

4.11.2 Participants: what constitutes public involvement?

While many would agree that the public should participate in democratic decision-making, few would agree on 'who' the public is or on how the public is best represented. Salter and Slaco (1981, 178) demonstrate that while some informants resented the 'impedences' of public participation, others argued over who the public is. Just as in the discourses on new communications technology, one commission stereotyped the public in terms of proponents (normal responsible citizens) and opponents (young activists). There is no cut and dry answer to this question. Still, we do reject the liberalist answer that the public is a bunch of dispersed individuals. The identification of the public in and through consensus decisionistics is a complex and nuanced science in itself, recognizing the dynamics and variations of participant groups and their representatives. Certainly, however, such participants indentification must go beyond head counting on the one hand or the reductionism of stratified groups such as union or political parties on the other. What is also very important is that the communicational and politico-economic conditions must be propitious for group formation, realignment and re-identification. The resources must also be available for groups to get informed communicational and technical expertise in order to be on an equal footing with their corporate and governmental counterparts.

One of the duties of public inquiries should be to 'attract participation from "missing" sectors of the public (other advocate and interest groups, volunteers, and individuals).' (Salter and Slaco 1981, 216)

Moreover, while consensus-oriented participation does not mean head counting we must also recognize the difficulties involved in rendering it operative. We are not suggesting that the task is easy or even attainable in any definitive sense. Rather, consensual participation and reevaluation of laws and norms of social order is a constant process and involves a constant readjustment of the social order including the alliances of groups within it. Still, these operating difficulties must be recognized for what they are, operating difficulties, and not misconstrued as some fundamental weakness in the democratic ideal itself.

We refer here at such length to the Salter and Slaco report because we cannot but agree that one of the corner stones of any form of participatory democracy is a fully interactive and publicly accountable decision-making process anchored in genuinely public inquiries. Their study is very pertinent to our theme here in that it looks at what conditions, above and beyond the availability of a forum/feedback channel, would be required to assure democratic participation in decision-making.[21]

Based on the criteria that they have outlined as necessary for democratic public participation at the community level in public hearings, we would propose to examine whether or not new communications technology fulfills these same criteria or whether more than a feedback channel is required. In this vein, we would suggest the following test of public inquiry.

4.12 Pilot project or dare: the electronic public inquiry

Let us begin by letting the public participate in decision-making concerning new communications technology. And let us use this new technology as the means by which to do so. If it succeeds in encouraging and legitimating full-dimensional public participation in decision-making, then the claims must be believed and new communications technology should continue to develop those discursive procedures that made the participation feasible. However, should public participation be found wanting, then such a pilot project would cast into doubt both the technology and the claims.

Prior examples of public participation in decision-making about communicational and cultural matters can be found, although new communications technology was not the medium for them. In West Germany, decisions concerning the implementation of new electronic technology in the workplace were taken in consultation not only with managers and scientific experts but also with workers (Friedrichs/UNESCO 1981). Canada may also qualifiedly be praised, despite deficiences in subsequent parliamentary accountability, for its public inquiry system whereby hearings about current issues are more or less open to public intervention (Salter and Slaco 1981). The Applebaum/Hébert Commission on culture in Canada is yet another recent example of an attempt at public participation. However, these public voices, though they may even find their way in to the commission's reports seldom manage to become legislation. The decision-making or legislative step is still relatively closed to participation by public interests as Salter and Slaco have shown. They advocate inquiries to implementation of their recommendations rather than to procedures. For an inquiry to be truly participatory, policy-making must be accountable to it.[22]

Thus, we would propose a field trial. One in which the stakes are high. The wired community will be used to solicit feedback of a very complex nature to the politicians concerning precisely the implementation and development of new communications technology in society. The participatory exercise will be carried out as a sort of 'electronic' public inquiry with special effort being made to hear from as many groups of the public sphere as possible. Where public consensus can be achieved (if indeed it can), then the government which declares new communications technology to enhance democracy will be committed to act upon the consensually and 'electronically' achieved decisions, to act, that is, in the form of concrete policy-making concerning new communications technology. However, where the degree of interaction and participation is found by both public and researchers to be inadequately facilitated by this new technology, then the government will also be committed to act accordingly. It would be committed to admit the limited degree of participation permitted by various procedures of new communications technology, to reevaluate its possible contribution to democracy and to look seriously at what other forms of discursive, social and epistemic change would be required in order to realize electronic society's claims to emancipatory, democratic society.

In short, what we are proposing is a feasibility study to evaluate the capacity of actualizable communications technology to contribute to public participation in the political sphere, and we are using as an issue for discussion the public perception of new communications technology. We would evaluate the degree to which the currently 'realizable' forms of new communications technology could lead to a full dimensional (including structural and functional feedback) participation and the degree to which such participation is allowed to penetrate into the political sphere, as is claimed.

For example, this field trial could be said to lead to structural and functional feedback providing the electronic exchange for public input into the very structure and design both of the system used in the trial, and of subsequent systems, as well as public expression concerning the ultimate functions and limitations of the technology. Salter/Slaco have shown how public inquiries are often concerned with regulation in the design phase and during the development and application of new technologies (1981, 208). This involvement should apply to new communications technology as well. In the design of a home videotex service, the public would not only be consulted concerning certain types of desired content, it would also be consulted about accessing patterns, content management, command languages, and inter-terminal links.

Participation in the sphere of decisionistics proper to problems of new communications technology would at least partially realize what we earlier called dialogical, structural and functional dimensions of interactivity.

The main advantages of this pilot project are that it: (1) uses new communications technology as a medium and (2) discusses as its subject new communications technology. This two-fold nature will make the results of the experiment all the more irrefutable in that (a) it is setting up discursive practices in the very medium that the proponents of new communications technology say is democratic in order to discuss and test the degree of democratic participation permitted; (b) the results will say something about new communications technology as discourse and about discourses on new communications technology, including not only those which we have already examined but perhaps also an alternative public discourse; and (c) where the degree of public participation in and on new communications technology is found wanting, policy-making bodies should be instigated to make changes in both the technological and non-technological communicational environment, before laying claim to the democratization of society – following such a study we may take for granted a marked decline in futurological political speculation concerning new communications technology.

Such a study is implicitly called for by the Canadian federal department of communications (DOC) when it acknowledges the need for a forum for public awareness concerning issues of new communications technology:

> Closely related to accountability, then, is the question of public familiarity with the issues. The CCC subcommittee on the individual and society wishes to organize a forum for public awareness purposes. (Foote/DOC 1981, m.s., 16)

If we were to take DOC at its word, then it too would welcome a field trial to see whether the discursive and political relations promised by the proponents of new communications technology could actually be realized in the practices of this technology in the social sphere. It is also only fitting that the first political issue to be discussed in such an electronic forum would be new communications technology given that the results of the experiment to test how communication facilitates or discourages public participation in decision-making will influence the mode of communication used for the discussion of all political issues and subsequent decision-making.

Such a project to test the role of new communications technology in the political sphere has been hinted at earlier as the key issue of this technology by Gordon Galbraith in 'La programmation communautaire et la vie politique' (CRTC/Government of Canada 1974, 53–4). Other guidelines for how such a project might be undertaken as an issue-exploratory project involving the public sphere may be found in Harms' and Richstad's suggestions for 'Issue-exploratory dialogue' which dealt specifically with the right to communicate. For these researchers as well, issue exploration can only proceed if the type of communication practised is dialogical, multi-way interaction. They propose 'an alternative model of human communication (. . .) that begins with a foundation for a multi-way, interactive, participatory communication process.' (1975, in Harms et al. 1977)

In the US there are a few exemplary experiments in participatory decision-making from which to take our cue. Douglas Cater's work at the Aspen Program on Communication and Society and F.P. Chisman's 'Policy interest and FCC policy making' (in Pool, *Journal of Communication,* Spring 1977), both suggest projects for studying and for increasing the degree of public political participation via new communications technology, although neither elaborated criteria of evaluation at all close to those of Habermas, Bakhtin or UNESCO.

This pilot project would be somewhat empirically oriented in that it would study specific communication practices as realized in the public sphere. The interpretation of the results would, however, be based on a discursive critique. Specific test practices would be examined to see the degree to which they reiterate the procedures of classical, control-oriented (hence participatory) reason and/or the extent to which they allow for transformations that constitute an interactive, participatory, dialogical, and public-interest-serving practice of communication.

The test would be set up and analyzed with the following questions in mind:

1 What are the specifically technological (functional structural) requirements for participatory communication?
   (a) Can a very nuanced public exchange take place using the present state of the art, new communications technology?
   (b) If so, what are the procedures that favor this?
   (c) If not, what alternative procedures would be required and can technology realize them?
   (d) What is the feasibility of a genuinely participatory protocol, where all of the earlier mentioned dimensions of feedback would be included?
   (e) Can new communications technology itself expand the amount of diversity and public participation in what is so far a very hegemonic society of decision-making?

2 What are the non-technological, communication competences

or procedures required for interactive communication and participatory decision-making?

(a) Must participants from the public sphere receive education in other forms of communicational competence such as political rhetoric in order to participate on an equal footing?

(b) What role must educational institutions play in giving citizens not only the technological competence to dialogue with this new technology but also non-technological mastery of discursive procedures?

(c) What non-technological discursive procedures continue to encourage a hegemony over discourse and decisionmaking, i.e., to encourage monopolies of knowledge and political power?

3 What extra-discursive practices must change in order that participatory communication and decisionistics be facilitated?

(a) Economic changes? Changes in international and national legal relationships? A whole shift in, or dissolution of institutionalized regulatory bodies? (Mosco 1979)

4 Must the overall discursive and epistemological environment change before new communications technology would contribute to participatory communication and decision-making, i.e., must we have a totally new episteme?

5 If we have a totally new episteme, will new communications technology embody the same discursive procedures as it has been seen to embody today? Will technology still mean instrumentality – how to make? Or will technology perhaps change its procedures, its epistemology and its face to mean some form of interactive knowing? And in this case, should the word 'technology' still be employed to refer to that alternative set of practices.

4.13 A bit of irony

Our suggestion for an 'electronic' public inquiry of new communications technology employing new communications technology is not made here without a degree of self-irony. It is not entirely naive and idealistic. The proceedings of such an inquiry would be studied while in process from many of the same perspectives and with many of the same criteria in mind as put forward by the Slaco/ Salter study of non-electronic public enquiries. Our aim is not to suggest that such an inquiry could easily be run successfully. Rather we wish simply to question whether an electronic medium for public inquiry could overcome the non-democratic pitfalls pointed out by their study of non-electronic inquiries. If it could, then something could be said for the democratizing potential of new communications technology. For example, in light of the procedures of discourse on and of new communications technology, would this technology defuse the monopoly shared by science and politics, which Salter and Slaco so convincingly showed to militate against a developed and fully dimensional public role in decision-making? Would all of the other conditions for dialogue be met?

4.14 Communications/technological ombudsman

In closing this chapter, we would like to insist again that the preceding suggestions have two very clear limitations: (1) they are not put forward as commands from an individual scientist for society at large to obey. Rather, they are suggestions concerning ways in which the public sphere could be involved in communication policy decision-making. These suggestions must themselves be subject to public decision-making. However, there is here a small double bind in our favor for unless a means of involving public participation in decision-making over new communications technology already exists – and we contend that it does not, hence our project – then the public jury cannot begin to work. This is the logical paradox of autonomy that has plagued political philosophy since the time of Locke, Hobbes, and Rousseau. How do we establish an autonomous order without recourse to an initiating transcendental moment which dictates that the order will be autonomous? As yet, I see no exit from this paradox and suggest that we accept it and get on with the business of working to establish an autonomous social order. The decision about whether to establish such a trial will reside in the first instance with some transcendent, extra-consensual instance, i.e., with policy-making bodies such as departments of communication. Once in place though, this form of public participation and consultation would then be run by the public groups involved. In other words, if they wish to claim to be democratic, communication regulatory bodies would have little choice but to initiate such a forum for participatory decision-making. (2) there is in this study a penury of exact technological instructions for how to make these systems actually practise the various dialogical procedures suggested. For example, we are simply not trained to stipulate how to set up computer-communications which will respect what we have called structural and functional feedback, and 'knowledge dialogue,' or how to program 'command languages' which respect the user's worldview. And furthermore, we cannot decide now what needs to be decided consensually.

If it is altogether unfeasible though, as the 'experts' may well say, then at least the degree of interaction will be recognized for all its limitations. However, when the social scientist, or the public for that matter, admits to not being a technical expert, this does not imply that the power of decision-making over new communications technology should be thrown entirely back into the camp of the technical or bureaucratic expert. Ways must be found to link the public's very certain knowledge of public interests to scientific and technical expertise which bears the onus of answering these public interests. The Science Council of Canada's study on science and technology in decision-making, *Regulating the Regulators* (1982), provides the fascinating example of the Cambridge, Massachusetts community committee on recombinant DNA which refused to appoint scientific experts on the grounds that the lay members were perfectly capable of using the reports and testimony of the experts to form their own opinion rather than disenfranchising themselves and relinquishing, as in Weber's bureaucratic society, political power to the scientific expert.

Much of Habermas' recent work on communication can be seen as an attempt to come to grips with what he sees as a gap between the communicational practices of scientific experts and those of the public sphere. Habermas suggests that the reason why science has so much power today is that there is no longer any common ground on which the public sphere can speak with science in order to express its interests. The debate on new communications technology is a typical case in point. The experts work on programs in isolation from public concerns and public interests cannot adequately be formulated because of a lack of awareness about who the 'experts' are and could potentially be doing. Habermas says this problem can only be rectified if a go-between field of discourse is found, a form of communicational competence with which both parties can dialogue on an equal footing. The go-between language must allow scientific and public discourse to communicate without usurping or vulgarizing one or the other. The go-between is absolutely essential if the public is to play a role in decision-making concerning technology and society.

The go-between must act as an arbiter between the public and the experts, in such a way as to avoid giving priority to technically specialized knowledge. This mediator or go-between sphere of discourse could be called the 'communications/technological ombudsman' and would not be an appendage of the State but would instead be a referee sphere of discourse between the institutions of State, science, and business on the one hand and public interest groups on the other. It would be incorrect to qualify the ombudsman as a person or an agent. It is rather a site for a go-between practice of discourse in the public sphere. In that there would be an office of the ombudsman, this would merely be an institutionalized site for the mediation of scientific and public discourse although ideally this site would become ubiquitous as should the public sphere. The key to the practices of the ombudsman discourse would not be a vulgarizing discourse, i.e., one that translates scientific jargon back into the dominant discursive traits and themes of the day. Rather, it would be a new discourse in which consensually agreed upon rules are followed by all parties of the discussion. The ombudsman sphere of discourse would be the discursive site responsible for evaluating the discursive procedures of new communications technology in order to ascertain whether they conform to what the public consensually decides is communication in the public interest. Furthermore, the ombudsman sphere of discourse would be the discursive field in which to assess expressed public communication needs and make representations to the scientific experts for their realization.

Finally, the suggestions and criticisms in this study of new communications technology would have to be submitted to the ombudsman sphere of discourse which would in turn decide upon their degree of public desirability as well as upon their capacity for potential technical realization.

CHAPTER 5

Conclusion; Discursive critique: Socio-epistemological foundations for an alternative approach to communication and technology

The qualitative traits of rationality in society have not changed with new communications technology. They have expanded to become the dominant traits of all reason. They have moved us from 'laissez-faire' capitalism, towards late, State-run, bureaucratic capitalism whereupon the role of specialized technology in bureaucratic decision-making has become so pronounced as to merit a change in nomenclature to 'technocracy.' In other words, a quantitative expansion of the dominance of the traits of instrumentalist rationality does in the end amount to a sort of qualitative change in society. And, at this point, we might recall that Hegel once said that a sufficiently great quantitative change produced a qualitative change.

We might see in the proposed ubiquity of new communications technology whereby informatization will infiltrate not only every activity of the work place but also the kitchen, the tv room, the school, etc., a case in point of the expansion of the procedures of quantification, calculation, analysis, classification, organization, means-ends logic, efficiency-seeking, instrumentalist communication and knowledge. Non-rational procedures of knowledge and legitimation which accentuate contextualization, grey areas, ambivalence, relativity, interactive determination of value and meaning are not compatible with most new communications technology.

What is surprising, though, is that throughout history there have been other forms of knowledge, forms which directly contested the exclusivity and supremacy of classical, analytico-referentialist knowledge. The Vienna School and Anglo-American philosophy espoused rationalism and the analytico-referential in their attempts to build exact sciences, including logical structures of meaning and truth. Wittgenstein's *Tractatus Logicus Philosophicus* has been treated by Russel and Whitehead, among others, as an attempt to exclude everything from the realm of knowledge except formalizable relations of logical and atomistic variables. However, while Russel and his proselytes remained enchanted with the undertaking, Wittgenstein and many others, including the philosophers of everyday speech (Austin), became disenchanted with the capacity of such an approach to language and to knowledge. In the *Investigations,* Wittgenstein gave up his analytico-referential approach in order to concentrate on the more elusive and vague regularities of the language game and of ethics. His view of knowledge and communication moved from one that thought structured modes of communication and knowledge could be frozen towards one that renounced exactness in favor of a vaguer description of the interactive flexibilities and activities of communication as act.[1]

The age-old school of 'communication' known as hermeneutics has also insisted that knowledge consists of much more than systematic analytico-referential models. It is a process of contextual explanation or understanding. Indeed, as we have shown, a socio-historical form of hermeneutics is important for understanding how and why the discursive procedures of systems theory became so dominant in society in the first place and what its social underpinnings and consequences are.

The systems model of communication that seems to penetrate all spheres of communication via new communications technology is nothing new. It is, however, more extended and more restrictive in the 'information society.' The vast percentage of present day job advertisements for systems engineers would seem to testify to this imbalance.

At this stage it would be mistaken to interpret the above considerations as proposing some form of true knowledge which should replace systems theory and models. Our aim in this discussion was merely to suggest that if there is any change in the episteme of new communications technology, it is in the degree of closure it presents to any form of knowledge other than systems analysis.

5.1 Alternatives in discovering, isolating, and encouraging irregularities

While it has been suggested that the forms of reason underpinning industrialism have not changed very much, certain irregularities or transformations have emerged. Both Foucault and Weber insist that while there is a dominant episteme or form of rationality, it is possible for certain other practices or procedures to appear which violate the regularities of the dominant episteme.

Weber describes rationality as being punctured by the irregularity of charisma. When bureaucracy's rationality seems threatened, then an irrational, discontinuous moment occurs. Charisma is described by Weber as the moment when the routine of mass life in bureaucratic society is opposed by the personality and creativity of the entrepreneur. However, for Weber this irregularity is recuperatory since, strangely enough, the very trend of rationality is maintained, indeed, saved, by the occasional accidental moment of charisma.

After years of treating the madman as an object, Foucault sees Freud's introduction of a discursive practice which not only speaks about but also speaks to the madman as an irregular transformation. We move from an episteme which incarcerated Sade to one which heralded Artaud, Dostoyevsky, and Nietzsche.

Were we not to allow for irregularities and transformations, we would be guilty of reducing the universe to fixed and universal laws which insisted upon continuity at the expense of any possible discontinuity. This would be a trait of classical, liberalist discourse. For liberalism, no discontinuity is allowed to disturb causalist links between events, objects or states of society. (Kumar 1978, 57)

To conclude, we would like to suggest that the discursive approach to new communications technology elaborates an alternative epistemology from which to consider both communication and technology.

We have adopted the discursive approach due to a certain dissatisfaction with traditional approaches not only as applied to new communications technology but also as applied to communication in general. These traditional approaches exhibit certain methodological weaknesses and epistemological tenets that place into question not only their epistemo-theoretical consistency but also their social and political legitimacy. The discursive approach avoids these failings and provides a coherent epistemological foundation for a critical study of communication and technology that also legitimates social practice in terms of political and technological intervention in a more convincing and constructive manner.

5.2 The definition of the object of study

5.2.1 Disintegration

Cees Hamelink describes mainstream administrative communication research as:

> a long series of empirical-analytical fragmentary, piecemeal studies, guided by the dichotomy of facts and values, directed by the interest to steer social technology for status quo purposes and epistemologically handicapped by the Kantian tradition of confining reality to predefined categories which are applied to it. (1980, 17)

Policy-making bodies in the developed world have to date (1987) already undertaken or sponsored many research projects and studies concerning the social implications of new communications technology, the gamut of which is quite representative of the piecemeal studies mentioned by Hamelink. These studies have treated everything from 'access' through 'legislation,' 'information providers,' 'satellite earth stations,' 'services for the handicapped,' 'community service organizations,' to needs or tastes for specific new communications technology services. After ploughing through the lot one cannot quite situate the major problems. Still less can one suggest on what basis a coherent and global guideline for understanding and intervening in the horizon of communication and technology might be. What is lacking is a coherent way of integrating not only all of these topical studies but also the various objects and issues (idea-objects) that are raised by each study. We find a study on smart toasters, another on transborder data flow or content-carrier regulation, followed by one on pay tv, with no attempt to see how all of them might be linked to the socio-political horizon of communication and technology.

Disintegration exists because up till now most knowledge concerning communication and technology has been organized

NEW TECHNOLOGIES NEW SERVICES / ISSUES **Technology**	**Services**	access	surveillance/ privacy	diversity/ leisure/ creativity	impact on other types of comm. privatization	symmetry/asymmetry bi-directionality/ unidirectionality	institutional impact/ e.g. disemployment	homogeneity/ diversity	concentration content/carrier	Fed./Prov. regional/cen-tral concerns	national/ cultural identity	national/ sovereignty	information rich information poor	Military-industrial complex
Cable fibre optica coaxtal cable (increased channel capacity) (versati-lity of delivery mode)	videodisc/ cassette	•		•	•	•	•	•	•		•		•	
	off-air broad-casting pro-gramming	•		•	•	•		•	•	•	•	•	•	
	pay T.V.	•		•	•			•	•	•	•		•	
	vertically appealing programming	•		•	•	•		•	•	•	•		•	
	closed cir-cuit moni-toring	•	•		•	•	•		•	•		•	•	•

Telidon Videotex	storage and retrieval	•	•	•	•	•	•	•	•	•	•	•	•	•
	electronic mail and merchandizing	•	•		•	•	•		•	•		•		•
	2-way communication	•	•	•	•	•	•		•	•	•		•	•
Satellite	specialized info. retrieval and transmission	•	•			•	•		•	•	•	•	•	•
	surveillance systems	•	•		•	•			•	•		•	•	•
	fast long distance comm.	•	•	•	•	•			•	•	•	•	•	•
	direct broadcast to home receivers	•	•		•	•		•	•	•	•	•	•	•

Figure 4.2 Disintegrated issues and referents

according to the procedure of referentiality. Machine-objects such as smart toasters or satellites have been piled onto each other. Alternately, that knowledge has been strewn across isolated and fragmented issues (idea-objects) ranging from free flow to content. For example, publishers tend to sponsor anthologies of unlinked articles on various aspects of new communications technologies, with no attempt to establish correlations or to point out incompatibilities (Forester 1981; Dertouzos and Moses 1980; Robinson 1978).

Where certain authors do list integrated priorities hierarchically, the reader is forced to choose blindly the set of objects he or she likes best. Gotlieb (1980, 43) makes licensing all important, whereas Nora and Minc (1978) stress sovereignty above all else. Governmental priorities concerning a valid integrating principle or directive for research can also seem to change as often as the executive director.

The horizontal and vertical axes in Figure 4.1 indicate respectively the disintegrated list of techniques and issues that dominate discourse on new communications technology. Mainstream approaches to communication and technology make no attempt to show the relationship between the various media/techniques and issues, or to integrate, organize or hierarchize in any way these concerns. The social impact of this is the following: where choices between various technologies and interventionist policies must be made, there exists no common guiding principle. It does not suffice to multiply studies on smart toasters, satellites, and robots without an integrating link which allows us to situate them all within our present social context. Both the issues and the regulation of technology, including communications technology, are currently in a state of utter confusion.

But what is wrong with an absence of focus? An absence of some coherent integrating principle for the discussion concerning new communications technology and society would seem to give way either to a certain type of 'laissez-faire' telos or to a reaffirmation of the status quo. Where we concentrate on fabricating and duplicating lists or techniques with no attempt to integrate them, one might suggest that the guiding principle is, by default, liberalist 'laissez-faire.' Where informatized gadgets are placed on an equal ontological footing with transborder data flow, then the integrating principle can only be one of commodity consumerism.[2]

Furthermore, the absence of integration makes for a form of side-tracking or paralyzing confusion. If one concentrates all of one's energies on separating categories of machines ranging from kitchen appliances to computers and omits the more global perspective of how all of the objects are a form of knowledge in a common social environment, one might be said to have already implicitly made a choice of integrating principles.

If one can not find solid ground upon which to see how both pay tv and the coupling of electronics and genetics arose out of this very society, then one must accept to be depoliticized, i.e., to have no means of suggesting technological and political intervention. In the event of a conflict of issues of the development of one technology over another, decision-makers must have an explicit integrating principle by which to formulate a set of socially pertinent priorities in relation to technology.

5.2.2 Integrative framework

A discursive study of communication views all of the various machines and issues as an integrated whole in order to see what types of relations exist between them and what principle might possibly link them. Such an integrative framework permits the formulation of directives for research and policy in the areas of communication and technology.[3]

A discursive approach would suggest the following integrative principle: in that all of these objects/techniques/referents/issues ranging from two-way cable through democratic participation are forms of social communication and in that they are a kind of knowledge (as technology), they should be treated as social discourses of knowledge. Furthermore, technology seen as a complex of discourses of knowledge may be socially contextualized by examining how it relates to other practices of knowledge and discourse in society such as economic or educational practices. This same integrative framework would apply not only to techniques of new communications technology but also to issues whereby specific issue treatment was seen to have been made possible by a particular configuration of social discourses. Our means of integration was then to point out the regularities and irregularities of communications technology as procedures of social discourses of knowledge. This technological complex of discursive procedures of knowledge was then related to other discursive practices of knowledge both in this society and in other societies. And at other times discursive procedures of knowledge were then related as a means of broadening the socio-historical context of analysis.

Whereas traditional referentialist approaches to communication and technology have often defined and elaborated their models as object-specific,[4] one would suggest an approach which can apply to *any* practice of technology or communication in any medium.

The discursive approach finds common procedures of social interaction and knowledge production in various types and media of communication and technologies. It allowed us to uncover the specific procedures that permitted the development and use of a certain technological medium, the production of a certain communicational content, the creation of a certain effect, etc. Thus, a discursive approach to communications and technology is not medium- or referent-specific, but rather context-specific. It is more likely to revise its methods not on the basis of a change of referent-object, but on that of a change of actual technological and communicational practices in time and space, i.e., a change in the historical context of communication and knowledge practices. The discursive approach manages to look beyond specific media in order to see continuities and social patterns which otherwise might not be apparent. A discursive approach can isolate not only continuities but also historical change. Unlike structuralism or effects studies it does not reduce all discourse throughout history to some fixed set of categories such as the 'semantic square' (Greimas 1966) or the 'gratification effect' (Blumer and Katz 1974). A discursive approach recognizes that the procedures it may uncover at one time and place are very likely to change and be transformed throughout history. Such an approach not only postulates the possibility of social and technologial change, it also provides a means of analyzing whether or not such change has actually occurred.

5.3 The need to surpass positivist empiricism

5.3.1 The empiricist's paradox: confidence in referentiality at all costs

Mainstream studies of communication and technology tend to denigrate what they call metaphysical or speculative studies on the grounds that they do not respect the exigencies of empirical epistemology, namely that objects can only be known through our indubitably trustworthy sensory experience of them.

Given this epistemological dictate, it was indeed curious to find that most of these die-hard empiricists resorted to two discursive procedures that belied empiricism:
1) fictive narrative scenarization, and 2) futurism. Where, then, is the sense experience upon which the empirical claim to scientificity so heavily relies? These discursive procedures were seen to indicate a loss of the empirical object-referent. Yet, as the history of the quarrels between the Columbia School and Frankfurt School would attest (Jay 1973), mainstream communication studies are always quick to denounce other critical discourses as non-empirical, as lacking factual evidence, as speculative, and, consequently, as unfounded.

But the mainstream use of futurology and fiction actually reveals its incapacity to refer to actual, factual experiences and objects such as 'new communications technology/society/revolution.' Empirical studies seemed to have lost their scientific legitimation, their referent-object. Mainstream communication research is forced to use the discursive procedures of futurism and fiction to modify its results in order to uphold its empirical doctrine that:

> Reality coincides with its actual factuality as a system of quantifiable instrumental relations that could not transcend the one-dimensional space of their being-as-such. (Hamelink 1980, 13)

These procedures reveal something problematical about the status of the referent to which empiricism so dearly clings.

5.4 Epistemology based on a recognition of the 'discursive formation of the object'

Ever since the discoveries of quantum mechanics, as documented in Heisenberg's *Physics and Philosophy* (1958–62), even the pure sciences have had to face up to what might be termed the 'crisis of referentiality' of our models of knowledge. Our scientific representations of the world were placed into doubt by some of the contradictions and inconsistencies of quantum mechanics' experimental results. Less 'pure' science, such as literary criticism (before semiotics came along), had long ago admitted of such a crisis. Mainstream communications studies and traditional approaches to technology have a lesson to learn from Heisenberg's attempts to account for what happened in his experiments (cf McClean). The traces left in the cloud chamber of elementary particles were not observable reality. They were productions of the interaction of 1) the object – elementary particle, 2) the measuring discourse or instrument – other elementary particles, and 3) the discourse or cognitive knowledge filters of the experimenter.[5] In other words, the results of our experiments are a production of the interaction of the object under study and the discourse and knowledge of the scientist or scientific community.

So again the referentialist question is posed: 'do the referents "the communications society" of "the technological revolution" exist or not?' Machlup, Bell (1981), Porat (1978), and in Canada of course, Valaskakis (1981), were all seen to have declared unequivocally that there is a referent and that they had discovered it, e.g., GNP statistics, whose adequacy in the world of communication is undoubted. But what, then, does one do with documents that also talk about 'the communications revolution' but which quite embarrassingly for the new gurus date back as early as the 1930s? (Albion 1930, 718–22) Are the 'newness' of the referent technology and the 'referent' itself to be placed into doubt here as does Weizenbaum in his somewhat facetiously entitled text, 'Once more the computer revolution' (1981).

5.5 The industrial revolution – a discursive constitution

Raymond Williams, in *Culture and Society* (1958) and *The Long*

Revolution (1961), perhaps provides us with a solution to the above epistemological and ontological dilemma. He states that the discourses on the industrial revolution 'were defining society rather than reflecting it.' Speaking of Carlyle's *The Condition of England* and of Engels' *The Condition of the Working Class in England* (1845), as well as of the industrial novels of Dickens, Williams suggests that these writers were contributing to the image, the insight of what the industrial revolution was more than they were giving scientific representation of it. These texts were not reporting the industrial revolution. They were exhorting it. Their words were full of perspective, bias, pressure and constitutive interpretation in relation to the supposed object of their discourse, namely, the industrial revolution.

Indeed, as Krishan Kumar (1978, 48) argued, these authors started off with a pre-conceived model of modern industrial society which even the referentialist discourse of economists and demographers did not necessarily confirm.[6] For example, 'urbanization' – the demographic flood into the cities – is said to have occurred much earlier than demography would confirm (Kumar 1978, 53ff). The redundancy of the semantic categories into which the industrial revolution was fit, namely, the 'terrible beauty of it all' and the 'hope and despair of the world' (Kumar 1978, 53ff), would indicate that rather than referentiality what is in operation here is 'the rhetoric of the technological sublime.' These discourses on technology were actually seen to have played a role in constituting the society of new technology. They constructed the discursive categories in which technology and technological society were both developed and enunciated. In other words, the discursive approach showed that the discourses of technological society are perhaps not simply reflective of the fact but are before the fact and are co-constitutive of it.

5.6 The discursive formation of the object

How then does a discursive approach avoid the paradoxes of referentiality and yet speak about communications technology and society? Discursive critique deviates from the referentialist basis of language by suggesting that what is important is not so much the content of a discourse, i.e., the semantics, as the rules, procedures or means by which the content is said. Foucault in *Discourse on Language* (1971–72), suggests that traditional studies in the social sciences, including communications studies, have tended to treat discourse as something transparent to the referent and consequently as something inconsequent. Too often, he says, Western science has seen to it that discourse be permitted as little room as possible between thought or knowledge and its object. Discourse was understood as a mere interjection mediating knowledge and the world without interference. Discourse was reduced to thought clad in signs or else discourse was understood as structures of language that were brought into being merely to name certain referents or to have a certain effect of meaning. This referentialist conception of language, itself an example of a discursive procedure, is referred to by Foucault as 'the theme of original experience' wherein ideas, things, and thoughts were believed to pre-exist discourse and to float around somewhere just waiting to be named (Foucault 1972, 228).

Therefore, to study communication and technology in society, be it an 'information society' or not, it was not the content of discourse that needed to be looked at, but rather the 'ways of saying,' the rules, procedures, assumptions of discourse and knowledge that make saying one thing and not another possible.

The epistemology and ontology of a discursive approach to new communications technology/society solved the problem of referentiality. Discursive critique showed how the communication revolution and society are brought into being as objects for our knowledge via certain discursive practices. For example, the 'communication revolution' was seen to be a statement permitted by liberalist discursive procedures of discourse which considered 'change' to be both inevitable and progressive. Technology, more specifically new communications technology has the ontological status of a discursive formation as defined by Foucault:

> a complex group of relations that function as a rule; it lays down what must be related, in a particular discursive practice, for such and such concept to be used (. . .) To define a system of formation in its specific individuality is therefore to characterize a discourse or a group of statements by the regularity of the practice. (Foucault 1969–72, 74)

Thus, discourse was seen to be not a perfect imitation (mimesis) or representation of the object technology as the empiricists would have us believe. Rather, it formed or constituted the object:

> (. . .) in discourse something is formed (according to well-definable rules); that this something exists, subsists, changes, disappears (according to rules equally definable) in short, that, along with all that a society may produce (. . .) there is the formation and transformation of 'things said.' (Foucault 1969–72, 9)

The discursive approach demonstrated that Bell's schema of the evolution of society from 'industrial' to 'post-industrial,' 'service' and 'knowledge' societies and now, finally, to the 'information' society, involved merely a reclassification of many traditional professions as 'knowledge-,' 'service-' or 'information-' related jobs. For example, preachers who existed in industrialist and feudal times and who have not become more numerous have simply been reclassified as 'information workers.' Garbage collectors have become 'sanitation engineers' and hence also a part of the 'knowledge sector.' Nor is this word play innocent for Bell uses it to support his declaration of 'the end of ideology.'

Thus the question concerning the existence of a technological or communicational revolution in society underwent a major shift in focus. Instead of the question: does there or does there not exist a 'communications revolution/society'? one asked how do the discourses of the Daniel Bells and even the Karl Marxes form or constitute the technological or communications revolution/society? One then turns one's attention away from discourses on new communications technology and toward new communications technology as discourse only to find the same discursive formation of the object to be operative.

The important thing to study about new communications technology *per se* was seen to be not the 'what' – content – of its data banks or hardware objects, but, the discursive procedures of new communications technology as discourse. These procedures – the 'how' – include such rules as the organization of networking, the discursive patterns of accessing, the logical 'tree' paths of association of information which are either permitted or not permitted, and the relationships of inclusion and exclusion of the various users, information providers and programmers. All of these rules and many more were seen to constitute the discursive procedures or the pragmatic rules of new communications technology as a social discourse that is practised today.

How then does a discursive analysis avoid the paradoxes of the empirical approach to communication? It does so by refusing to attach ontological validity to a referent-object as the exclusive justification and legitimation of the scientificity of discourse. Discourse analysis is a recognition of the impossibility of empiricist epistemology and a renunciation of any attempt to occult the failure of the empirical, i.e., the crisis of referentiality. Discourse analysis recognizes that, intervening between the object to be known and the knowing subject, there are always rules of discourse and knowledge which perturb the results and cloud the transparency of the representational model to the world. The one materially graspable object is discourse itself, but even this is subject to the perturbations of meta-discourse. Discursive analysis avoids the paradoxes of empiricism by means of a statement and practice of what might be called the 'epistemological modesty' which recognizes the discursive mediation of the objects of scientific study.[7]

5.6.1 The specific pertinence of a discursive approach to the study of new communications technology

Recognition of the discursive formation of the object serves to justify even further a discursive approach not only to new communications technology but to communication and technology in general.

Indeed, it is the group of 'gurus' of the 'communications age' who themselves, perhaps unwittingly, enunciate the discursive heart of technology. By speaking of the 'information age,' the 'knowledge society' and the 'communications revolution' such writers as Bell, Toffler and Porat are, at least semantically, strictly linking technology to knowledge and discourse. But, these authors establish these links only occasionally and only thematically, i.e., as content. Their method has not yet awakened to the necessity of an alternative approach once discourse or communication becomes the centre of things. Rather than trying to demonstrate the 'information society' discursively, they revert to empirical methods and argue its existence on the basis of 'economic referents,' namely the percentage of GNP devoted to certain sectors of the economy. They are, so to speak, inconsistent in their arguments and methods with the content of their findings. Their method is empirico-referential, i.e., based on what they believe to be true reflections of society, that is economic statistics. Their findings are that society is first and foremost a communicational or informational activity. To be consistent they would have to study this communicational activity not economically but discursively.

5.7 The dynamic historical contextualization of the 'object' of study

5.7.1 A-historicity – technological amnesia

Another weakness of much theory of communications technology is its negligence of history. Sören Kierkegaard (1944–71) once warned that the unhappiest man was one who could not live in the present and who was always fleeing it for the past or future. It is perhaps the case that the *futurist* tendencies of discourses on new communications technologies risk making society the 'unhappiest man.' To the degree that the past is mentioned at all it is done so in an abstract way. The past is the progressive, evolutionary causalist way to the future which in turn is really all that interests futurists. Events such as the development of telematics are spoken about as though they were just about to occur or as though they had fallen from the sky. Most documents appeared to speak only about the future of computers while extremely few mentioned the history of its costs, applications and successes or failures. In other words, *the history* of the interdiscursive relationships that were responsible for the formation of computer-communications links was repressed in the Freudian sense of the term.

5.7.2 Historical approach

While not wishing to fall back into the trap of cliché, it still needs to be said: the man who loses his memory is condemned to repeat the past. A historical positioning of the relationship of new communications technology to society is essential if we are to learn at all from the mistakes and successes of the past. This study of the history of discursive relationships that have constituted technological revolutions and societies at least since industrialism has allowed us to discern that which is genuinely 'new,' 'revolutionary,' and 'transformational.' A comparison of the procedures of communications technology with other procedures in contemporary and in past societies has enabled us, rather than talking about communications technology as some unknown but utopian promise of further social change, to evaluate the actual degree of this technology's revolutionariness. We can assess revolution in terms of the degree of discontinuity of discourses of new communications technology in relation to discourses of the past. The discursive approach meets, we believe, the challenge of dealing with technology in history:

> One question that comes up in my mind is whether what we are trying to do is not to study technology, but to, in a sense, expand and develop the study of history, cultural change, political choice, human goals, to include technology. Really, the problem is that these subjects have not known how to deal with technology and what we are seeing in this complexity is that it is not simple to deal with it from the point of view of history. (Macoby 1980, 8)

Furthermore, we are not seeking the kinds of causality and continuity that underlie traditional history. As Paul Veyne writes in his article 'Foucault révolutionne l'histoire' (1978), Foucault abandons the *a priori* assumptions of causality, continuity, progress, or even 'epistemological rupture' (Bachelard 1971), in favor of an intimate look at concrete practices of discourse and the dispersed and complex relations between these practices. An 'archeology' of these practices uncovered not only similarities but also irregularities: hence Foucault's insistence upon both the continuities and discontinuities of discourses throughout history, quite independently of any master pattern such as the dialectic.

The episteme – the sum of the discursive procedures that we looked at in chapters one and two – is equivalent to the historical context that situates contemporary practices of new communications technology and discourse about it, and shared many traits with the context of industrialism. The discursive archeology enabled us to argue against the futurists in the following manner: in order for one to be convinced that we are indeed up against a 'technological revolution,' a 'future shock' or a 'new kind of information society,' one would have to show that the discursive procedures that constitute the dominant episteme and that organize our society have indeed suffered radical transformations in relation to their predecessors.

Consequently, our task is to unearth various discourses on new communications technology as well as historical discourses on the general relation of technology to society in an attempt to get a picture of their relationship to the dominant episteme(s).

5.8 The idealization of technology as an immanent essence abstracted from context

'Technology is inherently good or evil.'
'Technology is inherently neutral.'
'It all depends on how you use it.'

Many of the discourses of new communications technology seen in this book are variations on these themes. All three amount to an abstraction of technology seen as either some sort of 'ideal type' or as something entirely empty and neutral waiting to take on the forms and values given by its use. In his essay 'Technique,' Cornelius Castoriadis suggests that these two main approaches or themes tend to abstract from a particular, historical context, the context in which technology is realized in each of its various forms. We must avoid essentializing technology or abstracting an ideal 'technology' from specific socio-historically situated practices. Treating technology as specific discursive practices shows that technology is not neutral or inherently anything.

> But there is nothing different with the global attitude towards technics: most of the time, contemporary opinion, be it popular or knowledgeable, remains stuck in the antithesis of technics as a simple man-made instrument (perhaps today misused) and of technics as an autonomous factor, fatality or 'destiny' (benevolent or malevolent). As such, thought is continuing its ideological role: give society the means with which to avoid thinking about its real problem and with which to elude its responsibility in the face of its own creations. (Castoriadis 1978, 222, author's translation).

Technology exists as specific practices or procedures that take on meanings and values within concrete social contexts.[8] The discursive approach is an attempt to meet Mumford's challenge against a theory or an economy of technology that is based not on organic needs, historical experience, human aptitudes, ecological complexity, but upon a system of empty abstractions such as power, mobility and growth.' (Mumford 1963, page) A discursive approach by virtue of concentrating on actually practised technologies in concrete social situations has tried to avoid what Carey and Quirk call abstract idealization of fetishism. Treating technology as discourse is an attempt to begin to dismantle the fetishes of 'communication for the sake of communication, and of decentralization and participation without reference to content or context.' (Carey and Quirk 1970, 423)

5.9 Contextualization

The discursive approach to technology elucidated what the abstractions failed to recognize, namely that technology not only changes its uses or values within a context but that technology, understood as a particular discourse, is formed – designed and built – within a particular context. The formations of technology, i.e., its way of existing as a practice or as an event, are inextricably linked to the context. Technology, like any discursive formation, is made possible, is produced, and has meaning by virtue of a set of discursive and non-discursive procedures that constitute the epistemic environment of technology's production and reception as a particularized discursive practice. Technology like any other discursive practice is made possible and structured by its episteme.

Such a contextualized theory of technology makes it impossible to mouth the variations upon the cliché 'technology is neutral.' The only way to characterize technology is as a form of knowledge and discourse characterized by specific and inherent values, structures and designs that are determined within and in turn determine the context that situated it and the uses intended for it. What is more, technology itself, as an integral part of that social context, also contributes to the set of procedures and their transformations that make up that context. For example, we have seen how computerized voice recognition technologies are both determined by and determinate of a larger social context in which surveillance is valorized.

In order to answer some fundamental questions about the uses, values, structures and design of new communications technology in our society, we have sought to identify the actual episteme that positions it and makes it possible. We hope to have shown that not only the host of discourses' that herald the coming of the 'communications revolution but also specific communications technologies are themselves made possible not by some individual genius, author, or inventor but by a set of discursive rules that, as episteme, condition both what are considered to be valued statements about and practices of technology.

By studying both technology and its social context as sets of regularized discursive practices, we have sought to avoid *a priori* universalizations about either. In terms of possible intervention in the sphere of politics and technology regulation, the avoidance of such abstract universalizations is essential because universals or ideals are *not* potentially alterable in and through social praxis.

5.9.1 A priori context formation

While traditional discourses on communication and on technology may not always pretend to have achieved the feat of pure, objective a-contextuality, they often assume an *a priori,* inflexible, and unreflective view of context; for example, the automatism of the capitalist/socialist dyad. We have seen that the discursive procedures of usurpation of voice, stereotyping of publics, appeal to authority and exclusivity are all a form of *a priori* context formation.

5.9.2 The changing of context(s)

The discursive procedures that constitute the episteme – context – have been shown to change throughout history. The procedures that constitute context must constantly be revised by any study that seeks to contextualize communication and technology. In our re-contextualization of new communications technology, we sought to look at the context anew in order to answer the following questions: why new communications technology here and now? In relation to what other discourses? How would it be different in another epoch? Was, is, and will new communications technology be contextually possible? And, if possible, at the expense of what other kinds of discourse practice?

The contextualization of discourse by first considering the episteme as a sum of the various discursive rules that make a statement possible is described by Foucault as an attempt:

> to reveal discursive practices in their complexity and density; to show that to speak is to do something (. . .) other than to express what one thinks; to translate what one knows, and something other than to play with the structures of a language; to show that to add a statement is to perform a complicated and costly gesture, which involves conditions (and not only a situation, a context and motives) and rules (not the logical linguistic rules of construction); to show that a change in the order of discourse does not presuppose 'new ideas,' a little invention and creativity, perhaps also in neighbouring practices, and in their common articulation. (Foucault 1972, 209)

Instead of attributing the discovery of the communications revolution to the 'gurus' of the information society, we have looked to the discourses of the technological build-up between the two world wars, traces of which we found in the discourses of Dessauer (1927) and Schumacher (1930), among others. We also invoked the context of the domination of the specialist's discourse in interdiscursive relation to practical economic discourse which tries to grapple with the recession and crisis of late, State capitalism (Gotlieb and Zeeman 1980, 'Introduction'). We then added to that context the discourses of educational institutions where various disciplines are re-evaluated and resituated on the hierarchical scale of 'financial' priority. Only by building up a picture of interrelations between the various types and fields of discourse practices that environ new communications technology can one avoid idealizing it or its social context.

5.10 Power/Knowledge – the focus for studying the social impact of new communications technology

5.10.1 Out of focus

Be it due to omission or occultation, it is often difficult to find some form of explicit, common conceptualization behind the way in which the issues are raised in most discourses on new communications technology. Such a dispersion of foci is often excused on the grounds of either the complexity of the problem or of the multi-disciplinarity of the approach. Unless we can identify the extant foci and provide alternative ones it seems unlikely that new tacks will be taken on the problems of the social impact of new communications technology. Indeed, the failings of the old approach will not even come to light. Furthermore, any study must make its own focus explicit in order to provide some perspective on the host of verbiage surrounding this 'fad,' and in order to establish certain priorities for political and technological intervention.[9] We have seen that the focus of traditional studies of new communications technology is often not simply *missing* but is actually *occulted,* and with good reason.

5.10.2 In focus

The focus of this study, and of any discursive critique, stemmed from the position that the crux of any social order is the relations of power and control that characterize it. Indeed, the very fabric of any social order is relations of control and power. Without them there would be no social organization. There would be only anarchic amorphous mass. The social impact of communication and technology is strictly linked to its accompanying relations of power in society. A discursive analysis allowed us to focus in on relations of power and social control by virtue of the way in which it was able to demonstrate how throughout history a monopoly over certain discursive procedures is strictly associated with a monopoly of knowledge which in turn legitimates the exercise of social control and political power.

5.11 How to study power and social control?

> But if power is in reality an open, more or less coordinated (in the event no doubt, ill-coordinated) cluster of relations, then the only problem is to provide oneself with a grid of analysis which makes possible an analytic of relations of power. (Foucault 1980, 199)

Power is not only made visible as traces left in discourse; power and domination are concomitant with and constituted as institutions of discourse and knowledge which enjoy a hierarchical and exclusive superiority over other discourses. Power lies in this hierarchical exclusivity of interdiscursive relations. No empirical study of decision-making or non-discursive behavior could see such relations.

Discourse or communication practices were perceived as traces throughout history of the practices of power and knowledge, and were also seen to be the conditions of possibility of the exercise of power and knowledge in society. If Hamelink once criticized mainstream American communication studies for insisting too much on the answers and voting patterns of hog farmers in Ohio (Hamelink 1980, 6), this is perhaps because, at least since the critical theory of the Frankfurt School, power and domination have been understood as surreptitious. Foucault's description of power as 'discourse's latent nature and its brutality' would meet with almost total incomprehension on the part of most empiricist students of mainstream communication:

> right is (. . .) the instrument of this domination (. . .) right (not simply the laws but the whole complex of apparatus, institutions and regulations responsible for their application), transmits and puts in motion relations that are not relations of sovereignty, but of domination. Moreover, in speaking of domination, I do not have in mind that solid and global kind of domination that one person exercises over others, or one group over another, but the manifold forms of domination that can be exercised within society. (Foucault 1980, 85–96)

The fundamental difference regarding the empirical and discursive approaches to power here lies in the discursive approach's insistence that power is related to all forms of discourse and control at even the most minute and dispersed levels of society. What is more, power is not a question of conspiracy on the part of certain groups of discourse users. It is not simply a case of capitalists exercising power while workers do not. No discourse is exempt from power relations; power is the constant companion of knowledge and discourse. In terms of a strategy for technological intervention, the power focus does not posit utopia as the absence of power but rather asks how alternative discourses favoring alternative relations of power may be developed. The discursive approach to technology:

> entertains the claims to attention of local, discontinuous, disqualified, illegitimate knowledge against the claims of a unitary body of theory which would filter, hierarchize and order in the name of some true knowledge and some arbitrary idea of what constitutes a science and its objects. (Foucault 1980, 83)

5.12 The relationship of knowledge to value, of theory to practice

5.12.1 Positivist separation of knowledge and values, of theory and practice

Mainstream administrative communication research has pretended to be objective and value-free. It 'describes' rather than passing judgment or prescribing remedies. Such a framework for research (a partial off-shoot of the positivism of the Vienna Circle, but primarily once again of classical discourse's confidence in referentiality) placed great confidence in the capacity of science to represent the world, free of perturbation by individual or social values. Communication study was, in certain circles, reduced to 'scientific,' statistical measurement of observable effects (Blumer and Katz 1974). Often these experiments pretended to have no connection to prescriptions for the practice of communications in society.[10] (Ellul 1954, 'Préface')

If we were to suggest the major underlying assumption of this confidence in 'scientificity' it could be stated as the exact opposite of the position held by 'discursive critique' that no discourse of knowledge, be it object-level, theoretical, or meta-theoretical, can be separated from the legitimation of power and from the relationship of domination and exclusion within the various communities of science.

The question of why one chooses one's epistemological stance is an elusive one. Nevertheless, some anthropology (Geertz 1973), some interpretations of quantum mechanics, and certain Third World reactions against Western science would all seem to make it difficult to hang on to the theory of objectivity of representation, science, and technology. If one does accept that objective, value-free knowledge is quite impossible, then the only alternative, besides silence, is perhaps the unocculted affirmation of the interests of particular scientific and technological practices. Many theorists of technology and of communication have countered the value-free school with a call for the overt recognition that various theoretical perspectives lead to various interests and their affirmation in political practice. What then are the practical interests of mainstream vs discursive approaches to communications technology?[11]

5.13 Empiricist epistemology as a reaffirmation of the 'factual' status quo

'All we can do is tell it like it is.' Such might be, in a nutshell, the expression of studies based on an epistemology of empirical referentiality. Such an epistemology has consequences not only for how one purports to know reality but also for if and how one purports to change reality. If science is based entirely on factual referents then all one can do is to study the facts and accept them. No prescriptive moment of duty or possibility is added to the observation of factual, descriptive analysis.

De Sola Pool and Solomon's study of policy concerns surrounding the question of transborder data flow typifies this stance. The fundamental message of their paper for the OECD is the following: 'Don't try to change, stop or interfere with technological progress since the technology is already in place and is neutral; all you can do is describe it, and, where it might be nefarious to social relations, there is still no need to worry since things will balance themselves out if left to their own devices.' (Pool and Solomon, OECD No.3, 1980) Such is their plea for deregulation of transborder data flow. In other words, an epistemology which limits itself to 'fact' or 'referent' is fundamentally non-interventionist and pro-deregulation in that one can only 'laissez-faire' and describe.

5.14 An epistemology of the possibility of change

Earlier in his career, Pool contradicted his own above-mentioned position when he suggested that policy research not only describes and projects possible impacts of new technology but also determines what technologies will occur.[12] Here, Pool is admitting that discourses on communications technology are value-laden and affect the practice and development of specific technologies. He is as much as admitting the discursive formation of the object.

A discursive analysis, by conceding that reality and meaning are at least co-constituted by various types and practices of discourse, allows that the 'reality' of technology is not a static absolute reality restricted to the 'facts' *hic et nunc*. Discourse plays an active role in the construction of these 'facts.' Consequently, discourse on new communications technology can potentially change that technology. Intervention via discourse is possible in order to change technology considered as discourse. Discourse on new communications technology can constitute technology in certain ways rather then merely 'saying it like it is.' Given theoretical discourse other than the positivist, 'laissez-faire' approach to factuality, it became possible to imagine and constitute new communications technology as it 'ought' to be (according to certain interdiscursively defined interests) and not merely to resign oneself to how things are, to the way it 'is.' Thus, the potential for change is both an epistemological and a social possibility. We are not arguing supreme Fichtean idealism here. We cannot put the moon in our backyards simply by thinking it possible. However, the production or disappearance of scientific objects, such as the circulation of blood or the ether, is a co-production involving scientific and cognitive discourse as much as the real world that the former mediate. Especially as concerns socio-discursive reality such as new communications technology and society, the active co-constitutive role of scientific discourse is enormous.

We have tried to show how the 'fact' – practices – of new communications technology shift with the shifts in the discursive contexts – the episteme – that situate it. Consequently, once we have properly understood the current context which situates communication and technology, we can understand not only how the 'facts' – practices – were constituted but also how an alternative set of practices or 'facts' of communications technology could be possible. In the last chapter we have tried to conceive potential transformations of the discursive procedures that make up that technological context. The search for an alternative context of new technology in society was inspired by Habermas' argument that for an alternative society we need an alternative pragmatics, i.e., an alternative set of discursive procedures, an alternative set of rules of communicational competence.

Such an epistemology of potential asks not only 'in what context is new communications technology practised as it is?' but also 'in what context could new or alternative communications technology possibly be designed, developed, and deployed in order to favor certain socially advantageous interests or possibilities?' Discursive practices can and do change. In keeping with critical theory, and in the name of what Adorno called 'negative dialectics,' we have refused to accept the status quo of the referent. Such would be resignation to false utopia. An epistemology of potential would refuse to limit itself to an extant state of affairs. It holds out rather for a more utopian form of communications technology in society, for less hegemonic, less repressive procedures of discourse. It holds out for a discursive communicational context in the form of a discursive community which would crown a certain type of communication as discourse of knowledge but in such a way as to avoid the exclusions and hierarchies of classical, referential, liberalist discourse:

> the existence of a communicative community is the precondition for all knowledge in the subject-object dimension and that the function of this communication itself can and must become the theme of scientific knowledge as an intersubjective metadimension for the objective description and explanation of data. (Apel 1971, 7–44)

Ultimately, we do not prescribe any definite forms of technology. Instead, we suggest which discursive procedures would have to exist in the sphere of decision-making about technology for both technology and politics to become more democratic. Placing discourse at the heart of technology and of society demonstrates the enormity of the task of social and technological intervention while simultaneously insisting that intervention is possible.

5.15 The acceptance or rejection of the pre-given logical, discursive space from within which to talk about new communications technology

5.15.1 Myth as overcoding

The redundancy of the treatment of new communications technology is puzzling. In scores of texts, each aware of the others' treatment of the subject, we constantly run across the same lists of techniques, issues, clichés, etc. The constant use of these points is an attempt to 'naturalize' certain attitudes and beliefs about communications technology while obscuring the relativity and contextual divergencies of the meaning of technology for society. This process of 'naturalization' is referred to as myth by Barthes (1957) and is associated with ideology. 'Naturalization' is a term which Barthes uses to mean the association of a meaning with a phenomenon in a way that purports to be inevitable, natural, automatic, and unquestionable. The 'technological imperative' is a perfect example of 'naturalization' whereby the necessity of technology, regardless of which technology, is taken for granted. It is a 'false imperative.' For this reason Barthes associates naturalization with myth and ideology. 'Technology is inherently neutral' is but one such extrapolated universalization, one such overcoded myth.

5.12.2 Epistemic relativization

A discursive approach, by virtue of looking at the pragmatic conditions of all and every meaning, seeks to avoid universalization in favor of exposure and relativization of procedures of semantic association. Showing how the procedure of 'progress,' for example, is associated with a specific episteme demythifies its claims to universality. Discourse analysis seeks to 'de-automatize' semantic associations of technology by illustrating the dependence of these upon particular contexts of discursive procedures. In so doing, a discursive analysis points out the possibility of a transformation of these procedures, and, hence, of a change in the social meaning of technology for society.

> A demystification avoids a premature fossilization of the meaning surrounding a term or terms with new communications technology; reflection replaces automatism (. . .) (Seeley 1980, 47)

In short, discursive criticism has the same task as that which Bacon, Locke, Descartes, et al., gave themselves at the beginning of the rise of 'new' science. These thinkers sought to think their way out of an old logical space – discursive episteme – and into a new, more adequate one to meet the needs of the times. Discursive criticism, not content to remain within it, criticizes the old space. It then sets itself the aim of perceiving how transformations toward an alternative new space might be achieved.

On the basis of the above epistemological concerns for an alternative approach to the social implication of new comunications technology practical guidelines for further research into communications and technology can be made. Communication researchers and administrative bodies should attempt to see what conditions of speaking and knowing have made possible the specific yet redundant ways of currently conceiving of the social importance of communication and the ways of designing, developing, and deploying communication. Furthermore, to alter aspects of communication and society in order to constitute an alternative or better communication for society will, to a large degree, require certain 'transformations' in the whole context of society's rules of discourse and science as well as in their links to power legitimations.

Finally, the lacunae of traditional studies of communications technology and the respective compensation for these on the part of the discursive approach can be schematized as follows:

Table No. 1

Traditional approach	Alternative approach
Disintegrated: lack of integration of techniques, functions and issues;	*Integrated*: view based on common denominator of discourse, i.e., a satellite and a smart toaster are both practices of social discourse;
Empiricist's paradox: insistence upon the criterion of empirical observation and its adequacy as a representation of the world – objectivity of science;	*Epistemology based on discourse*: the recognition of the perturbation of empirical results; and the belief that 'facts' and 'events' are discursively manifest and co-constituted;
A-historical: futurological bent and evolutionary view of history which ignores the past and refuses to answer to the specific demands of the present context of communication practice;	*Historical contextualization*: attempt to fulfill the need for a historical study which contextualizes new communications technology in relation to the past ways of conceptualizing it as well as in relation to the demands made by current social conditions – i.e., a form of discursive materialism;

Table No. 1 Continued

Traditional approach	Alternative approach
Idealization/abstraction: idealization of technology to an abstraction from the concrete context which is formative of both the uses and the very structure of new communications technology;	*Contextualization*: attempt to develop a contextual theory of technology and society whereby technology is neither 'inherently neutral' nor 'inherently value-laden' but where it is studied as what grows to be, to be used, and to be judged within specific contexts;
A priori assumption of context: an *a priori* view of what the context of new communications technology is once and for all, be it one of social responsibility or exploitation;	*Changing context*: constant reconstitution of contexts based on a study of the traces of context left by past and present discourses;
Out of focus: no theoretical framework by which to judge the validity and consistency of statements and positions re host of new communications technology;	*In focus*: bringing to light the dominant focus of power and social control; eventual search for an alternative focus: democracy and participation;
Claim to objectivity: the positivist belief that the theory of communication is neutral and objective, descriptive as opposed to prescriptive;	*Affirmation of interests*: all theory, including positivism, is value-laden or linked to ideology and therefore these values should not be occulted but rather exposed and defended as discursive and knowledge guide-lines for action, i.e., for technological intervention and the legitimation of alternative configurations of power;
Reaffirmation of status quo: the acceptance of the status quo because it is 'fact' and because the *possible* is not a viable object for empirical knowledge;	*Potential*: the criticism of existing practices of communication in view of postulating the possible practices in a possible state of affairs;
Closure within an accepted discursive space: no attempt to reflect upon the logical space within which their positions are enunciated; no misery due to extreme cliché of existing texts; no attempts to overcome the limits of their own critical discourse.	*Relativization of logical space*: a constant attempt to dejargonize the clichés surrounding the debate, to expose the mythification as well as to find discourse with which to talk about this phenomenon; to recognize the constraints or procedures of one's own discourse while seeking to transform those which deny potential alternatives.

Contextualization **DOMINANT EPISTEME**

INTEGRATING PRINCIPLE **DISCOURSE**

CENTRAL FOCUS Emancipatory interaction/democracy vs Social control/concentration of power

TYPOLOGY: #1 Quality of life — Socio-economic — Regulation — Freeflow — Dependence and development

Technology	Services	access	surveillance/ privacy	diversity/ leisure/ creativity	impact on other types of comm. privatisation	symmetry/asymmetry bi-directionality/ unidirectionality	institutional impact/ e.g. disemployment	homogeneity/ diversity	concentration content/carrier	Fed./Prov. regional/central concerns	national/ cultural identity	national/ sovereignty	information rich information poor	Military-industrial complex
Cable fibre optica coaxtal cable (increased channel capacity) (versatility of	videodisc/ cassette	•		•	•	•	•	•	•		•		•	
	off-air broadcasting programming	•		•	•	•		•	•	•	•	•	•	
	pay T.V.	•		•	•			•	•	•	•		•	
	vertically appealing programming	•		•	•	•		•	•	•	•		•	
	closed cir-													

	data													
Telidon Videotex	information storage and retrieval	•	•	•	•	•	•	•	•	•	•	•	•	•
	electronic mail and merchandizing	•	•		•	•	•		•	•		•		•
	2-way communication	•	•	•	•	•	•		•	•	•		•	•
	specialized info. retrieval and transmission	•	•			•	•		•	•	•	•	•	•
Satellite	surveillance systems	•	•		•	•			•	•		•	•	•
	fast long distance comm.	•	•	•	•	•			•	•	•	•	•	•
	direct broadcast to home receivers	•	•		•	•		•	•	•	•	•	•	•

Figure 5.1 is the conclusion to the epistemological introduction. It illustrates the move away from a study based on referent-objects and toward a study which seeks to find a focus, an integrating principle and to contextualize both new communications technology and discourses about new communications technology.

NOTES

Chapter 1 Procedures of discourses on new communications technology: the episteme

1 Most of the groundwork for this section was undertaken in a seminar given in the Graduate Program in Communications at McGill University, Montreal. I wish to acknowledge the participation of the students in this seminar as regards both the unearthing of many of the documents cited here and the collective analyses of these texts. I can only thank them for having contributed to making the seminar so enjoyable and for having ensured a broad representation of works in the corpus.

2 For a justification of the non-exhaustivity in establishing a corpus while maintaining a degree of representativity, Greimas suggests that one must have a random selection of texts, analyze them, and then test other texts from a larger sample in order to ensure that the results of the previous analysis hold true for the remainder. A.J. Greimas, *Sémantique structurale,* Paris, Larousse, 1966.

3 To date, the most flagrant case of speakers being locked into a discursive procedure in the debate on new communications technology has been the letters to the editor following the 1983 New Year's issue of *Time* which declared the computer 'Man of the Year.' Whether the writers were for or against the computer, all fell into the trap of discussing it in terms of the oppositional procedure on a metaphysical level: man vs machine.

4 For a discussion and critique of the concept of 'scopophilia' in Metz as derived from Lacan, see Paisley Livingston, 'Discipliner la communication: sémiologie et cinéma,' in *Communication/ Information,* vol. 5, no. 2/3, 1983.

5 A recent study on access for DOC also interprets problems of communication in terms of access to information objects: 'The key issue was how data banks might be used to make information that is of interest and use to the public more widely and cheaply available.' (Dornan and Wells/DOC 1981, 32) Even ministerial reports discuss communication in terms of the possession and exchange of bits of information or of pieces of hardware that store and organize information:

> it is basic to the government's thinking that coming improvements in data communications should be distributed equitably throughout all of Canada if critical disparities between regions are to be avoided. (*Green Paper*/DOC 1973, 5)

6 The now renowned protectionism of its microcomputer industry is a case in point of protecting hardware and economic exchange.

7 The fact that, in Canada, the Department of Communication is

currently concentrating on seeding hardware industries and that software and information provision industries are far behind the development of hardware (Kurchak/DOC 1981, 74) is further evidence of this. Indeed, it has been suggested that videotex systems such as Prestel and Telidon have been somewhat less than commercially successful largely due to the absence of attractive and pertinent software.

8 For a discussion of the referentialist assumptions that constitute the episteme of the seventeenth century and of its links to power, see Timothy J. Reiss, 'Cartesian Discourse and Classical Ideology,' in *Diacritics,* 19–27, Winter 1976.

9 For a discussion of the very intricate and powerful functioning of the procedures of order and analysis in classical science, see: Michel Foucault, *Les mots et les choses,* Paris, Gallimard, 1971.

10 Karl Marx, *Capital,* New York, International Publishers, I, 3.

11 Computer programming might be capable of reflective knowledge, but it is still primarily concerned with calculation, problem-solving, and decision-making at the object level of knowledge.

12 For a discussion of the concept of a semiotic square, see: A.J. Greimas and Fr. Rastier, 'Jeu de contraintes sémiotiques,' in Greimas, *Du Sens,* Paris, Seuil, 1970.

13 Louis Marin also refers to the theme of eternal mediation as constitutive of what he calls the 'ideology of representation.' In his *Critique du Discours,* Marin suggests that, since the grammarians of Port Royal, theories of language have depended upon the theme of eternal mediation in order to justify representation. The first case of communicational mediation is the eucharist, whereby the word mediates between man and God: 'This is my body.' Louis Marin (1975), in J.-Fr. Lyotard (1977, 32–33).

14 For a discussion of the concept 'subsumes,' see Greimas 1966. Greimas suggests that when one semantic subsumes another there exists an axiological relationship of domination of the subsuming over the subsumed level. We have extrapolated this notion to apply to enunciative voices which subsume each other much like the Russian dolls which contain each other. For a discussion of the logical power of hierarchical subsuming levels in art, music, and mathematics, see Douglas Hofstadter, *Godel, Escher, Bach,* New York: Vintage Books, 1982.

15 The American insistence on free flow is borne out by its record of intervention at UNESCO as well as by V. Mosco's research on policy-making in the FCC (1979). Bloom's study (1982) on the legal and policy definitions of communications also reinforces such an interpretation of the American approach to communication regulation.

16 Karl Marx, *Capital,* Vol. 1, Chapter 3, New York: International Publishers.

17 The Irving Brothers and other owners of press and mass media conglomerates in Canada testified before the 'Kent Commission' on

monopoly because they were 'simply following the indisputable right and desire for growth, a natural desire as well.' (Irving Brothers on *The National,* CBC, Spring 1981)

18 For a discussion of post-marxist reception of Marx's theories of control-oriented works vs non-control-oriented interaction, see Albrecht Welmer's article on Habermas' insistence of interaction in Marx. Welmer insists that Habermas' work is a reminder that Marx insisted not only on work but also on interaction as a sensuous human activity. Most marxists ignored interaction concentrating only on work. Habermas seeks to rectify this neglect by concentrating on the social formative activity of communicational interaction. Albrecht Welmer, 'Communication and Emancipation: Reflection on the "linguistic turn" in Critical Theory,' polycopy of an unpublished manuscript, Montreal, McGill University, McLennan Library, 1974.

19 That control is a primary interest of many of the discourses on computer-communications technology is further borne out by such advertising as the CNCP television ads and the Telidon publicity where we are confronted with a constant and unocculted call for reduced risk-taking, greater predictability, and, in short, greater control over the environment in which decision-making and problem-solving must operate.

20 For example, a recent consulting report done for the Montreal cable television companies by TAMEC (Ouimet 1979) is a flagrant testimony to the desire of the *Subject* (corporations), to tip the scales of legislation in their favor in order to exclude 'unregulated competitive companies' and the telephone companies from the active role of *Subject* or *Sender*. The role suggested for the public is still that of *Receiver,* receiver of information on a cable that has two-way capacity. Here we might quote McLuhan's remarks on such a role:

> Freedom to Listen: in a world where effective expression (. . .) is reserved only for a tiny minority is freedom to put up or shut up. (1967, 21)

21 It costs $1000 Cdn to be a voting member of VISPAC. INFORMAT is a cartel of Torstar and Southam, the two press monopolies in Canada.

22 Observation of Apple users was facilitated by an invitation to join the Apple Club at the telematics laboratory, Department of Communication, Université du Québec à Montréal, 1980–81.

Chapter 2 New communications technology as discourse: techne

1 Toshiba Oka, speaking about *Data Network Developments and Policies in Japan* (OECD 1980, 69ff), lists several of the positive uses of contributions to the quality of life that new communications technology is said to have provided in Japan: A) enterprise systems; B) systems related to social welfare and national life (for which there is a low demand); C) personal usage systems including automatic seat

reservation, nationwide banking data (MARS), automated meteorological data; acquisition systems and data communication systems for market information; D) scientific and technological retrieval systems; E) resource sharing networks among universities; F) shared hospital information systems (SHIS); G) a Health Care network system.

2 It remains to be seen how much protection Canada's recently passed Freedom of Information Bill (C–52) will provide. Citizens' rights lawyers have already expressed reservations about its effectiveness. For more information on European and Scandinavian protection in this area, see Vandeberghe/OECD, No. 1 1979.

3 This individualist ethos cuts both ways, however. Conservative members of parliament have long argued against precisely such hook-ups because they infringe upon the individual's ownership of himself. This contrary position, far from weakening our argument, merely reinforces it by demonstrating the centrality of individualism as it operates a class division in the social environment of new communications technology.

4 Mark Graf and Roger Johnson, 'Computer Angst,' in G.E. Lasker (ed.) *Applied Systems and Cybernetics,* vol. 5, Systems Approaches in Computer Sciences and Mathematics, Proceedings of the International Congress on Applied Systems Research and Cybernetics, December 1980, 2467–2471.

5 For more discussion of Peirce's triadic semiotics based on an epistemology of the interactive constitution of meaning and knowledge throughout an ever-expanding sign-field see my Ph.D. dissertation: Marike Finlay, 'The Potential of Irony: From a semiotics of irony towards an epistemology of communication praxis.' Diss. Université de Montréal 1981, Part III.

> there is only one mental law, namely that ideas have a tendency to continually extend themselves and to affect certain others which are situated in relation to them in a particular relation of affectability. (Peirce, *Collected Papers,* 6.04)

6 The penury of qualitatively pertinent and attractive protocol has since been cited as a primary contribution to Telidon's lack of commercial success.

7 One apologist who does not argue that centralization will occur is James Taylor (1981). Cf Finlay-Pelinski (1982b) where we reply to some of the idealizations of technology made by Professor Taylor.

8 The American firm Times/Mirror has recently purchased Telidon as its standard videotex system.

9 We may quote a few more examples of the overtly expressed desire for control in business magazines' coverage of business communications:

> The parts of an electronic device are essentially tools of control, and they are unique in that they operate at speeds in millionths of seconds, and when linked together organize kinds of matter that had previously defied control. And it is the capacity of electronic

> gear to organize vast quantities of data and present it for human inspection almost instantaneously that is the basis of the industry's greatness. ('The Electronic Business,' *Fortune,* April 1977, 138)
> But EDP (electronic data processing) is designed to do much more for American business than increase the productivity of clerks. What modern business needs even more than atomated record-keeping tools are techniques to establish control (. . .) More and more companies are looking for that elusive thing called control. (Bello, 'The War of Computers,' in *Fortune,* October 1959, 130)

10 However, to be fair, certain projects such as the Ida Grassroots projects for supplying pertinent information to farmers demonstrate a discursive irregularity which respects regionality despite this background of centralization, something for which Canada has a very positive reputation around the world, e.g., the 'Challenge for Change' program.

11 The debate over whether the 'privatique' or the 'télématique' systems will be accepted still rages. Several specialists in organizational communication argue that 'privatique' will be adopted and that it will defuse centralization as well as encourage creative communications by individuals (Taylor 1981). As in this chapter, we argue elsewhere (Finlay-Pelinski 1982b) that it will take more than horizontal networking to ensure such qualities in an information system.

Chapter 3 Powermatics: focusing in on power and social control

1 For a very complete discussion of the meaning of the term 'critique' as it comes down to us from critical philosophy, see Trent Schroyer's *Critique of Domination* (1973).

2 For an excellent summary of the association of technology and science for control since the Enlightenment, see William Leiss' *The Domination of Nature* (N.Y.: Braziller, 1972).

3 For an excellent criticism of both leftist and rightist theories of technology based on simple or symptomatic causality, see Jennifer Daryl Slack's *Communication Technologies and Society: Conceptions of Causality and the Politics of Technological Intervention* (Norwood, N.J.: Ablex, 1984).

4 For a more adequate discussion of the implications and interpretation of quantum mechanics for a theory and epistemology of communication, see Marike Finlay's 'The Potential of Irony: from a semiotics of irony toward an epistemology of communication praxis,' Dissertation (Université de Montréal, 1981).

5
> Throughout the period of the Second and Third International, there was a powerful tendency toward what Russel Jacoby has termed 'automatic Marxism,' a determinist view of history as proceeding unerringly toward its final goal, governed inexorably by the 'laws' of dialectics. (Leiss 1974, 337)

By referring to automatic marxism, I wish to insist, of course, that there are non-reductionist readings of Marx, such as those of Bakhtin and Habermas, to whom I refer at length in the conclusion.

6 Courts ruled against attempts to reveal lists of subscribers to a pornographic film distributed by QUBE. See Robert Block, *Canadian Lawyer,* November 1982, 10.

7
> 1) There is a progressive mathematization of experience and knowledge, a mathematization which, starting from the natural sciences and their extraordinary successes, extends to the other sciences and to the 'conduct of life,' itself.
> 2) There is the insistence on the necessity of rational experiments and rational proofs in the organization of science as well as in the conduct of life.
> 3) There is the result of this organization which is decisive, namely, the genesis and solidification of a universal technically trained organization of officials that becomes the absolutely *inescapable* condition of our entire existence. (Weber 1920, 1ff, in Marcuse 1968, 204)

8 Richard Gilpin, while trying to absolve the West of any guilt for the underdeveloped world, claims that Western nations became so rich not by exploiting the resources of Third World countries but because of an increase in Europe's capacity for economic transactions, i.e., because of a communicational advantage. Gilpin's theory, which he derives from D.C. North, R.P. Thomas, and Harold Innis, is of interest not because of its position on development problems but because it shows the great advantage that yet further communications transactional capacity in the form of new communications technology will bestow on developed countries at the expense of underdeveloped countries. It is, says Gilpin, important to entertain the possibility that economic exploitation of the Third World depends first and foremost on communicational exploitation.

> While it would be foolish to deny the fact or the importance of colonial exploitation, more efficient transactions have been of much greater and more lasting importance in the political supremacy of the West. (North and Thomas 1973, *The Rise of Western Europe,* in Gilpin 1980, 234)

Gilpin goes on to state that this historical explanation is still pertinent to an understanding of the great advantage that the West continues to enjoy over the Third World, an advantage rendered even greater by the greater speed and lower costs of new communications technology:

> The jet airplane, satellite communications and computer have created the necessary technological means for the integration of the world economy by Western corporations, but economic integration could not have proceeded as far as it has without the underlying major decrease in the costs of conducting global business enterprises. (1980, 235)

9 We will discuss in more detail the conditions under which this new

technology could possibly facilitate non-hegemonic social order (Conclusion). We may already warn against viewing such an attempt as restating a relativist position toward technology since a non-hegemonic technology would be a fundamentally different form of technology than what we have so far described. It might be so divorced from instrumentalist procedures that it might no longer be pertinent to label such an alternative 'technology.'

10 One partial explanation for the survival of capitalist social relations, despite the most dramatic advances in production forces, has to do with the nature of modern engineering, the source of those technological advances. (Noble 1977, xxiv)

11 In modern management, these engineers who had been trained in science and weaned upon large-scale corporate enterprise fused the imperatives of corporate capitalism and scientific technology into a formal system. Representing a shift in engineering focus from the natural to the social realm, from production forces to social relations, modern management constituted a deliberate attempt to ease the tension between the two, making both fit within the confines of corporate society. Moreover, as these engineers became managers in industry, private capital itself began to assume the appearance of modern technology, the management experts lending to the power of capital the sanction of objective science. Not alone the actual machinery of production but the entire bureaucratic operation of corporate enterprise took on the guise, of an efficient, well-oiled machine – against which individual opposition could not but appear 'irrational.' (Noble 1977, xxvi)

12 George Grant explains Weber's notion of a value-free science within its historical content. Weber was too aware of the socialization of knowledge to believe that a completely objective form of rationality was possible. However, Weber did want the university to function independently of institutional, technique-laden values which the administrative power of the time wished to force upon knowledge. Grant suggests that by value-free knowledge, Weber meant that thought should be other than technique, in the sense of instrumentalist, bureaucratic reason. Grant interprets Weber as holding out for the separation of instrumentalist rationality – techne – and forms of reason – episteme. George Grant, 'Ideology in Modern Empires,' in John Flint and G. Williams (eds), *Perspectives of Empire* (Longman, New York, 1973), 191–197.

13 A slip of the tongue by Canada's Department of Communications betrays such a surrender to instrumentalist reason, whereby technology is a subject acting upon a passive object-society. It advocates research and policy 'to foster a positive social adaptation to altered technological and economic realities.' (Foote/DOC 1981, 'Research Directives Paper,' 4) Thus, it is up to society to adopt and to adapt to technology which unilaterally acts upon society in the name of efficiency and rationality. The remark is yet another sign of

the entrenchment of instrumentalist rationality in the bureaucracy that underpins new communications technology in society.

14 Elsewhere, Weber elucidates these types of legitimation:

> 1) Rational grounds – resting on a belief in the 'legality' of patterns of normative rules and the right of those elevated to authority under such rules to issue commands (legal authority).
> 2) Traditional grounds – resting on an established belief in the sanctity of immemorial traditions and the legitimacy of the status of those exercising authority under them (traditional authority).
> 3) Charismatic grounds – resting on devotion to the specific and exceptional sanctity, heroism or exemplary character of an individual person or of the normative patterns of order revealed or ordained by him (charismatic authority). (Weber 1947–64, 328)

These types of legitimation, then, lead to various types of representation:

> 1) Appropriated representation
> 2) Representation on the basis of socially independent grouping
> 3) 'Instructed' representation – elected representatives or representatives chosen by rotation or lot or any other manner of exercise of powers of representation which are strictly limited by an imperative mandate and a right of recall, the exercise of which is subject to the consent of those represented.
> 4) Free representation – not bound by instruction/in a position to make one's own decisions. (Weber 1947–64, 417)

In view of this contradiction between rationality's claim to democracy and its obstinate attachment to social control, Weber rejected the liberalist dicta of democracy as based on some intrinsic claim to equal rights. He concentrated instead on types of representation in social organization as well as on means or relations of control:

> he did not believe in democracy as an intrinsically valuable body of ideas: 'natural law,' 'the equality of men,' 'their intrinsic claims to equal rights.' He saw democratic institutions and ideas pragmatically: not in terms of their 'inner worth' but in terms of their consequences in the selection of efficient political leaders (. . .) in modern society such leaders must be able to build up and control a large, well disciplined machine, in the American sense. The choice was between a leaderless democracy or a democracy run by leaders of large party bureaucracies. (C. Wright Mills (ed.), in Weber 1946–78, 42)

Given the above remarks, it would be foolish to try to make of Weber a mouthpiece for the democratic ideal. One should note, however, that Weber shows the incompatibility of 'rationality' with democracy. This distinction will prove indispensable when the question of democracy is addressed in the conclusion.

15 It would have been quite feasible to isolate yet another corpus of discourses on new communications technology, discourses which could be referred to as 'neo-liberalist' and whose content would have been far more critical of the potential social effects of new

comminications technology. The Berkman (1982) text is a typical example. Still the discursive procedures of neo-liberalism would coincide with many of those of liberalist rationalism by merely providing a negative transformation of them. For example, in the double binds brought to light in chapter one, the negative sides would be the neo-liberalist argument that 'it is dangerous to adopt,' 'damned if you do,' 'free-flow is imperialist,' 'the workers will be exploited,' etc. All are skeptical statements made possible only by the discursive transformations of liberalist confidence in individualism and equilibrium. These transformations move towards neo-liberalist cynicism about the State's and other institutions' commitment to the rights of individuals, and all of them maintain the same themes and forms of reasoning. Neo-liberalism is merely the negatively determined face of liberalist procedures of discourse.

Chapter 4 Alternative procedures of discourse: epistemologies of technology

1 Hugh Faulkner, a former Canadian Minister of State for Science and Technology, echoes the call for an alternative conceptualization: 'We are all "developing" nations in the new information society and solutions to these new problems cannot always be found in tradition or precedent.' Opening address for the International Federation of Information Processing Congress, August 1977, Toronto, reprinted in *Computer World,* 15 August 1977.

2 Here, a word of caution is in order. One of the greatest risks of a discursive critique which seeks to transcend the episteme it criticizes is that of regressing in its own discursive practice to the very procedures criticized. Alternatives come about by virtue of gradual transformation as opposed to leaps and bounds. Therefore, it is not surprising to find the beginnings of an alternative to be constantly slipping back to the old episteme. Nor is this book innocent of such lapses. Often, we fall prey to contradiction and to a practice of classical procedures which we sought to supplant.

3 In *Gutenberg Two,* we find a typical example of policy recommendations designed only to meet the criteria of regulation of abuse and ignoring the inherent propensity toward social control by the technology itself:

1) Creation of a corps of bonded professional computer operators bound by strict oath of secrecy.

2) A total ban on discursive of personal information from the files except to the individual named.

3) Mandatory security measures, both physical and electronic for the protection of data.

4) Severe criminal penalties, including mandatory jail sentences, for breaches of privacy on the part of the 'keepers of the files' and

for those, government and company officials, etc., who incite, condone or benefit from such breaches.
5) Civil redress for those who may have been damaged by improper disclosure of the content of their files.
6) Mandatory destruction of certain types of information might also be required so that certain types of master files cannot be created. (Godfrey & Parkhill 1980, 89)

4 Rowat, a political scientist studying freedom of information in Canada, writes in 'Le secret administratif' (1973, 48) that 'Information Canada,' the government agency which oversees information dissemination for the government, exercises certain forms of domination by controlling information:

> Nous prenons conscience qu'il s'agit d'une tradition entretenue par les gouvernements et les fonctionnaires à leur seul profit. La Connaissance confère le pouvoir et ils répugnent à abandonner leur mainmise sur l'information. Les pouvoirs publics peuvent en effet manipuler les informations, en hâter ou retarder la communication pour favoriser le parti en place; ils peuvent même lancer des campagnes publicitaires pour masquer les erreurs commises.

5 Before examining this alternative to the liberalist doctrine of individualism we can see in the American objection to these discursive transformations that the plane of conflict is reduced to whether or not the discursive traits of liberalism and 'laissez-faire' should be abandoned.

> The UNESCO proposal has caused a certain disappointment among the largest television companies in the US as well as Europe. In fact, at an assembly of regional television associates which took place in Rome last March, North American and European television enterprises declared the proposal unacceptable. In their opinion, what represents the greatest danger of such a declaration is that one day or other it could serve as the basis for regulating, under the pretext of international convention, the conduct of nations and companies in the question of space transmission. In order that the point of view of the free traffic information should prevail, we must ask ourselves if it is possible from a cultural and educational point of view to discuss the concept of free-flow of information without giving the impression of imposing our own political philosophy on other countries (. . .) The proposal presented by UNESCO reflects the opinion of a large number of underdeveloped countries that a form of control and regulation is essential for the development of their nation and who wish to arrive at a balance between control and freedom. But on the other hand, commercial advertising is an integral creation of our system of free enterprise which other countries must recognize. (US Government Memorandum to Samuel de Palma from John E. Upston, US National Commission for UNESCO, Advisory Recommendations – typewritten copies, 1972, in Mattelart 1978, 95)

6 a) radio frequencies are public property (. . .)
b) the Canadian broadcasting system should be effectively owned and controlled by Canadians so as to safeguard, enrich and strengthen the cultural, political, social, and economic fabric of Canada (. . .)
d) the programming provided by the Canadian broadcasting system should be varied and comprehensive and should provide reasonable, balanced opportunity for the expression of differing views on matters of public concern (. . .)
e) all Canadians are entitled to broadcasting service in English and in French (as public funds become available) (. . .)
g) the national broadcasting service should:
i) be a balanced service of information, enlightenment and entertainment for people of different ages, interests, and tastes covering the whole range of programming in fair proportion (. . .)
(Broadcasting Act 1968, Information Canada)
Of like spirit are the UN 7 Bill to create the National Film Board of Canada and the Canada Council Act 1957 (Ottawa 1970).

7 Peirce may often be quoted as affirming that truth is a consensual establishment and not an absolute referential one:
'thought, belief and its fixation are established in the community and vice versa' (*Collected Papers,* 5.378).

> The real, then, is that which, sooner or later, information and reasoning would fianlly result in, and which is therefore independent of the vagaries of me and you. Thus, the very origin of the conception essentially involves the notion of a community without definite limits and capable of a definite increase of knowledge. (1972 (ed. Moore), 115)

'so that reality depends on the ultimate decision of the community.' (*Ibid.,* 115)
See also, *Collected Papers* (8.012) and (7.319).

8 Of course, it would be quite tenable, though not within our present competence, to argue that non-Western societies have not always inferred a principle of emancipation as the ground of their social theory and legitimation. Nevertheless, I think it fair to agree with Habermas here, at least as regards the West. Scholars of other traditions inform us that the emancipatory principle is quite remote from the centre of their political and religious teachings.

9 Kevin Wilson has shown that privacy legislation is virtually useless to citizens because of the enormous a-symmetry between corporate and natural individuals in terms of capacity to bring a case to court and to mount a legal proceeding. (Wilson 1984)

10 Karl-Otto Appel, 'From Kant to Peirce: the semiotical Transformation of Transcendental Logic,' *Proceedings from the Third International Kant Congress* (Lewis, White & Beck, eds, 1970).

11 I owe this bit of facetiousness to Kevin Wilson.

12 In Canada, the DOC and the Quebec Ministry of Communication are

to be praised for having sponsored research in the area of services to the handicapped. At the Université du Québec à Montréal, Michel Cartier has organized a research project of Telidon and the handicapped as part of the Agora group. Projects were also planned for study of possible uses that could be made by ethnic communities until budget cuts made them unfeasible.

13 le discours de notre vie pratique est plein de mots appartenants à d'autres, il en est dans lesquels nous fondons notre voix, oubliant à qui ils sont, il en est dont nous nous servons pour renforcer nos mots à nous, considérant qu'ils font pour nous autorité, il en est enfin que nous peuplons de nos propres aspirations qui leur sont étrangers ou hostiles. (Bakhtin, *Problèmes de la poétique de Dostoievski,* trans. Guy Verret, Lausanne, Editions l'Age d'Homme, 1929–1970, 211ff)

For Bakhtin, quasi-direct discourse exhibits far more social interactivity and receptivity to the discourses of others than does free, indirect discourse. Quasi-direct discourse is the active relation of one discourse to another within the mobile constructional activity of language itself. Language is no longer subsumed under authorial jurisdiction. Quasi-direct discourse is a sort of 'steeping' of the reporting context within the reported context of speaking. Quasi-direct discourse is 'polyphonic.'

> La multiplicité de voix et de consiences indépendantes et non confondues, l'authentique polyphonie de voix pleinement valables est effectivement la particularité profonde des romans de Dostoievski (*Ibid,* 10–12).

Bakhtin suggests that free-direct discourse belongs to an episteme of the 'classical inviolabilities of the boundaries of authorial dominance.' He describes this discourse as a linear style, as an ideological assurance and as dogmatism where there are clear-cut external contours for the speech of the other. He associates quasi-direct discourse with the authoritarianism and dogmatism of the middle ages and the seventeenth and eighteenth centuries. Quasi-direct discourse, on the other hand, is a plethora and flux of types and levels of discourse. All speech and values flow freely into one another in an open field or circuit. Bakhtin relates this discursive type to the 'realistic and critical individualism in its pictorial styles and tendency to permeate reported speech,' and to 'relativistic individualism beginning with the development of this interaction and its decomposition of the absoluteness of the authorial context.' (Bakhtin/Volosinov, *Le marxisme et la philosophie du langage,* trns. M. Yaguello, Paris, Minuit, 1977, 127ff)

14 The illocutionary force of a speech act, which brings about an interpersonal relationship between consensually interacting participants, arises from the binding force of acknowledged norms of action: to the extent that a speech act is part of consensual interaction it actualizes an already established value-pattern.

(Jürgen Harbermas, 'Some distinction in universal pragmatics,' *Theory and Society,* 3, No. 2, Summer 1976, 158)

See also Jürgen Habermas, 'Preparatory remarks to a theory of communication competence,' in Habermas and Luhman, *Theorie der Gestellschaft oder Sozialtechnologie* (Frankfurt am Main, Suhrkamp, 1971), 101–41.

15 To a very limited degree this was the aim of the latest Apple desk computer, Lisa, which boasts of allowing businessmen to interact with it without having to learn a complex language and by using only slight variations on their own lingo.

16 Whether or not they have an explicitly linguistic form, communicative actions are related to a context of action, norms and values (. . .) Without the normative background of routines, roles, forms of life, in short, conventions – the individual actions would remain interdeterminate. All communicative actions satisfy or violate normative expectations or conventions. (Habermas, *Communication and the Evolution of Society,* trans. Thomas McCarthy, Boston, Beacon Press, 1979, 35)

validity claims are at least partially and implicitly related to discursive vindication. (*Ibid,* 23)

they (validity claims) must be in accordance with existing norms and must merge with value patterns. (*Ibid,* 158)

Universal pragmatics aims at a reconstitution of the rule systems over which adult speakers must have mastery in order to use sentences in utterances at all, regardless of the specific natural language to which the sentence belongs or the context in which it happens to be embedded. (Habermas, 'Some distinctions in a universal pragmatics,' *op. cit.,* 156).

17 The ideal speech situation derives from the attempt to construct the universal ethics of speech where it is recognized that communicational interaction holds the key to ethical and political practice.

The ideal speech situation entails:

symmetrical distribution of opportunities for the selection and execution of speech acts that relate to propositions as propositions, to the relation of the speaker to his utterances, and to the observance of the rules. (K-O Appel, in Habermas 1979, 205)

Habermas is quite specific about the procedures that are included in the ideal speech situation. One is obliged to : 1) represent something; 2) express one's intentions truthfully; and 3) establish legitimate interpersonal values:

1) To choose the propositional sentence in such a way that either the truth conditions of the proposition stated or the existential proposition of the propositional content mentioned are supposedly fulfilled. ('Wahr')

2) The speaker must want to express his intentions truthfully so that

the hearer can believe the utterance of the speaker. ('Wahrhaftig') 3) To perform the speech act in such a way that it conforms to recognized norms or to accepted self-images so that the hearer can be in accord with the speaker in shared value orientation. (Verstaendlich') (Habermas 1979, 29)

The final rule of the ideal speech situation is the most important. It is the right to discuss the rules of communication at a meta-communicational level, at the level of 'discourse' as opposed to the level of communication (discourse here is unrelated to Foucault's use of the term):

> 'discourse' as that form of communication which is free from the constraints of the very process of action and experiences, and which allows for an exchange of arguments on hypothetical validity claims (whereby truth and legitimacy may count as discursively redeemable validity claims, while veracity can only be subject to a test of consistency over a period of continued interactions).
>
> It is only with the transition to 'discourse' that the validity claims of an assertion or the claim for legitimacy of a command viz. the underlying norm, can explicitly be questioned and topicalized in speech itself. (Habermas, 'Some distinctions . . .,' *op. cit,* 163)

18 We are of course arguing here for a communication policy that transgresses the monopoly of the State or of the corporate and bureaucratic oligopoly. For this to occur practices in a much broader political sphere must also change but this discussion, however crucial, is too vast for our current task.

19 Robert A. Alfred, 'Paradigms of relations between State and society,' in Lindberd (ed.), *Stress and Contradiction in Modern Capitalism: public policy and the theory of the State*, Toronto, Lexington: 1975.

20 We have since participated in a project led by Salter, Leiss and Levy on how science is used to legitimate decision-making in areas ranging from dangerous chemicals to communications, concentrating especially on the role of the public in establishing what is good science and what is rational decision.

21 It is interesting that Salter's own work in this area has tended more and more toward a discursive analysis (Conference, Graduate Program in Communications, McGill, Fall 1983).

22 This raises the whole question of legal control of the right to communicate. It has been suggested that the right to communicate would be less susceptible to legal control than the right to free flow. This is primarily because free flow was often regulated by commodities-oriented laws, e.g., the law for the protection of intellectual property, etc. There remain many questions concerning the legalization of the right to communicate: is it confusing moral and normative laws? Should corporate law or space law or cultural law faculties deal with it? Can an international law be formulated before individual nations formulate such a law? See Green, Mikulowski

Pomorski, and LeDuc, in Harms et al. 1977, 298, 89ff, 160 respectively, and Bishop and Blumenwitz, in Fisher 1982, 37 and 13 respectively.

Our own personal fear over the legality is that we will revert to rigid communicational constraints imposed from above as opposed to following rules of communication decided upon consensually at the level of 'discourse' within the public sphere. Perhaps only the exigency of 'discourse' in the sense of public discussion and decisions upon rules of communication to be followed within a community should be fixed into statutes of (international) law.

Conclusion: Discursive critique: socio-epistemological foundations for an alternative approach to communication and technology

1 For an excellent discussion of this shift, see Toulmin and Janik (1973) and Shorske (1981). Indeed, the former make a very good case for considering even the *Tractatus* as evidence of a disillusionment with an analytico-referential paradigm of knowledge.

2 V. Mosco (1979) suggests much the same notion when he states that in the case of the American Federal Communications Commission the absence of action guided by an integrating principle of public interest has played into the hands of much better organized and integrated corporate interests and hence of a principle of 'laissez-faire.'

3 An integrative approach has been found lacking and has been called for by several government research and regulatory bodies ranging from the CRAB report (1979, 9) through the CRTC report (1980, 79), studies by the Institute for Research in Public Policy (Russel 1978, 40), to a recent symposium on new technology (Smith in *Daedalus* 1980, 9)

4 The continental semiotic model, for example, comes up short in communication studies because it could not deal with discourse both as an act and as an interaction. It could only treat communication as an object that filled the categories of syntax and semiotics. The effects model can only deal with discourse as an impact, ignoring that there is a symbolic and culturally contextualized production of signs involved in all communication as well as an effect. (Carey in Blumer and Katz 1974)

5 Donald McClean, 'Taking the Revolution Seriously: Intellectual Obstacles to Information Policy Making,' proceedings of the annual meeting of the Canadian Communications Association, 26 May 1981, Halifax, m.s.

6 Krishan Kumar, *Prophesy and Progress* (New York: Penguin, 1978, 48):

> While it was left to sociology to turn that image into a 'model,' it is clear that both in the popular consciousness and in the

sociologist's model these were ideas and feelings that had roots in artistic and literary representation, rather than in scientific investigation.

7 I develop the problem of reference at the level of meta-discourse (critical discourse) at length in Marike Finlay-Pelinski, '"Criticism" in the Margins: Barbara Herrnstein Smith's "On the Margins of Discourse",' *Canadian Review of Comparative Literature* (March 1982), 56–75.

8 I deal at length with the tendency to abstract technologies from social context and to abstract an ideal 'technology' from specific practices of technologies in 'William Seiss on Technology,' and 'Technology as Practice,' in *Canadian Journal of Political and Social Theory*, vol. X, N. 1–2 (1986), pp. 174–95; and vol. XI, N. 1–2 (1987), pp. 198–214.

9 Some policy studies concur with this call for focus suggesting that it is necessary for a reinterpretation and modification of development in the social reception of new communications technology as it now stands. Such a focal point should:

> provide mechanisms whereby the many different policies and policy alternatives could be related to each other and analyzed with respect to their total impact and from which the results of such analyses could be fed back to the responsible development bodies for interpretation and, if necessary, augmentation or modification. (*Green Paper,* DOC 1973, 7)

10 A typical statement of this positivist confidence might be reproduced here from Ellul's work on technology:

> il ne s'agit pas de porter dans ce travail un jugement éthique ou esthétique sur la technique. Evidemment, dans la mesure où l'on ne reste pas purement photographique, et dans le mesure où l'on est homme, une certaine prise de position peut transparaître. Mais elle n'est pas si éminente qu'elle empêche une objectivité plus profonde. (Ellul 1954, 'Préface')

11 Since Peirce, through the Frankfurt School's 'critical theory,' and more recently to Halloran (1974, 3), Weizenbaum (1975, 264–5), and Hamelink (1980, 3), many communications theorists reaffirm that value has an undeniable and inextricable place in communications research. Since the mid-sixties, policy-making bodies such as UNESCO, despite many euphemisms, have been pushed by Third World countries to recognize certain values of regional and cultural divergence from Western culture. For years, Carey has been stating that the interest of 'good' democracy should take precedence over such futile debates on centralization vs decentralization.

More recently, the MacBride Commission has, if somewhat cautiously, followed suit in the reaffirmation of a communication research and policy that would reaffirm certain democratic and regional interests:

> Due attention should be given to the fact that it is the concepts themselves that should be changed. One of the main aims of communications policies must be to correct and to adjust existing structures in order to meet the needs for broader flows and for the democratization of communication. (MacBride 1980, 20)

12 Ithiel de Sola Pool, 'The Rise of Communications Policy Research,' *Journal of Communications,* 33.

BIBLIOGRAPHY

1 Mass media coverage of the electronic revolution (1919–1983)

ARTANDI, Susan (1979). 'Computers: Cause of the Curse For Problems.' *New York Times*, 5 August, Sec. 11, p.24.

BAKCSY, de Alex (1979), in 'The Information Utility An Alternative to Newspapers?' *Production News*, November, p.15.

BARRET, Tom (1980). 'Home Computers: Buyer Resistance May Be High; A Dream Whose Time Hasn't Come?' *Vancouver Sun*, 9 July, Sec. 1, p. 5.

BATES, William (1978). 'Home Computers So Near and Yet . . ' *New York Times*, 26 February, Sec. 3, p. 3.

BELLO, Frances (1959). 'The War of the Computers.' *Fortune*, October, pp. 128–11

'Now Factory Lighting Influences Income Life.' (advertisement) (1919). *Scientific American*, 12 July, p. 25.

BERGER, Ivan (1976). 'Home Sweet Computerized Home.' *Popular Mechanics*, September, pp. 112–113.

BERKELEY, Edmond C. (1949). *Giant Brains of Machines That Think*. New York: John Wiley and Sons

BERKMAN, Dave (1981). 'The Video Revolution.' reference unavailable.

BLENSTEIN-BLANCHET, M. (1974). 'Supprimer les sondages,' *Le Monde*, Sec. 5, p. 23.

BLOCK, Robert (1982) 'Here's Looking at You: the pure seductive convenience of two-way T.V. will persuade us to tell the tube our most intimate secrets.' *Canadian Lawyer*. November, pp. 8–39.

BLUMENTHAL, Ralph (1978). 'Electronic Fraud Accompanies Tellerless Banking.' *New York Times*. 26, March, Sec. 1, p. 1.

BRANSCOMB, Lewis (1979). 'Information: The Ultimate Frontier.' *Science Magazine*, 12, January.

BROW (1980). 'The Manners of a Machine.' *Saturday Night*. September. pp. unavailable.

BRUGGE, Peter (1982). 'The Dark Side of Computerization.' *Der Spiegel* in *World Press Review*, November, pp. 28–30.

BURCK, Gilbert (1964). 'Management Will Never Be the Same Again,' *Fortune*, August, pp. 125–126.

BURCK, Gilbert (1964). 'On Line in Real Time.' *Fortune*, April, p. 144.
BUSHELL, Don D. (1966). 'For Each Student a Teacher.' *Saturday Review*, 23 July, pp. 31.
BURTON, Harold M. (1975). 'That Incredible Economy South of the Border.' *Fortune*, September.
BUSIGNIES, Henri (1972). 'Communication Channels.' *Scientific American*, September, pp. 99–113.
BYLINSKI, Gene (1975). 'Here Comes the Second Computer Revolution.' *Forum*, 6 November.
'Can Computing Machines be Used to Formulate Major Decisions of Policy?' (1953). *Fortune*, October, p. 129.
CAREY, Frank T. (1979). 'Computer Terminals and their Utility.' Editorial. *New York Times*, 6 June, p. 26.
CHASE, Stuart (1954). 'Machines That Think.' *Reader's Digest*, January, pp. 143–146.
CLINES, Francis X. (1979). 'About New York; A Layman's Trip into the Mega-Mega Land of Computers.' *New York Times*, 7 June, Sec. 2, pp. 1 and 8.
'The Communications Revolution' (1980). C.B.C. Radio – 'Ideas' Series. October.
'The Computer Age: Automating the Boss' Office' (1965). *Business Week*, April 7, p. 59.
'Computer Fraud.' (1982). *Time*. 1, February, pp. unavailable.
'Computers Getting Off Track?' (1957). *Fortune*, October, p. 118.
'A Computer in Every Home.' (1979). *Newsweek*, 2 April, pp. 79–80.
COOK, J.S. (1973). 'Communications by Optical Fibre.' *Scientific American*, 229, no. 5, November, p. 28.
COUSINS, Norman (1966). 'The Computer and the Poet.' *Saturday Review*, 23 July, p. 42.
DAGENAIS, Angèle (1980). 'L'information à domicile.' *Le Devoir*, 9 November, pp. 6–9.
'Dark Screen Future.' (1951). *Time*. July, pp. unavailable.
'The Datamation 100: The Top 100 Companies in the DP Industry.' *Datamation*, 26, No. 7, July 1980, 87–182.
'Data Transmission and the Real-Time Systems.' (1965) *Duns Review*, September, p. 161.
'Diagnosis by Computer.' (1959). *Scientific American*, September, p. 113.
DIEBOLD, John (1966) 'The New World Computering: Tomorrow's Computer Will Revolutionize Business, Education, Communications, Science in Ways Only Dimly Foreseen.' *Saturday Review*, 23 July, pp. 17–18.
DITLEA, Steven (1979). 'When a Computer Joins the Family.' *New York Times*, 30 August, Sec. 3, pp. 2 & 5.
DURIEUX, Claude (1977). 'La Revolution de la Vidéo Communication.' *Le Monde*, March, pp. 20–21.
'Electric Calculating Machine.' *Science Newsletter*, No. 40, 14 September.
'Electricity is used to Solve Crime Mystery.' (1927). *Popular Mechanics*, April 27.

'Electronic Calculator: Uses 18,000 Tubes to Solve Complex Problems.' (1946). *Scientific American*, June, p. 248.
'Electronic Intricacy in Compact Design.' (1956). *Life*, 21 May, pp. 108–109.
Electronics, 6. III (1975). 17, April.
EMERSON, Thomas (1972). 'Communication and Freedom of Expansion.' *Scientific American*, September. pp. 163–172.
ENGEL, Leonard (1953). 'Electronic Calculators: Brainless but Bright.' *Harper's Review*, April pp. 84–90.
'Le'évènement: video-casettes et video-disques.' (1974). *Le Monde*, Sec. I, p.8.
'Expanded Ocean Surveillance Effort Set.' (1978). *Aviation Weekly and Space Technology*, 10 July.
FANO, R.M., and F.J. CORBATO (1966). 'Time-Sharing on Computers.' *Scientific American*, September, p. 129.
'Feats of Science help movies give vivid picture of a world ruled by machines.' (1927). *Popular Mechanics*. March, pp. 424–425.
'Le FBI enquête sur l'espionage industriel aux USA.' *Le Devoir* 24 September.
FEDER, Barnalsj J. (1981). 'AT&T Sets Video Standards.' *The New York Times*, pp. 27 and 30.
'Electric Calculating Machine Devised for Complex Problems.' (1940). *Science Newsletter*, 14 September. pp. 163–164.
Fortune's Discovery of the 500 Largest Industrial Corporations, May 1975.
FRIEDBAR, Richard (1958). 'Where is Science Taking Us?' *Saturday Review*, 3 May, p. 47.
FULFORD, Robert (1980). 'How Telidon May Change Your Life.' *Saturday Night Review*, September, pp. unavailable.
GOLDEN, Marcia (1981). 'High Tech at Home: Making Friends with a Talking Toaster.' *Ms.*, October, pp. 12 ff.
GOLDMARK, Peter C. (1972). 'Communication and Community.' *Scientific American*, September, pp. 143–150.
GOLDSTEIN, Tom (1979). 'Business and the Law; On Protection and Privacy.' *New York Times*, 8 June, Sec. 4, p. 4.
GOTTFRIED, John (1979). 'Home Entertainment; A Man's Best Friend is His Computer.' *Saturday Review*, 13 October, pp. 63–66.
GREENBERG, Martin, Ed. (1953). *The Robot and the Man*. New York: The Gnome Press.
GEENBERGER, Martin Ed. (1964). 'The Computers of Tomorrow.' *Atlantic*, May, pp. 63–67.
'Gronk. Flash. Zap. Video Games are Blitzing the World.' (1982). *Time*, 18 January.
HENLON, Michael (1979). 'Sex, money, politics – Computers know all.' *Toronto Star*, 17 March, p. 4.
HAYNES, Dave (1979). 'Computer Ready to Run Your Life.' *Winnipeg Free Press*, 25 August, Leisure sec., p. 1.
HELLER, Yione (1979). 'Well Done Garielle Computer Tells Student.' *Toronto Star*, 13 March, p. 11.

HELLMAN, Martin E. (1979). 'The Mathematics of Public-Key Cryptography.' *Scientific American*, 241, No. 2, August.
'High Tech at Home.' (1982) *Ms.*, 12 October.
'High Technology – Wave of the Future or a Market Flash in the Pan?' (1980). *Business Week*, 10 November, pp. 80–96.
'Hommage to Orwell.' (1956). *Nation*, 6 October, p. 278.
'How Computers Are Changing Your Life.' (1969). *U.S. News and World Report*. 10 Nov, pp. 96–98.
HUMPHREYS, David (1980). 'Picking up on the Micro Revolution.' *Toronto Life*, November, pp. 66ff.
IMOSE, Heroshi (1972). 'Communication Networks.' *Scientific American*, September, pp. 117–128.
'Information' (1955). *Business Week*, 30 July, pp. unavailable.
'Intelligence Amplifier.' (1956). *Time*. May, pp. unavailable.
'L'intelligence artificelle, nouvel enjeu du défi japonais.' (1981). *Le Monde*, December.
Inuktitut (1980). Special Issue on Communication. Ottawa: Ministry of Indian and Northern Affairs Government of Canada, November.
Jane's Book of Weaponry (1980s).
JASTROW, Robert (1978). 'Toward an Intelligence Beyond Man's.' *Time*, 20 February, p. 47.
JOHNSTON, David (1982). 'Customer Feels Hounded By Computer Watchdog.' *The Gazette*. 21 December.
KAMEN, Hope (1979). 'Canada May be Computer McDonald's.' *Winnipeg Free Press*, 29 September, Sec. I.
KEDDY, Barbara (1979). 'Communications in the Home: Where It's Going.' *Toronto Globe and Mail*, 23 July, Sec. 1, p.17.
KEDDY, Barbara (1979). 'Got A Question? Will Give You Everything from Race Results to Recipes.' *Toronto Globe and Mail*, 30 January, Sec. 1, pp. 1 and 2.
KIERAN, Michael (1980). 'Humans Live in Technological Soup.' and 'TV to Come: Catering to the Individual.' *The Globe and Mail*, 8 December, p. 19.
KLASS, P.J. (1974). 'UN will face broadcast satellite use?' *Aviation Week and Space Technology*, 2, IX.
KRON, Joan (1978). 'When the Computers Come Home.' *New York Times*, 12 January, Sec. 3, p. 8.
KRUGER, Alice (1979). 'Women vs. Computers.' *Winnipeg Free Press*, 5 Nov, Sec. I, p. 7.
'Labor of Year is Done in 400 Hours.' (1957). *Life*, 18 February, p. 92.
LANGER, Richard (1977). 'Computers Find a Home (Your's).' *New York Times*, 25 August, Sec. 3, pp. 1 and 4.
LEAR, John (1966). 'Whither Personal Privacy?' *Saturday Review*, 23 July, pp. 36 and 41.
LESSING, Lawrence P. (1954). 'Computers in Business.' *Scientific American*, January, pp. 21–25.
'Light chart of "L" system shows flow of Power.' (1927). *Popular Mechanics*. p. 747.

LIPETZ, Ben-Ami (1966). 'Information storage and retrieval.' *Scientific American*, September, pp. 224–242.
MARSH, Peter (1980). 'Communications for the Information Age. ("System X").' *New Scientist*, 23 October, pp. 235–238.
McCARTHY, John (1966). 'Information.' *Scientific American*, September, p. 65–72.
McELHANY, Victor (1977). 'Technology: From Revolution to Evolution.' *New York Times*, 29 June, Sec. 4, p. 5.
MACY, John, W., Jr. (1966). 'Automated Government.' *Saturday Review*, 23 July, pp. 23–25.
MARTIN, Linda Grant (1975). 'The 500: A Report on Two Decades,' *Fortune*, May.
MATHEW, Walter M. (1980). *Monster or Messiah*? Jackson: University Press of Mississippi.
MATHEW, Clide (1979), 'Stopping Computer Chomping of Privacy.'
'Men and Machines and TNEC.' (1940). *Business Week*, 4 May. *New York Times*, 28 April, Sec. 1, p. 25.
'Microelectronics.' (1977). Special edition of *Scientific American* No. 237 September.
MILLER, Vernon F. (1966). 'The Town Meeting Reborn.' *Saturday Review*, 23 July, p. 345.
MORGAN, Thomas (1961). 'The People-Machine.' *Harper's*, January, pp. 53–57.
MOSKOWITZ, Milton, Michael KATZ, and Robert LEVERING (1980).
'The Computer People.' In *Everbody's Business*: *An Irreverant Guide to Corporate America*. San Francisco: Harper & Row, 429–455.
'Toronto Star Home Computer Program Detailed.' (1979). *Winnipeg Free Press*, 18 February, Sec. I, p. 8.
'Nato Defense Technology Outlined.' (1978). *Aviation Week and Space Technology*. 17 July, p. 62.
'Nehru Finds U.S. like Soviets in Part.' (1953). *New York Times*. June.
NEUMANN, John von (1955). 'Can We Survive Technology?' *Fortune*, Vol. 55, June, p. 152.
'Never Stumped' (1950). *New Yorker*, 4 March, pp. 20–21.
'The New Age.' (9 special issues of *Saturday Review* devoted to the potentiality of the automation revolution and its implications for our society) (1966). *Saturday Review*, 23 July, pp. 15–42.
'New Electrical Control.' (1930). *Literary Digest*. 25 October.
'New eyes for computers: Chips that see.' (1982). *Popular Science*. January.
'New Nationwide T.V. Net. . . .' (1951). *Life*, Vol 31(2).
'Nur Idioten gegen technischen Fortschritt.' (1981). *Die Zeit*. June.
'Office Robots.' (1952). *Fortune*, January, pp. 82–87.
'Oppenheimer calls Computer Just as Vital to Science as Invention of Microscope.' (1953). *New York Times*, April.
OSTRY, Bernard (1979). 'A Revolution is Brewing in Your TV Set.' *Toronto Star*, 24 February, Sec. 3, p. 4.

'Our Gray, 21-Inch Lives.' (1956). *The Nation*. December.
OWEN, Ken (1981). 'Teaching Computers to Learn.' *The Times* in *World Press Review*, May.
'Parade of the Business Machines.' (1935). *Business Week*. October.
PEBBEL, John (1966). 'Book Publisher's Salvation?' *Saturday Review*, 23 July, pp. 32–33.
PIERCE, John R. (1966). 'The Transmission of Computer Data.' *Scientific American*, September, pp. 145–156.
POURY, Volta (1963). 'Computerizings.' *Atlantic*, December, pp. 140–143.
PROLL, John (1954). 'The Thinking of Men and Machines.' *Atlantic*, July, pp. 62–65.
'Queen Bee.' (1935). *Scientific American*. October, p. 204.
Radiotelevisione, informazione democrazea, convegno nazionale del PCI, (1973). Roma 9–31 Marzo. Rome: Editori Ruiniti.
RENSBERGER, Boyce (1976). 'Low-Cost Computers Beginning to Move Into the Home.' *New York Times*, 4 May, Sec. 7, pp. 1 and 23.
'La Rentrée des ordinateurs. (1981). *L'Express*, 25 September, pp. unavailable.
'The Robot Einstein.' (1945). *Newsweek*, 12 November, pp. 94–95.
RIESER, Carl (1962). 'The Short-Order Economy.' *Fortune*, August, p. 91.
ROSE, Sanford (1970). 'Making the turn to a peacetime economy.' *Fortune*, September, pp. unavailable.
ROSE, Stanford (1975). 'This misguided furor about investments from abroad.' *Fortune*, May, pp. unavailable.
ROSS, Irwin (1973). 'Labor's big push for productionism.' *Fortune*, March.
'Sacred Electronics.' (1956). *Time*. December.
SALISBURY, David F. (1980). 'The Third "Industrial Revolution": Robot Factories and Electronic Offices.' *Christian Science Monitor*, 8 October, pp. 1 and 9.
SARNOFF, Eric (1966). 'No Life Untouched: By the End of the Century Computers Will Affect Every Field in Innumerable Ways; Some Specific Predictions.' *Saturday Review*, 23 July, pp. 21–22.
SARNOFF, Robert W. (1974). 'The electronic revolution: dawn of a new era.' *Economic Impact*. Washington: No. 7.
SCHUYTEN, Peter (1978). 'Technology, the Computer Entering Home.' *New York Times*, 6 December, Sec. 4, p. 4.
SCHUYTEN, Peter J. (1979). 'Don Regan of Merrill: Bullish on His Computer.' *New York Times*, 25 May, Sec. 4, pp. 1 and 6.
SCHUYTEN, Peter J. (1979). 'Home Computer: Demand Lags.' *New York Times*, 7 June, Sec. 4, p. 2.
SCHUYTEN, Peter J. (1979). 'Technology, Computer Show: Big Numbers.' *New York Times*, 31 May, Sec. 4, p. 2.
SCHUYTEN, Peter J. (1979). 'Technology; Turning TV Set into a Computer.' *New York Times*, 28 June, Sec. 4, p. 2.
'Scientist Says Man Outpaces Himself.' (1953). *New York Times*, April.

SEARS, Val (1979). 'The Robots are Here. As You Read This Your Life is Changing.' *Toronto Star*, 10 March, pp. 1 and 18.
SHEPPARD, Robert (1979). 'Computer Goofs Anger Shoppers; Study Shows Many Prefer Cash to Cards.' *Toronto Globe and Mail*, 6 March, Sec. 1, p. 8.
SHIELS, Merrill et al. (1980). 'And Man Created the Chip.' *Newsweek*, 30 June, pp. 50–56.
SHRAGE, Michael (1980). 'Goodbye "Dallas", Hello, Videodiscs.' *New York Magazine*, 17 November, pp. 38–56.
SMITH, Delbert D., (1971). 'Educational Satellite Telecommunication: The Challenge of a New Technology.' *Bulletin of Atomic Scientists*. April, pp. unavailable.
SMITH, William D. (1977). 'I.B.M. vs A.T. & T. – The Coming Struggle.' *New York Times*, 9 January, Sec. 7, p. 23.
DE SOLLER PRICE, Derek (1959). 'An Ancient Greek Computer.' *Scientific American*, June, pp. 60–67.
'Space Surveillance Deemed Inadequate.' (1980). *Aviation Week and Space Technology*. 16 June, pp. unavailable.
'Spacecraft Survivability Boost Sought.' (1980). *Aviation Week and Space Technology*. 16 June, pp. unavailable.
'Spécial Informatique.' (1974). (a special issue on informatics), *L'Expansion*, July–August.
'Spy Satellites, somebody could be watching you.' (1978). *Electronics and Power*, August, p. 573.
STEWART, Walter (1968). 'The Next Big Revolution Will Happen in your Livingroom.' *Macleans*. December. pp. 14ff.
'Stop the Export of Data to U.S. Sytems, M.P. urges.' (1979). *Toronto Globe and Mail*, 26 Nov, p. 5.
'Super-Ape Needs Intelligence to Best Machine.' (1954). *New York Times*, April, pp. unavailable.
SUPPES, Patrick (1966). 'The Uses of Computers in Education.' *Scientific American*, September, pp. 207–220.
SUPPES, Patrick (1966). 'The Uses of Computers in Education.' *World Press Review*. May, pp. 21–23.
'There is no "maybe" in the world of digital communication.' (1968). *New York Times*. 30 June, pp. unavailable.
'Tinkertoy T.V. (1954). *Newsweek*, pp. unavailable.
'Today's Computer Technology. . .' (advertisement for Fujitsu). (1982). *Japan Echo*. pp. unavailable.
TOFFLER, Alvin (1970). *Future Shock*. New York: Bantam.
TOFFLER, Alvin (1980). *The Third Wave*. New York: Bantam Books.
'Touch-down, Columbia.' (1981). *Time*. 27 April, pp. 14–15.
'Unarmed Bandit.' (1979). (Editorial). *The Globe and Mail*, 22 May, p. 6.
UNGER, Arthur (1980). (series of articles on new communications technology). *Christian Science Monitor*, 4, 5, 6, and 7 November.
WELLECH, Henry (1966). 'Big Brother Computer.' *Newsweek*, 25 July p. 80.
'Vivre avec l'ordinateur.' (1978). *Express*. 23 December. p. 67.

WALKER, E.M. (1981). 'Matching Bits with the Computer.' *Psychology Today*, June, p. 108.
'Wanted – Genuine Free Enterprise in T.V.' (1959). *The Canadian Forum*, No. 39, September, pp. unavailable.
'Watson Cites Limitations of Computer Capabilities.' (1968) *New York Times*. 12 November, pp. unavailable.
'What Life May be Like in Our "Wired World"' (1979). *Toronto Star*, 10 March, p. 181.
'Window on the World: The Home Information Revolution.' (1981). *Business Week*, 29 June, pp. 74–83.
WIRBEL, Loring (1980). 'Somebody is Listening – There's a Computer on the Line.' *The Progressive*, November, 16–22.
'Wired: P. Schwob's Apartment.' (1979). *New Yorker*, 21 May, pp. 26–29.
WOODBURY, David O. (1956). *Let "Erma" Do It: The Full Story of Automation*. New York: Harcourt, Brace & Co.
'The Word Processing Supplement' (1977). (a special section devoted to word processing). *Fortune*, 46, October.
'X-Rac.' (1954). *Life*, 24 May, pp. 109–110.
'You'll Talk To Your Briefcase.' (1979). *Toronto Star*, 13 March, p. 1 and 11.

II Research works dealing specifically with new communications technology and with policy

ABELSON, Philip H. Allen Hammand (1977). 'The Electronic Revolution.' *Science*, p. 1087–91.
ALBION, Robert G. (1931–32). 'The Communication Revolution.' *American Historical Review*, 37, pp. 718–720.
'Appelbert Report.' (1982). Applebaum. Louis, Hébert, Jacques. et al. Report of the Federal Cultural Policy Review Committee. Ottawa: Government of Canada.
D'ARCY, Jean (1961 and 1969). 'Direct Broadcast Satellites and the Right to Communicate.' *EBU Review*, No. 118. pp. 14–18.
BABE, R. (1975). *Cable Television and Telecommunications in Canada: an Economic Analysis*. Michigan: Michigan State University.
BAER, Walter S. (1973). *Cable T.V.: A Handbook for Decision Making*. Santa Monica, California: Rand Corp.
BAER, Walter S. (1978). Telecommunications Technology in the 1980's.' In *Communications for Tomorrow; Policy Perspectives for the 1980's*. Ed. Glen O. Robinson. New York: Praeger Special Studies. pp. 3–55.
BARAN, Paul (1975). 'Interactive Broad Band Cable Systems.' In *IEEE. Transactions on Communications*. Com. 23, January.
BARAN, Paul (1971). *Potential Market Demand for Two-Way Information Services to the Home, 1970–1990*. Institute for Future, R–26, December.
BARUCH, J.J. (1969). *Interactive Television*. Mediacom, October.

BARNOW, Erik (1966). *A Tower in Babel: A History of Broadcasting in the U.S. to 1973*. New York: Oxford University Press, 105–107.

BARRON, Ian, and Ray CURNOW (1979). *The Future with Microelectronics*. London: Frances Pinter Ltd.

BELL, Daniel (1981). 'The Social Framework of the Information Society.' In *The Microelectronics Revolution*. Ed. Forrester. Cambridge, Mass.: MIT Press, pp. 500–549.

BERNADET, R.J. (1979). 'Macro-Economie de la Société.' Montreal: Gamma Group.

BERNSTEIN, J. (1979). *Micro-Economics of Information. Structural and Regulatory Aspects*. Montreal: Gamma Group.

BERRIGAN, Frances, Ed. (1977). *Access: Some Western Models of Community Media*. Paris: UNESCO.

BERTRAM, Raphael (1976). *The Thinking Computer: Mind Inside Matter*. San Francisco: W.H. Freeman and Co., p. 186 and 194.

BLOOM, Michael (1982). 'Problems of Definition in International Communication Law: Internal Trade Aspects of Transborder Data Flow; Legislation in Canada and the United States.' Montreal: McGill m.s.

Blue Paper (1979–80). Ottawa: Department of Communications/ Government of Canada.

BLUMENWITZ, Dieter (1978). 'Freedom of Information in the Light of International Law.' In *Freedom of Information – a human right*. Munich: Hans-Seidel-Stiftring, pp. 13ff.

BODEN, Margaret (1981). 'The Social Implications of Intelligent Machines.' In Forrester, Ed. *op. cit.*, pp. 439–452.

BORCHARDT, Kurt (1978). *Actors and Stakes: A Map of the Communications Area.* Computers and Communications Working Paper W–78–8. Cambridge, Mass.: Harvard University, Program of Information Resources Policy.

Branching Out: Report of the Canadian Computer Communications Task Force, Ottawa: Department of Communications, Government of Canada. (1972).

BRANSCOMB, Lewis (1980). 'Computer Technology and the Evaluation of World Communications.' *Telecommunications Journal*, April.

BROWNSTEIN, Charles N. (1978). 'Interactive Cable T.V. and Social Services.' *Journal of Communications*. (Spring, 1978) Vol, 28, No. 2, pp. 142–48.

C.B.C. (1980). 'CBC 2. A proposal for national, non-commercial, satellite delivered CBC television services.' A Report prepared for the C.R.T.C., August. Ottawa/Hull: Government of Canada.

C.B.C. (1980). *Canadian Broadcasting and Telecommunications: Past Experience, Future Options*: A Report prepared for the C.R.T.C. Ottawa/Hull: Government of Canada

C.B.C. (1982). *Quarterly Report*. June 6.

CCTA (1979). *The Clyne Committee: A Critique*. Ottawa. Government of Canada, July.

Canada Council Act (1970). Ottawa: Government of Canada.

'Canadian Broadcasting Act.' Ottawa: Government of Canada.

C.R.T.C. (1980). 'Public Hearings on Communication in the North' (Fall). Ottawa: Government of Canada.

C.R.T.C., (1980). *The 1980's: A Decade of Diversity; Broadcasting, Satellites, and Pay TV*. Ottawa: Government of Canada.

CHAGY, Gideon, Ed. (1971). *The State of the Arts and Corporate Support*. New York: Paul S. Erikson Inc.

(Report of the) Clyne Committee (1979). *Canada and Telecommunications: the Report of the Consultative Commission on Telecommunications and Canadian Sovereignty*. Commissioned by the Department of Communications. Ottawa: Government of Canada.

COMPAINE, Benjamin (1981). 'Shifting Boundaries in the Information Market Place.' *Journal of Communication*, No. 7, Winter.

COVVEY, Dominic H., and Neil Harding McALISTER (1980). *Computer Consciousness: Serving the Automated '80's*. Reading, Mass.: Addison-Wesley Publishing Co.

CRAB (1979). Department of Communication. Ottawa: Government of Canada.

Canadian Videotex Consultative Committee's Individual and Society Sub Committee (1971). *Telecommission Study*. Ottawa: Government of Canada.

Communications Satellite Act of 1962. Washington, D.C.: Government Printing Office.

Communications Study Advisory Council Briefing, April 17, 1973. New York: Government Printing Office.

Computer/Communications Policy – A Position Statement by the Government (1973). Ottawa: Government of Canada, April.

Consultative Committee on the Implications of Telecommunications for Sovereignty (1979). *Telecommunications and Canada*. Ottawa: Department of Communications, Government of Canada.

CUNDIFF, W.E., and Mado REID, 1978. *Issues in Canadian/U.S Transborder Computer Data Flows*. Proceedings of a Conference held in Montreal sponsored and published by Institute for Research on Policy, Montreal: September 6.

'Davey Committee (1970). *Special Senate Committee on Mass Media*. Ottawa: Government of Canada.

DECKER, Horst (1976). *Die Massenmedien in der post industriellen Gesellschaft. Konsequenzen neuertechnischer u. wirtschaftlicher Entwicklungen fur Aufgaben u. Struckktunen d. Massenmedien*. In D.B.R.D., Gotlingen: Scwartz.

DENICOFF, Marvin (1980). 'Sophisticated Software: The Road to Science and Utopia.' In Moses and Dertouzos, Ed. *The Computer Age: A Twenty Year Survey*. Cambridge, Mass.; M.I.T. Press. pp. 367–391.

DERTOUZOS, Michael (1980). 'Individualized Automation.' In Moses and Dertouzos, Ed.,*op. cit.*

DERTOUZOS, Michael L. and Joel MOSES (1980). *The Computer Age: a twenty-year view*. Cambridge: MIT Press.

DICKSON, Edward M. with R. BOWERY (1973–4). *The Video Telephone: Impact on a New Era in Telecommunications*. New York: Praeger Publishers.
DIEBOLD, John (1969). *Man and The Computer: Technology as an Agent of Social Change*. New York: Praeger.
DIETSKIE, Fred I. (1981). *Knowledge and the Flow of Information*. Cambridge, Mass.: M.I.T. Press.
DOC (1972). *Community Communications in Five Ontario Cities*. Canada: Supplies and Services, p. 49.
DORNAN, Chris and Diane WELLS (1981). 'Videotex Availability to Users.' Ottawa: Department of Communications, Government of Canada.
DORSEN, N., and S. GILLERS, Eds. (1974). *None of Your Business: Government Secrecy in America*. New York: Penguin Books.
DREYFUS, Hubert L. (1972–1979). *What Computers Can't Do: The Limits of Artificial Intelligence*. New York: Harper Colophon.
DUNN, D.A. (1971). 'Cable Television Delivery of Educational Services.' IEEE Eascon Conference, Washington, October.
DURLACK, J., and P. ROOSEN-RUNGE, Eds. (1980). *Communications, Computers and Human Settlements*. Proceedings Symposium. Toronto: York University.
EDELSTEIN, Alex S., John E. BOWES, and Sheldon M. HARSEL (1978). *Information Society: Comparing the Japanese and American Experience*. Seattle: International Communications Centre.
EISENSTADT, S.N. (1955–56). 'Communication Systems and Social Structure: An Exploratory Comparative Study.' *Public Opinion Quarterly*, 19, pp. 153–167.
ELLIS, David, (1968). *Evolution of Canadian Broadcasting System: Objectives and Realities, 1928–1968*. Ottawa: Government of Canada, Dept. of Supply and Services.
ENGLISH, E. (1973). 'Canadian Telecommunications: Problems and Policies.' In *Telecommunications for Canada: An Interface of Business and Government*. Ed. English. Toronto: Methuen.
ETZIONO, Amitai (1968). *The Active Society*. New York: The Free Press.
ETZONI Amitai and Eugene Leonard, (1971). 'Minerva': a participatory technology system.' *Bulletin of Atomic Scientists*, November.
EVANS, Christopher (1979). *The Mighty Micro*. London: Victor Gollancz Ltd.
FARQUHAR, John A. (1972) *Education and library services for community information utilities*. Santa Monica, California: Rand Corporation.
FAULKNER, Fernande (1979). *Vista/Telidon Seminar*. December. Transcript. (A summary available from Jeff Campbell of Bell Canada, Montreal).
FAULKNER, Hugh (1977). 'Introductory Remarks to the International Federation of Information Processing Congress.' Toronto, August. 1977. In *Computerworld*, 15 August.

FEDOSEYEV, P.N. (1981).'A philosophical interpretation.' in *The Social Implications of the Scientific and Technological Revolution*. Paris: UNESCO, pp. 103–123.

FEELEY James, (1980). 'The Canadian Telidon Field Trials.' Ottawa: Department of Communications/Government of Canada.

FINLAY-PELINSKI Marike (1982b). 'Nouvelle technologie des communications' émancipation ou contrôl social?' in *Communication/Information*. Automne, Vol. V, no 1, pp. 147–177.

FINLAY-PELINSKI, Marike (1983). 'Technologies of Technology: A Critique of Procedures of Power and Social Control in Discourses on New Communications Technology.' Program in Communication, Working Papers Series: McGill University, Montréal.

FISHER, Desmond (1982). *The Right to Communicate: A Status Report*. Reports and Papers on Mass Communications, No. 94. Paris: UNESCO.

FOIDART, Donald (1980). 'Headingley Before and After the IDA Project. Phase I (Before).' Ottawa: Department of Communications/Government of Canada.

FOOTE, John (1981). Social Aspects of Videotex Services: Proposed Research Directives, Ottawa: Department of Communications/Government of Canada. November, m.s.

FORESTER, Tome (Ed.) (1981). *The Microelectronics Revolution*. Cambridge: MIT Press.

FOX, Frances (1979–80). 'Annual Report.' Ottawa: Department of Communications/Government of Canada.

FOX, Frances (1982). 'Introductory Address to the Canadian Communication Association Meetings held at the Assembly of Learned Societies.' Ottawa: Department of Communications/Government of Canada.

FRENCH, Michael B. (1980). 'The Semi-conductor Industry: An Overview.' *Datamation*, 26, No. 2, February, 164–170.

FRIEDRICHS, G.V. (1981). 'Automation, technological, social and economic change.' In Ed. Cohen *The Social Implications of the Scientific and Technological Revolution*. Paris: UNESCO. pp. 274–284.

FRITKIN, A. (1973). 'Space communications and developing countries.' In *Communications, Technology and Social Policy*. Eds. G. Gerbner, L. Gross, and H. Melody. New York: Wiley Interscience.

FURHOFF, Yars (1974). *Communications Policies in Sweden*. Paris: UNESCO.

GALBRAITH, Gordon (1974). 'La Progammation communautaire et la vie politique.' CRTC/Government of Canada. pp. 53–4.

GANELY, Oswald H. (1979). 'Communications and Information Resources in Canada.' *Telecommunications Policy*, December.

GANLEY, Oswald H. (1980). *The United States-Canadian Communications and Information Resources Relationship and its Possible Significance for Worldwide Diplomacy*. Cambridge, Mass.: Harvard University Program on Information Resources Policy.

GARDINER, W.L. (1980). *Public Acceptance of the New Information*

Technologies: The Role of Attitudes. Montreal: Gamma Group.
GARRETT, John and Geoff WRIGHT (1981). 'Micro is Beautiful.' in Forrester, Ed., *op. cit.*, pp. 497–499.
GASSMAN, H.P. (1980). 'Introduction by the Secretariat.' In OECD No. 3, *op. cit.* pp. 11–19.
'The GE Project.' (1971). In *Behind the corporate image: What General Electric did not say in its annual report.* Cambridge, Mass.:
GERNER, G. (1973). 'Cultural Indicators: The Third Voice.' In *Communications Technology and Social Policy*. New York: John Wiley.
GERBNER, G., L. GROSS, and Wm. MELODY (1973). *Communications Technology and Social Policy: Understanding the New Cultural Revolution*. New York: John Wiley.
GILPIN, Robert (1980). 'The Computer and World Affairs.' in Dertouzos and Moses, Ed., *op. cit.* pp. 229–253.
GODFREY, David and D. PARKHILL, Eds (1980). *Gutenberg Two*, Toronto: Press Porcépic.
GOLDHABER, Michael (1980). 'Politics and Technology: Microprocessors and the Prospect of a New Industrial Revolution.' *Socialist Review*, 10, No, 4, p., 9–32.
GOLDHAMMER, Herbert (1971). 'The Social Effect of Communication Technology.' *The Processes and Effects of Mass Communication*. Revised. Eds W. Schramm and D. Roberts. Chicago: University of Illinois Press.
GOTLIEB, A., and R. GWYN (1972). 'Social Planning of Communications.' In *Communications in Canadian Society* by B.Singer. Toronto: Copp Clark.
GOTLIEB, C.C. (1978). *Computers In the Home*. Occasional Paper No. 4. Montreal: Institute for Research on Public Policy.
GOTLIEB, C.C. and Z.P. ZEEMAN (1980). *Towards a National Computer and Communications Policy: Seven National Approaches*. Toronto: The Institute for Research on Public Policy.
GRAF, Mark G. and Roger B. JOHNSON (1980). 'Computer Angst.' in Ed. G.E. Lasker. *Applied systems and Cybernetics*, vol. V, Systems Approaches in Computer Science and Mathematics. Published Proceedings of the International Congress on Applied Systems Research and Cybernetics. Acapulco, December, pp. 2467–2471.
GREES, Harold (1967). 'The New Technological Era: A View from the Law.' Monograph 7, Program of Policy Studies.
Green Paper (1973). Ottawa: Department of Communications/ Government of Canada.
GREENBERGER, Martin, Ed. (1962). *Computers and the World of the Future*. Cambridge, Mass.: MIT Press.
Grey Paper (1975). Ottawa: Department of Communications/ Government of Canada.
GUITE, Michel (1971). *New Technology for Citizen Feedback to Government*. Unpublished. By Computer Task Force. Ottawa: Department of Communication/Government of Canada, December.

GUITE, M. (1977). *Requiem for Rabbit Ears: Cable Television Policy in Canada*. Stanford: Stanford University Press.

GURSTEIN, Michael (1979). *Social Impacts/Social Uses of Telidon/Vista*. Status Report, Prepared for Bell Canada, December.

HAIGHT, Timothy R., Ed. (1979). *Telecommunications Policy and the Citizen*. New York: Praeger.

HALLMAN, Eugene S., et al. (1977). *Broadcasting in Canada*. London: Routledge & Kegan Paul.

HALLORAN, James (1974). 'Mass Media and Society: the challenge of research.' Leicester: Leicester University Press.

HAMELINK, C.J. (1977). *The Corporate Village: The Role of Transnational Corporations in International Communications*. Rome: 1 Department of Communications International.

HAMELINK, Cees J. (1979). 'Informatics: Third World Call for New Order.' *Journal of Communication*, 29, No. 3, Summer.

HAMELINK, Cees J. (1980). 'International Finance and the Informatics Industry.' A Paper prepared for the panel on Transnational Data Systems at the Conference on World Communications: Decisions for the Eighties, May 12–14. Philadelphia.

HASLAM, Gerald (1980). 'Information Provider Activities in Canada.' Paper presented at the Online Conference on Viewdata Services. London, England, March, m.s.

HINLEY, M.P., G. MARTIN, and J. McNULTY (1977). *The Tangled Net*. Vancouver: J.J. Douglass.

HOGNEKE, Edmund F.M. (1981). 'Digital Technology: The Potential for Alternative Communication.' In *Journal of Communication*, 31, No. 7, Winter.

HOWE, J. Louis (1979). *La Poussée Technologique et les coûts unitaires décroissants en télématique*. Montreal: Gamma Group.

HUGHES, P., and R. SASSON (1980). 'The Usage of International Data Networks in Europe.' In OECD No. 3, *op. cit.*, pp. 25–34.

Hutchens Commission (1947). *A Free and Responsible Press: A General Report on Mass Communication*. Chicago: University of Chicago Press.

In Search, Ottawa: Department of Communications/Government of Canada.

Instant World (1971). A Report on Telecommunications in Canada. Ottawa: Government of Canada.

JACOBS, Ira and Stewart E. MILLER (1971). 'Optical Transmission of Voice and Data.' IEEE *Spectrum*, 14, No. 2, February, p. 32.

JAKOBSON, R.E. (1979). 'Who Gets What in The Information Society Distributes: Aspects of Communications Policy-Making.' In *Telecommunications Policy and the Citizen*. Ed. T. Haight. New York: Praeger, pp. 29–54.

JANCO, Manual, and Daniel FURJOT (1972). *Informatique et Capitalisme*. Paris: Maspero.

JANKY, J.M. (1971). 'Optimisation in design of mass-production microwave receiver suitable for direct reception for satellites.' Diss. Stanford University.

JOHNSON, Nicholas (1970). *How to talk back to your TV set*. Boston: Little Brown.

JONES, Trevor, Ed. (1980). *Micro-electronics and Society*. Stony Straford: The Open University Press.

KAHN, Herman, and P. WIENER (1967). *The Year 2000*. New York: Macmillan Co.

KAPRON, Felix P., and Ake Koichi (1979). 'Physics of Optical Fibres of Communications.' *Physics in Canada*. Pt. I, 35, No. 2, March, p. 53; and Pt. II, 35, No. 4, July, p. 75.

KATO, Hidetoshi (1978). *Communication Policies in Japan*. Paris: UNESCO.

KATZ, Elihu, et al. (1978). *Broadcasting in the Third World; Promise and Performance*. London: Macmillan.

KEDDY, Barbara (1979). 'Transborder Flow of data.' *In Search*. Vol. No. 4, pp. 12–17.

'Kent Commission' Royal Commission on Newspapers (1981). Ottawa: Government of Canada.

KENT, Tom (1982).'The Kent Commission and the Press in Canada.' Conference presented at the Annual Meetings of the Canadian Communication Associations. Ottawa: May.

KIDDER, Tracy (1982). *The Soul of a New Machine*. New York: Avon Books.

KLING, Rob (1978). 'Value Conflicts and Social Choice in Electronic Funds Transfer System Developments.' *Communications of the ACM* (Association for Computing Machinery), 21, No. 8, August, 642–656.

KOCHEN, Manfred, Ed. *(1967). The Growth of Knowledge: Readings on Organisation and Retrieval of Information*. New York: John Wiley and Sons Inc.

KRIPPENDORF, Klaus, Ed. (1979). *Communication and Control in Society Networking*. New York: Gordon and Breach Science Publishers.

KURCHAK, Marie (1981). *Telidon: the information providers*. Ottawa: Department of Communications/Government of Canada.

LAVINGTON, S.A. *Early British computers*. Manchester: Manchester University Press.

LEMELSHTRICH, Noam (1974). *Two-Way Communication: Political and Design Analysis of a Home Terminal*. Beverly Hills, California, London: Sage Publications.

LEMPEN, Blaise (1980). *Information et Pouvoir: essai sur le sens de l'information et son enjeu politique*. Lausanne: L'Age d'Homme.

LENK, Klause, Ed. (1976). *Informations – Rechte – Kommunikationspolitik – Entwichlungsperspektiven d. Kabelfernsehens u. d. Breitbandkomm.* Darmstadt: Treche-Mittler.

LEONARD, Ross (1974). *Economic and Legal Foundations of Cable Television*. Beverly Hills: Sage Publications.

LERNER, Max (1939). *Ideas and Weapons*. New York: Viking Press, p. 13.

LICKLIDER, J.C.R. (1980). 'Computers and Government.' Eds, Moses and Dertouzos. In *op. cit.* pp. 87–126.

LINK, (1981). 'Videotext: a Worldwide Evaluation.' In *Videotex Monitoring Service*, 3, No. 1, January.

LOGUE, Timothy J. (1979). 'Teletext: Toward an Information Utility?' *Communications*, Autumn.

LOWI, Th., J. (1981). 'The Political Impact of Information Technology.' In Ed. Forrester, *op. cit.*, pp. 453–472.

LUSSATO, Bruno, and Jean BOUNINE (1979). *Télématique ou Privatique? Questions à Simon Nora et Alain Minc*. Paris: Editions d'Informatiques.

'The MacBride Commission' (1980). *One World, Many Voices*. Paris: UNESCO.

MacCLEAN, Donald (1981). 'Taking the Revolution Seriously: Intellectual Obstacles to Information Policy Making.' A Paper presented to the Annual Meeting of the Canadian Communications Association. 26 May.

MacHUGH, Joseph M. (1976). *Mass Media and Public morality. A Problem-law. Precept analyses with recommendations for self-regulation in the mass media*. New York: Vantage Press.

MacKINDER, Halford (1962). *Computer Efficiency Vs War: Democratic Ideals and Reality*. New York: Norton.

MACOBY, James (1980). 'Debate.' In *Daedalus*, *op. cit.*, pp. 9ff.

MADDEN, John C. (1980). 'Julia's Dilemma' and 'Simple Notes on a Complex Future.' In Eds., Godfrey and Parkhill. *op. cit.*, pp. 13–68.

MADDEN, John C. (1980). 'Telidon in Canada.' Paper at a conference entitled *New Developments in Canadian Communications Law and Policy*. Ottawa, 24–26 January, m.s.

MADDEN, John C. (1979). 'Videotex in Canada.' Discussion Paper. Toronto: 8 May, Delta Dialogue Series Seminar, No. 9.

MADDOX, Brenda (1974). 'A Brazilian Satellite – for what?' *New Scientist*, 25, No. IV.

MAHLE, Walter A., and P. RICHLER (1974). *Communications Policies in the Federal Rebublic of Germany*. Paris: UNESCO.

MANNING, Eric G. (1980). 'Computers, Telecommunications and Canada: Options for our Future.' The Gordon Lecture at l'Universite de Sherbrooke, 26 March. m.s.

MARSH, Peter (1981). *The Silicon Chip Book*. London: Abacus.

MARTIN, James (1978). *The Wired Society*. Englewood Cliffs, N.J.: Prentice-Hall Inc.

MASON, Richard (1979). 'The Role of Dialogue in the Measurement Process.' In Krippendorf. *op. cit.*

MATTELART, Armand and J. PIEMME (1980). *Télévision: enjeu sans frontières*. Grenoble: Presses Universitaires de Grenoble.

McHALE, John (1976). *The Information Society*. London: The Westview Press.

McHALE, John (1977). *The Changing Information Environment*. London: Paul Elek.

McCLEAN, J. Michael (1979). *The Impact of the Microelectronics Industry on the Structure of the Canadian Economy*. Montreal: The Institute for Research on Public Policy.

MELODY, William H. (1973). 'The role of advocacy in public policy planning.' In *Communications Technology and Social Policy*. Eds. G. Gerbner, L. Gross, and H. Melody. New York: Wiley Interscience.

MENZIES, Heather (1981). *Women and the Chip: Case Studies of the Effects of Informatics on Employment in Canada*. Montreal: Institute for Research on Public Policy.

MICHTHEIM, George (1963). *The New Europe; Today and Tomorrow*. New York: Basic Books.

Microelectronics (1977). *A Scientific American Book*, San Francisco: W.H. Freeman and Co.

MINC, A., (1980) 'The Informatization of Society.' In OECD, No. 3, *op. cit.*, pp. 154–159.

MITCHELL, Heather (1980), *Access to Information and Policy Making: A Comparative Study of Sweden, Australia and the United States*. Toronto: Commission on Freedom of Information and Individual Privacy, Research Publication File.

MOSCO, Vincent (1979). *Broadcasting in the United States: innovative challenge and organizational control*. Norwood, N.J.: Ablex Pub. Corp.

MOWSHOWITZ, Abe (1968). 'The Conquest of Will: Information Processing.' In *Human Affairs*.

MURDOCH, L., J. HELLAND, and J. SMITH (1979). *Freedom of Information and Individual Privacy: A Selective Bibliography*. Toronto: Commission on Freedom of Information and Individual Privacy, Research Publication No. 12.

National Information Policy (1977). *Report to the President of the U.S.* Washington, D.C.: Government Printing Office.

National Science Foundation (1980). *Science Technology and Uses of Information*. Washington, D.C.: Government Printing Office.

NICOLS, Rodney W. (1979). 'For Want of a nail?' In *Technology in Society*, 1, No. 2. New York: Pergamon Press Ltd.

NORA, Simon, and Alain MINC (1980). *The Computerization of Society*. Cambridge, Mass.: MIT Press.

OECD, (1975). *Allocations des resources dans le domaine de l'informatique et des telecommunications*. IIIième partie, rapport de base, préparé par S. Gill. Paris: OCDE.

OECD. No. 1 (1979). *Transborder Data Flows and the Protection of Privacy*. Proceedings of a Symposium held in Vienna, Austria, September 20–23, 1977. Paris: OECD.

OECD, No. 2 (1980). 'The Usage of Internationa Networks in Europe.' *Information Computer Communications Policy, No. 2*. Paris: OECD, 1980.

OECD, No. 3, (1980). *Policy Implications of Data Network Developments in the OECD Area*. No. 3– Information, Computer, Communications Policy, Paris: OECD.

OETTINGER, Anthony (1971). 'Communications in the National Decision-Making Process.' In *Computers, Communications, and the Public Interest*. Ed. M. Greenberger. Baltimore: Johns Hopkins Press, 74–113.

OETTINGER, Anthony G., P.J. BERMAN, and C.H. READ (1977). *High and Low Politics: Information Resources for the '80's*. Cambridge, Mass.: Ballinger Publication Co.

OKA, Joshisada. 'Data Network Development and Policy in Japan.' In OECD No. 3, *op. cit.*, pp. 44–76.

OSTRY, Bernard (1980). 'La Revolution des Communications.' In *La Revue Silencieuse de la Communications*. March, pp. 5ff.

OSWALD, H. Ganley (1979). *The Role of Communications and Information Resources in Canada*. Cambridge, Mass.: Harvard University Program in Information Resources Policy.

OUIMET, Alphonse (1980). 'The Communications Revolution and Canadian Sovereignty.' In Eds. Godfrey and Parkhill. *op. cit.*, pp. 131ff.

PARKER, Donn (1979). *Crime by Computer*. New York: Scribners.

PARKER, E.B. (1972). *Assessment and Control of Communications Technology*. Stanford: Stanford U. Press.

PARKER, E.B. (1975). 'Incidences Sociales des systemes de téléinformatique.' Conference on Policy of Data Processing and Telecommunications, 4–6 February. Paris: OECD.

PARKHILL, D.F. (1966). *The Challenge of the Computer Utility*. Reading, Mass.: Addison-Wesley Publication Company.

PARKHILL, D.F., (1980). 'The Necessary Structure.' In Eds. Godfrey and Parkhill, *op. cit.*, pp. 71ff.

PARKHILL, D.F. (1980). 'An Overview of the Canadian Scene.' Paper presented at the Online Conference on Viewdata Services. London, England, March, m.s.

PARKHILL, D.F. (1976). 'Society and Computer Communications Policy.' Proceedings of the Third International Conference on Computer Communications Toronto, 3–6 August.

PELLETIER, G. (1973). *Communications: Some Federal Proposals*. Ottawa: Government of Canada.

PERGLER, P. (1980). *The Automated Citizen*. Montreal: Institute for Research on Public Policy.

PIERA, F. (1979). 'Activities for the IBI in the Field of Transnational Data Flows.' OECD, No. 1, *op. cit.*, pp. 236 and 7.

PIERCE, J.R. (1977). 'Electronics: Past, Present and Future.' *Science*, March 18, p. 1092.

POOL, DE SOLA, Ithiel (1974). 'Direct Broadcast Satellites and the Integrity of National Culture.' In *Control of Broadcast Satellites Values in Conflict.* Palo Alto, Aspen Institute: Office of External Research, U.S. Department of State.

POOL, DE SOLA, Ithiel, Ed. (1973). *Talking Back: Feedback and Cable Technology*. Cambridge, Mass.: MIT Press.

POOL, DE SOLA, Ithiel (1977). 'The Changing Flow of Television.' *Journal of Communications*. Vol. 27, No. 2, (Spring, 1977). pp. 139–50.

POOL, DE SOLA, Ithiel, and R.J. SOLOMON (1980). 'Transborder Data Flows: Requirements for International Co-operation.' In OECD No. 3, *op. cit.*, pp. 79–139.

PORAT, Marc (1978). 'Communications Policy Man Information Society.' In Robins G.O. Ed. (1978). *op. cit.* pp. 3–60.

PORAT, Marc (1977). *The Information Economy*. Washingtom: U.S. Department of Commerce.

PORAT, Edwin (1976). 'The Social Implications of Telecommunications.' *Telecommunications Policy*, 1, No. 1, 1976.

The Prestel Users Guide and Directory (with Teletext Viewdata Magazine), 2nd Edn. Eastern Countries Newspaper Ltd. Rouen, Norwich, England: Prospect House.

Proclamation of the Canadian Radio-Television and Telecommunication Act. (1976). Ottawa: Department of Communication/Government of Canada.

PROULX, Serge (1982). 'La société informatisée,' presentation at the Annual Meeting of the Canadian Communications Association, Ottowa, May.

RANKIN, Murray (1977). *Freedom of Information in Canada: Will the Doors Stay Shut?* Toronto: Canadian Bar Association.

RABEAU, Y. (1980). *Tele-Informatics, Productivity and Employment: An Economic Interpretation*. Montreal: Gamma Group.

REDMAN, Kent and Thomas SMITH. *Project Whirlwind: the History of Pioneering Computer*. New York: Digital.

Regulating the Regulators: Science values and decisions. (1982). Ottawa: Science Council of Canada.

REID, A.L. 'Prestel, the British Post Office Viewdata Service.' In OECD No. 3, *op. cit.*, pp. 41–42.

A Resource for the active community developed by Broadcast Programmes and Research Branches of the C.R.T.C. Ottawa: Government of Canada.

RICHARDS, William, and George LINSEY (1979). 'Social Networking Analysis: An Overview of Recent Developments.' In Krippendorf. *op. cit.*

ROBINSON, Glen O. (1978). *Communications for Tomorrow: policy perspectives for the 1980's*. New York: Praeger.

RULE, James (1980). 'Privacy: Impact of the New Technologies.' In *Science, Technology and Uses of Information*. Washington, D.C.: National Science Foundation.

RULE, James, Douglas McADAM, Linda STEARNS, and David UGLOIO (1980). *The Politics of Privacy*. New York, London, Scarborough: Metor.

RUSSEL, R.A. (1978). *The Electronic Briefcase: The Office of the Future*. Montreal: The Institute of Research on Public Policy.

SALTER, Liora (1980). 'Public and Mass Media in Canada: Dialecticism in Innis, Communications Analysis.' In Ed. Melody Salter, and Meyer: *Communications and Dependency, The Tradition of H.A. Innis*. Norwood, NJ: Ablex.

SCHILLER, Herbert (1978). 'Computer Systems: Power for Whom and for What?' *Journal of Communications*, 28, No. 4, Autumn, 184–193.

SCHUTZ, Walter (1978). *Forschungsprojkte Kommunikationspolitische Kommunikationswisenschaftliche Bundesregierung (1974–1978)*. Bonn: Presse u. Informationsamt d. Bundesregierung.

Science Council of Canada. Committee on Computers and Communications (1978). *Communications and Computer: Information and Canadian Society*. Ottawa: Science Council of Canada.

Science Council of Canada (1982). *The Impact of the Microelectronic Revolution on Work and Working*. Ottawa: Science Council of Canada.

Science Council of Canada (1979). *A Scenario for the Implementation of Interactive Computer Communications*. Ottawa: Science Council of Canada.

SEELEY, Doug, (1980). 'The Ecology of Information or Avoiding the Pollution of Reality in the Information Age.' In *Proceedings of a York Symposium: Communications, Computer and Human Set.* March 19–21, p. 47ff.

SIMON, Herbert A. (1980). 'The Consequences of Computers for Centralization and Decentralization.' In Eds., Moses and Dertouzos, *op. cit.*, pp. 212–28.

SIMON, Herbert (1981). 'What Computers Mean for Man and Society.' In Ed. Forrester *op. cit.*, pp. 419–433.

SIMON, Herbert and H. NEWELL (1972). *Human Problem Solving*. Englewood Cliffs, N.J.: Prentice Hall.

SINDELL, P.S. (1979). *Public Policy and the Canadian Information Society*. Montreal: Gamma Group.

SLACK, Jennifer Daryl (1984). *Communication Technologies and Society: Conceptions of Causality and the Politics of Technological Intervention*. Norwood, N.J.: Ablex Pub. Corp.

SMITH, Anthony (1980). *The Geopolitics of Information*. Oxford: Oxford University Press.

SMITH, Anthony (1978). *The Politics of Information: Problems of Policy in Modern Media*. London: MacMillan.

SMITH, Delbert (1976). *Communications via Satellite: A Vision in Retrospect.* Leiden: Sitjthoff.

SMITH, Henry (1973). 'Goebels in space, Government use of Télécommunications.' Cinéaste, No. 4.

SMITH, Ralf Lee (1972). *The Wired Nation CATV: The Electronic Communication Highway*. New York: Harper Colophon Books.

STANTON, Frank (1975). *Report on Panel on International Information, Education, and Cultural Relations*. Washington, D.C.: Georgetown Center for Strategic and International Studies.

STERN, Nancy (1980). *From ENIAC to UNIVAC: an appraisal of the Eckhart and Mockley Computers*. New York: Digital.

STRACHEY, C. (1959). 'Time-sharing in Large Fast Computers.' In *Proceedings of the International Conference on Information Processing*. Paris: UNESCO, 336–341.

SUMMERS, Robert E., et al. (1966). *Broadcasting and the Public*. Belmont, Ca.: Wadworth.

SWANSKEY, (1971). 'Monopoly control of Canada's mass media.' *Communist Viewpoint*, Toronto. May-June.

Tamec Economic Consultants (1979). *Videotex Services: The Market Potential for Cable*. Montreal: unpublished study report for Videotron Cable.

TAYLOR, James (1982). 'Computer Aided Message Systems.' In *Office Information Systems*. Ed. N. Naffah. INRA: North Holland Publication Company, pp. 631–651.

TAYLOR, James (1981). 'Vues nouvelles sur le bureau de demain à partir des théories d'hier.' In *En Quête/In Search*, 8, 3. Ottawa: Department of Communications/Government of Canada, pp. 4–13.

TEHERANIAN, Majid, Ed. (1977). *Communications policy for national development*. London: Routledge and Kegan Paul.

Telidon Reports. Ottawa: Department of Communications/Government of Canada.

THOMPSON, G. (1980). 'Technology and the Information Society.' *In Search*, 26–31. Ottawa: Department of Communications.

TYLER, Michael (1979). 'New Media in the Information Economy: Prospects and Problems for Viewdata and Electronic Publication.' In *Proceedings of the Sixth Annual Telecommunications Policy Research Conference*. Ed. Herbert S. Dordick. Lexington, Mass.: Lexington Books, p. 227.

UNESCO (1968). *Communications in the Space Age: the use of the satellites by the mass media*. Paris: UNESCO.

UNESCO (1965). *World Radio and Television*. New York: UNESCO.

UNESCO (1981). Eds, Cohen, Prof. Robert S. *The Social Implications of the Scientific Implications of the Technological Revolution*. A UNESCO Symposium. Paris: UNESCO.

VALASKAKIS, K. (1979). *The Information Society: The Issues and the Choices. Integrating Report on Phase I.* Montreal: Gamma Group.

VALASKAKIS, K., and P.S. SINDELL (1980). *Industrial Strategy and the Information Economy: Towards a Game Plan for Canada*. Montreal: Gamma Group.

VANDENBERGHE, M. (1979). 'La vie privée et les banques de données.' In OECD No. 1, *op. cit.*, pp. 249–256.

VEITH, Richard (1976). *Talk-back T.V. -2-Way Cable T.V.* Blue Ridge Summit, Pa.: C/L Tab Books.

VOYER, R.D., and B.L. EDWARDS, 'The Wired Country.' (Unpublished). Statistics Canada.

WEIL, G.L., Ed. (1969). *Communicating by Satellite: an international discussion*. Report of an international conference sponsored by the

Carnegie Endowment for International Peace and the Twentieth Century Fund. New York: Carnegie Endowment Fund.
WEIZENBAUM, Joseph (1976). *Computer Power and Human Reason: From Judgement to Calculation*. San Francisco: W.H. Freeman and Co.
WEIZENBAUM, Joseph (1978). 'Once More – a Computer revolution.' *The Bulletin of the Atomic Scientists*. pp. 12–19.
WEIZENBAUM, Joseph (1981). 'Where Are We Going?: Questions for Simon.' In Ed. Forrester, *op. cit.*, pp. 434ff.
WESEL, Andrew E. (1976). *The Social Use of Information – Ownership and Access*. New York: John Wiley & Sons.
WESTIN, Alan F., and Michael A. BAKER Ed. (1971). *Databanks in a Free Society: Computers, Record keeping and Privacy*. New York: Quadrangle Books.
WHITTAKER, P.N. (1980). 'Satellite Business Systems (SBS): A Concept for the Eighties.' In OECD. No. 3, *op. cit.*, pp. 35–39.
WICKLAN, John (1979). 'Wired City U.S.A.: The Charms and Dangers of Two-Way TV.' *Atlantic Monthly*, February, p. 37.
WINOGRAD, Terry (1980) 'Toward Convivial Computing.' In Moses and Dertouzos Eds. *op. cit.*, pp. 56–75.
WINSBURY, Rex (1979). *The Electronic Bookstall: Push-button Publishing on Videotex*. London, England: International Institute of Communications.
WINSTON, Patrick Henry (1977). *Artificial Intelligence*. Reading, Mass.: Addison-Wesley Publishing Co.
WILLS, R. (1979). *Research and Developments in the Information Sector of the Canadian Economy*. Montreal: Gamma Group.
WILSON, Kevin (1985). 'The Social Significance of Home Networking: Public Surveillance and Social Management.' Doctoral Dissertation: McGill University.
WOODWARD, Charles C. (1974). *Cable Television: Acquisition and operation of CATV Systems*. New York: McGraw-Hill.

III Theories of communication, discourse, and technology

ADORNO, Theodor (1966–73). *Negative Dialectics*. Trans. E.B. Ashton. Boston: Beacon.
ADORNO, Th. W., Ed. (1972). *Der Positivismusstreit in des deutschen soziologie*. Darmstat: Luchterhand.
ALTHUSSER, Louis (1970). 'Ideology and ideological state apparatuses.' In *Lenin and Philosophy and Other Essays*, trans. by Ben Brewster. New York and London: Monthly Review Press.
APEL, Karl-Otto (1971). 'Szientistik, Hermeuneutik Ideologickritik.' In Ed. Habermas *Hermeneutik and Ideologiekritik*. Frankfort: Suhrkamp, pp 7–44.
APEL, Karl-Otto (1970). 'From Kant to Peirce: The Semiotical Transformation of Transformation of Transcendental Logic.' *Proc. from the Third International Kant Congress*. Ed. Lewis, White, and Beck.

APEL, Karl-Otto (1976). 'The Transcendental Conception of Language Communication and the Ideal of a First Philosophy.' In Ed. Herman Parret. *History of Linguistic Thought and Contemporary Linguistics*. Berlin: de Gruyter.

ARROW, K.J. (1963). *Social Choice and Individual Values*. New York: Wiley.

BAKHTIN, Mikhail (1977). *Marxisme et la philosophie du language*. trans. M. Yaguello. Paris: Minuit.

BAKHTIN, Mikhail (1929–70). *Problèmes de la poétique de Dostoievski*. Trans. Guy Vevret. Lausanne: Editions l'Age d'Homme.

BARTHES, Roland (1953). *Le degré zéro de l'écriture*. Paris: Seuil.

BARTHES, Roland (1978). *Leçon: Lecon inaugurale de la chaire de sémiologie littéraire du Collège de France*. Paris: Seul.

BARTHES, Roland (1957). *Mythologies*: Paris: Seuil.

BATESON, Gregory (1972). *Steps To An Ecology of Mind*. New York: Ballantine Books.

BATESON, Gregory (1979). *Mind and Nature: A Necessary Unity*. New York, London, Toronto: Bantam.

BELL, Daniel (1973). *The Coming of Post-Industrial Society*. New York: Basic Books.

BENVENISTE, Emil (1966). 'La nature des pronoms.' In *Problémes de linguistique générale II*. Paris: Gallimard, pp. 251–266.

Bill U.N. 7 Re: Creation of National Film Board (1970). Ottawa: Queen's Printer.

BOOKCHIN, Murray, (1971). *Post-Secondary Anarchism*. Berkeley: The Ramparts Press.

BOULDING, Kenneth (1962). *Conflict and Defence – A General Theory*. New York: Harper and Row.

BLUMER, Jay and Elihu KATZ, Ed. V (1974). *The Uses of Mass Communications: Current Perspectives on Gratification Research*. Beverly Hills, London: Sage Publications.

BRAVERMAN, Harry (1975). *Labor and Monopoly Capital: the degradation of work in the twentieth century*. N.Y.: Monthly Review Press.

BRZESINSKI, Zbignew (1970). *Between Two Ages: America's Role in the Technetronic Era*. New York: Viking Press.

CAMPBELL, Duncan, and Linda MELVERN (1980). 'America's Big Bar on Europe.' *New Statesman*, 18 July, pp. 10–14.

CARDOSO, F.M. and E. FALETTO (1979). *Dependence and development in Latin America*. Berkeley: University of California Press.

CAREY, James (1969). 'The Communications Revolution and the Professional Communicator.' In *The Sociological Review*, monograph 13. Ed. P. Holmes. Keele: University of Keele Press.

CAREY, James, and John QUIRK (1970). 'The Mythos of the Electronic Revolution I and II.' In *American Scholar*, 39, 1 and 2, Spring and Summer, pp. 219–241, and 395–424.

CAREY, James (1975). 'Canadian Communications Theory.' In Ed. G.J.

Robinson and D. Theall. *Studies in Canadian Communications*. Montreal: McGill University Press pp. 45ff.

CASTORIADIS, C. (1975). *L'institution imaginaire de la société*. Paris: Seuil.

CASTORIADIS, Cornelius (1978). *Les Carrefours du labyrinthe*. Paris: Seuil.

CASTORIADIS, Cornelius and Daniel COHN-BENDIT (1981). *De l'écologie à l'autonomie*. Paris: Seuil.

CATER, Douglas (1981). 'The Survival of Human Values.' *Journal of Communication*, 31, No. 1, Winter.

CHERRY, Colin (1978). *World Communication: threat or promise? A socio-technical approach*. New York: Wiley.

CLEGG, Stewart (1975). *Power, Rule and Domination: A Critical and empirical understanding of power in sociological theory and organizational life*. London and Boston: Routledge and Kegan Paul.

'Communication Research Third World Realities.' (1980). Workshop. Institute of Social Studies. The Hague.

COOPER, Barry (1981). 'Ideology, Technology, and Truth.' Calgary: M.S. unpublished.

Daedalus (1980). 'Some Issues of Technology,' Proceedings of a Conference, MIT, 30 April, No. 109.

DAHRENDORF, Ralph (1964). 'Recent Changes in the Class Structure of European Societies.' *Daedalus*, 92, No. 1, 225–270.

DARNTON, Robert (1975). 'Writing News and Telling Stories.' *Daedalus*, Vol. 104, No. 2.

DERRIDA, Jacques (1967). *L'écriture et la différence*. Paris: Seuil.

DESSAUER, Friedrich (1927). *Philosophie der Technik*. Bonn: Friedrich Cohen Verlag.

DOMENACH, DUMOUCHEL and DUPUY (1981). 'Projet de création d'un Centre d'Epistémologie des Sciences de l'Homme à l'Ecole Polytechnique – CREA.' Paris: Centre de Recherche sur l'epistémologie et l' autonomie, M.S.

DOUGLAS, J.D. Ed. (1971). *The Technological Threat*. Englewood Cliffs, N.J.: Prentice-Hall.

DRUCKER, Peter (1968). *The Age of Discontinuity*. New York: Harper and Row.

DUPUY, Jean-Pierre (1982). *Ordres et désordres: enquête sur un nouveau paradigme*. Paris: Seuil.

ECHO, Umberto (1976). *The Theory of Semiotics*. Bloomington: Indiana University Press

EDELMAN, J. Murray (1964). *The Symbolic Uses of Politics*. Urbana: University of Illinois Press.

ELLUL, Jacques (1977).*Le Système technicien*. Paris: Calmann-Lévy.

ELLUL, Jacques (1954). *La Technique ou l'enjeu du siècle*. Paris: A. Colin.

ENZENBERGER, Hans Magnus (1974). 'Constituents of a Theory of the Media.' Trans. Stuart Hood. In *The Consciousness Industry: On Literature, Politics and the Media*. Selected with postscript by Michael Roloff. New York: Seabury.

FEYERABEND, Paul (1975). *Against Method*. London: Verso.
FINLAY-PELINSKI, Marike (1982a). 'Semiotics or History: From Content Analysis to Pragmatics of Communicational Interaction.' *Semiotica*, (Winter), forthcoming.
FINLAY-PELINSKI, Marike. (1986–7). 'William Leiss on Technology,' and 'Technology as Practice,' *The Canadian Journal of Political and Social Theory*. Vol. X, N. 1–2, pp. 174–95: and Vol. XI, N. 1–2, pp. 198–214.
FINLAY-PELINSKI, Marike (1982c). '"Criticism" in the Margins: Barbara Herrnstein Smith's "On the Margins of Discourse".' *Canadian Review of Comparative Literature*. March, pp. 56–75.
FINLAY-PELINSKI, Marike (Ed.) (1983). 'Il était une fois la théorie . . .' special issue on theory of communication, *Communication/Information*, Vol. V, No. 2/3 (Winter/Summer 1983).
FINLAY-PELINSKI, Marike (1981). 'The Potential of Irony: From a Semiotics of Irony Toward an Epistemology of Communicational Praxis, A Study of Schlegel and Musil.' Diss. Université de Montréal.
FOUCAULT, Michel (1969). *L'Archéologie du savoir*. Paris: Gallimard. English translation is cited: Foucault, Michel (1972). *The Archeology of Knowledge*. New York: Pantheon.
FOUCAULT, Michel (1966). *Les Mots et les choses*. Paris: Gallimard.
FOUCAULT, Michel (1970). *The Order of Things*. London: Tavistock.
FOUCAULT, Michel (1971). *L'Ordre du discours*. Paris: Gallimard.
FOUCAULT, Michel (1977), *Language, Counter-memory, Practice*. Ed. & Trans. Donald Bouchard, Ithica, New York: Cornell University Press.
FOUCAULT, Michel (1971). 'Response to Steiner.' *Diactrics*, Winter, p. 60.
FOUCAULT, Michel (1975). *Surveiller et Punir*. Paris: Gallimard.
FOUCAULT, Michel (1980). *Power/Knowledge: Selected Interviews and Other Writings 1972–1977*. Ed. Colin Gordon. Trans. Colin Gordon et al. New York: Pantheon Books.
GALBRAITH, John Kenneth (1973). *Economics and The Public Purpose*. Boston: Houghton Mifflin.
GALTUNG, Johan and Maria HOLMBOE RUGAR (1970), 'The Structure of Foreign News.' In Ed. J. Tunstall. *Media Sociology*. Urbana, I11.: University of Illinois Press, pp. 259ff.
GEHLEN, Arnold (1980). *Man in the Age of Technology*. Trans. P. Lipscombe. New York: Columbia Univ. Press.
GEISSLER, Rainer (1973). *Massenmedien, Basiskommunikation & Demokratie: Ansatz zu einer normativ-epirischen Theorie*. Tubingen: Mohr.
GENETTE, Gerard (1972). *Figures III*. Paris: Seuil.
GOLHABER, Michael (1980). 'Politics and Technology.' *Socialist Review*, No. 52 (Vol. 10, No. 4), July–August, pp. 9–32.
GOSTSHALK, D.W. (1954). 'Needed for Liberalism: a New Metaphysics and a New Program.' In *Antioch Review*, Spring.

GOULDNER, Alvin W. (1976). *The Dialectics of Ideology and Technology: the origins, grammar and future of ideology*. New York: Seabury Press.

GRANT, George (1976). 'The Computer does not impose on us the ways it should be used.' In Ed. A. Rotstein. *Beyond Industrial Growth*. Toronto: University of Toronto Press, pp. 117–131.

GRANT, George (1973). 'Ideology in Modern Empires.' In Ed. J.E. Flint and G. Williams. *Perspectives of Empire*. New York: Longman. pp. 189–197.

GREIMAS, A.J. (1966). *Sémantique Structurale*. Paris: Larousse.

HABERMAS, Jürgen (1979). *Communication and the Evolution of Society*. Trans. Thomas McCarthy. Boston: Beacon.

HABERMAS, Jürgen (1971). 'Preparatory Remarks to a Theory of Communicative Competence.' In *Theorie der Gesellschaft oder Sozialtechnologie*. Frankfurt am Main: Suhrkamp, 101–141.

HABERMAS, Jürgen (1975). *The Legitimation Crisis*. Boston: Beacon Press.

HABERMAS, Jürgen (1976). 'Some Distinctions in Universal Pragmatics.' In *Theory and Society*, Summer, 3, 2, pp. 156–167.

HABERMAS, Jürgen (1971–73). *Theory and Practice*. Trans., John Viertel. Boston: Beacon Press.

HABERMAS, Jürgen (1968–70). *Toward a Rational Society*. Trans. Jeremy Shapiro. Boston: Beacon Press.

HABERMAS, Jürgen (1970). 'Toward a Theory of Communicative Competence.' Pt. I, in *Recent Sociology* No. 2, *Patterns of Communicative Behaviour*. Ed. Hans Peter Dreitzel. London: Macmillan, pp. 115–148.

HALL, E.T. (1966). *The Hidden Dimension*. New York: Doubleday.

HAMELINK, Cees J. (1980). 'New Structures of International Communication: the role of research.' Paper for the VIIth Assembly and the scientific conference of the International Association for Mass Communication Research, Aug. 25–29, 1980, Caracas, Venezuela.

HARDMAN, J.B.S. (1941). 'Free Press Authority.' In *Freedom of the Press*. Ed. Harold Ickes, New York: The Vanguard Press.

HARMS, L.S. and Jim RICHSTAD, Eds. (1977). *Evolving Perspectives on the Right to Communicate*. Honolulu: East-West Center, East-West Communication Institute.

HARMS, L.S., Jim RICHSTAD, and Kathleen A. KIE, Eds. (1977). *Right to Communicate: Collected Papers*. Honolulu: Social Sciences and Linguistics Institute, University of Hawaii at Manoa.

HEATH, Stephen (1972). *The Nouveau Roman*. London: Elek.

HEIDEGGER, Martin (1977). *The Question Concerning Technology and Other Essays*. Trans. Wm. Lovitt. New York: Harper and Row.

HOCKING, Wm. E. (1947). *Freeedom of the Press: a Framework of Principle*. Chicago: University of Chicago Press.

HOFSTADTER, Douglas (1980). *Godel, Escher, Bach: an Eternal Golden Braid*. New York: Vintage Books.

HORKHEIMER, Max. (1937–1972). 'Traditional and Critical Theory.' In *Critical Theory*. New York: Herder and Herder. pp. 188–243.
HOROWITZ, David, and David KOLODNEY (1969). 'The Foundations of Technology: Charity Begins at Home.' *Ramparts*, April.
INNIS, Harold (1951–71). *The Bias of Communication*. Toronto: University of Toronto Press.
INNIS, Harold (1950–72). *Empire and Communication*. Toronto: University of Toronto Press.
JAY, Martin (1973). *The Dialectical Imagination: a history of the Frankfurt School and the Institute of Social Research*. (1923–1950). Boston: Little Brown.
JENSEN, Jay (1976). *Liberalism, Democracy and the Mass Media*. Urbana: University of Illinois Press.
JANOWITZ, Morris (1967). *The Community Press in an Urban Setting*. Chicago: University of Chicago Press.
JUENGER, Fr. (1949). *The Failure of Technology*. Hinsdal, I11.: The Humanist Library.
JANIK, Allan and Stephen TOULMIN (1973). *Wittgenstein's Vienna*. New York: Simon and Schuster.
KLANE, Michael (1972). *War Without End*. New York: Vintage Books
KLAPPER, Joseph T. 'Mass Media and Engineering of Consent.' In Ed. Wilbur Schramm. (1964), *op. cit.* pp. 523–544.
KRANZBERG, Melvin and William M. DAVENPORT (1972). *Technology and Culture, an Anthology*. New York and Scarborough, Ont.: A Meridan Book.
KUHN, H.S. (1970). *The Structure of Scientific Revolution*. Chicago: The University of Chicago Press.
KUHNS, W. (1971). *The Post-Industrial Prophets: Interpretations of Technology*. New York: Harper and Row.
KUMAR, Krishan (1978). *Prophesy and Progress: The sociology of Industrial and Post-Industrial Society*. Harmondsworth: Penguin Books.
LACAN, Jacques (1966). *Ecrits*. Paris: Seuil.
LANE, Robert E. (1966). 'The Decline of Politics and Ideology in a Knowledgeable Society.' In *American Sociological Review*, 31, No. 5, pp. 649–662.
LA PIERRE, Richard T. (1954). *A Theory of Social Control*. New York: McGraw-Hill Book Co.
LASZLO, Ervin (1972). *Introduction To Systems Philosophy: toward a new paradigm of contemporary thought.* New York: Gordon and Breach.
LAUDON, Kenneth Craig (1977). *Communications Technology and Democratic Participation*. New York: Praeger.
KIERKEGAARD, Sören. 'The Unhappiest Man.' In *Either/Or*, trans, D. Swenson. Princeton: Princeton University Press, 1944–71.
LAZARSFIELD, Paul et al. (1964). 'Mass Communications.' In Ed. Schramm (1964). *op. cit.*, pp. 523–544.

LAZARSFIELD, Paul, BERELSON and GAUDET (1948). *The People's Choice*. New York: Columbia University Press.

LEISS, William (1974). 'Critical Theory and its Future.' *Political Theory*, vol. II, no. 3, August, pp. 330–349.

LEISS, William (1972). *The Domination of Nature*. New York: George Braziller, 1972.

LEISS, William (1971). 'Husserl's Crisis.' *Telos*, no. 8, Summer. pp. 109–121.

LEISS, William (1976), *The Limits to Satisfaction*. Toronto: University of Toronto Press.

LEISS, William (forthcoming) 'Nature as Commodity: Landscape Assessment and the Theory of Reification.' In *Boston Studies in the Philosophy of Science*. Ed. Cohen and Wantofsky. Dordrecht, Holland: Reidel.

LEISS, William (1970). 'Utopia and Technology: Reflections on the Conquest of Nature.' *International Science Journal*, Vol. XXIII, no. 4, pp. 576–588.

LENTRICCHIA, Frank (1982). 'Reading Foucault (Punishment, Labor, Resistance),' Parts I and II, in *Raritan*. Spring and Fall, vol. 1, no. 3 and vol. 2, no. 1, pp. 5–33 and pp. 41–70.

LERNER, Daniel and Wilbur SCHRAMM, Eds. (1967). *Communication and change in the developing countries*. Honolulu: East-West Center Press.

LEVI-STRAUSS, Claude (1970). *Tristes tropiques: an anthropological study of primitive societies in Brazil*. Trans. J. Russel. New York: Atheneum.

LEVITT, Kari (1970). 'The Hinterland Economy.' *Canadian Forum*, 50, July–August, 163.

LIPPMANN, Walter (1949). *Public Opinion*. New York: Macmillan.

LUHMAN, Niklas (in debate with Habermas) (1971). *Theorie der Gesellschaft oder Sozialtechnologie*? Frankfurt: Suhrkamp Verlag.

LUKES, Steven (1974). *Power: a radical view*. London, New York: Macmillan.

LYOTARD, Jean-Francois (1979). *La Condition Post-Moderne, Rapport sur le Savoir*. Paris: Minuit.

LYOTARD, Jean-Francois (1977). *Rudiments paiens*. Paris: Union Générale d'Éditions.

MACHLUP, Fritz (1962). *The Production and Distribution of Knowledge in the United States*. Princeton: Princeton University Press.

MACPHERSON, C.B. (1965). *The Real World of Democracy*. Toronto: CBC/Hunter Rose.

MALCUZINSKI, Pierrette (1982). 'Le dialogiome dans le roman latino - américain.' Conference presented at the founding meeting of the C.Q.F.D. - Centre Québécoise de l'Analyse du Discourse Social. November, Montreal.

MARCUSE, Herbert (1965–1968b). 'Repressive Tolerance.' In R. Wolff, B. Moore, H. Marcuse. *A Critique of Pure Tolerance*. Boston: Beacon. pp. 81–122.

MARCUSE, Herbert (1968). *Negations: Essays in Critical Theory*. Trans. Jeremy J. Shapiro. Boston: Beacon.

MARIN, Louis (1975). *La Critique de discours*. Paris: Minuit.

MARX, Karl (1967). *Capital Vol. I–IV*. Trans. S. Moore and E. Aveling, New York: International Publishers.

MARX, Karl and Frederick Engels (1959–1974). *The German Ideology*. Ed. and Intro. C.J. Arthur. New York: International Publishers.

MARX, Karl (1972). *Karl Marx: The Essential Writings*. Ed. and Intro. F. Bender. New York: Harper Torchbooks.

MATTELART, Armand (1979). *Multinational Corporations and the Control of Culture*. Trans. Michael Chanan. Sussex, New Jersey: Harvester/Humanities Press.

MATTELART, Armand (1978). 'Notes on the Ideology of the Military State.' In *Communications and Class Struggle*. Ed. Armand Mattelart and S. Seigelaub. New York: International General.

McCARTHY, Thomas (1978). *The Critical Theory of Jürgen Habermas*. Cambridge: MIT Press.

McLUHAN, Marshall (1967). *The Mechanical Bride*. Boston: Beacon.

McLUHAN, Marshall (1964). *Understanding the Media: the extension of Man*. New York: Signet.

McLELLAN, David (1976). *Karl Marx, his Life and Thought*. Frogmore, St. Albans: Paladin.

MELODY, SALTER, HEYER, et. al. (1981). *Communication, Culture and Dependency: The Tradition of H.A. Innis*. Norwood, N.J.: Ablex.

MUELLER, Claus (1970). 'Notes on the Repression of Communicative Behaviour.' In *Recent Sociology*, No. 2, '*Patterns of Communicative Behaviour*.' Ed. Hans Pert Dreitzel. London: Macmillan, 101–113.

MUMFORD, Lewis (1963). *Technics and Civilization*. New York: Harcourt, Brace, Jovanovich.

MURDOCH, G., and P. GOLDING (1984). 'For a Political Economy of Mass Communications.' In *Socialist Register*. Merlin Press.

NOBLE, David (1977). *American By Design*. New York/London: Harvester Humanities Press.

NOBLE, David F. (1982). 'M.I.T. – Whitehead Merger: The Selling of the University.' *The Nation*. Vol. 234, no. 5, (February 6), pp. 129–48.

NORTH, Douglas C., and Robert Paul THOMAS (1973). *The Rise of Western Europe*. Cambridge: Cambridge University Press.

OSBORNE, Adam (1979). *Running Wild: The Next Industrial Revolution*. Berkeley, California: Osborne/McGraw-Hill.

O'SULLIVAN, Ryan, and Mario KAPLAN (1981). *Communication Methods to Promote Grass Roots Participation in Latin America*. Communication and Society No. 6. Paris: UNESCO.

PEIRCE, Charles Sanders (1931–58). *Collected Papers of Charles Sanders Peirce*. Ed. Charles Hartshorne and Paul Weiss, (vols. 7 and 8 Ed. Arthur Burks). Cambridge, Mass,: Harvard University Press.

PEIRCE, Charles Sanders (1972). *Charles S. Peirce: The Essential Writings*. Ed. C. Moore. New York: Harper and Row.

PERRY, James M. (1968). *The New Politics, The Expanding Technology of Political Manipulation*. New York: Weidenfield and Nicolson.

PETERSON (1956). 'The Social Responsibility Theory of the Press.' In Ed. Siebert. *Four Theories of the Press*. Urbana: University of Illinois Press.

PYE, Lucien W. (1963). *Communications, Political Development and Social Change*. Princeton, N.J.: Princeton University Press.

PRIGOGINE, Ilya and Isabelle STENGERS (1979). *La Nouvelle Alliance: métamorphose de la science*. Paris: Gallimard.

RIESMAN, David (1950–77). *The Lonely Crowd*. New Haven: Yale University Press.

ROBBE – GRILLET, Alain (1963). *Pour un nouveau roman*. Paris: Minuit.

ROLING, N.J. ASHCROFT, and T. CHEGE (1976). 'The diffusion of Innovations and the Issue of Equity in Rural Development.' In Ed. Rogers. *Communication and Development: Critical Perspectives*. Beverly Hills: Sage Publications.

ROWAT, Donald (1973). *The Ombudsman Plan*. Toronto: McLelland and Stewart.

SALTER, Liora and Debra SLACO (1981). *Public Inquiries in Canada*. Ottawa: Science Council of Canada, Supply and Services.

SAURRAUTE, Nathalie (1956). *L'ère du soupcon*. Paris: Gallimard.

SCHELLING, Thomas C. (1960). *The Strategy of Conflict*. Cambridge, Mass.: Harvard University Press.

SCHILLER, H.I. (1976). *Communication and Cultural Domination*. White Plaines, New York: International Arts and Sciences Press Inc.

SCHILLER, Herbert I. (1969). *Mass Communications and American Empire*. New York: A.M. Kelley.

SCHILLER, Herbert I. (1973). *The Mind Managers*. Boston: Beacon Press.

SCHRAMM, Wilbur (1964). *Mass Media and National Development*. Stanford: Stanford University Press.

SCHROYER, Trent (1973). *The Critique of Domination: The Origins and Development of the Critical Theory*. New York: George Braziller.

SCHUMACHER, Fritz (1932). *Der 'Fluch' der Technik*. Hamburg: Boysen and Maasch Verlag.

SCUPHAM, John (1967). *Broadcasting and Community*. London: Watts.

SERRES, Michel (1968). *Hermes I: Ou la Communication*. Paris: Minuit.

SMITH, Anthony (1980). *The Geopolitics of Information*. Oxford: Oxford University Press.

SHORSKE, Carl E. (1981). *Fin-de-siècle Vienna*. New York: Vintage Books.

TOURAINE, Alaine (1971). *The Post-Industrialist Society*. New York: Random House.

TUNSTALL, Jeremy (1977). *The Media Are American*. New York: Columbia University Press.

UNESCO (1978). 'Interim Report on Communications Problems in Modern Society.' September, Paris: UNESCO.
'Vous avez dit culture technique?' (1982). Special Issue of *Esprit*. Fall, pp. 3–97.
WATZLAWICK, Paul (1977). *How Real is Real?: Confusion, Disinformation, Communication*. New York: Vintage.
WEBER, Max (1927–1964). *From Max Weber: Essays in Sociology*. Ed., and Trans. H.H. Gerth and C. Wright Mills. New York: Oxford University Press.
WEBER, Max (1924). *Gesammelte Aufsatze zur Soziologie and Sozialpolitik*. Tubingen: Mohr.
WEBER, Max (1921). *Gesammelte Politisch Schriffen*. Munchen: 1921.
WEBER, Max (1947–1964). *Max Weber: The Theory of Social and Economic Organization*. Ed. and intro. Talcott Parsons. Trans. A.M. Henderson. New York: The Free Press.
WELLMER, Albrecht (1974). 'Communication and Emancipation. Reflections on the "linguistic turn" in Critical Theory.' Polycopy of Unpublished Article. Montreal: McGill University.
WELLMER, Albrecht (1974). *Critical Theory of Society*. Trans. John Cumming. New York: Seasbury Press.
WHITE, David Manning (1950). 'The "Gate-Keeper" – Study of the Selection of News.' *Journalism Quarterly 63*, 27, No. 4, Fall, 383–390.
WILLIAMS, Raymond (1958–77). *Culture and Society*. Harmondsworth: Penguin.
WILLIAMS, Raymond (1976). *Keywords: a Vocabulary of Culture and Society*. Glasgow: Fontana.
WILLIAMS, Raymond (1961). *The Long Revolution*. Harmondsworth: Penguin.
WILLIAMS, Raymond (1974–8). *Television, Technology and Cultural Form*. Glasgow: Fontana.
WIRTH, Lewis (1964). 'Consensus and Mass Communications.' In Ed. Schramm (1964), *op. cit.*, pp. 523–533.
WITTGENSTEIN, Ludwig (1958–74). *The Philosophical Investigations*. Oxford: Blackwell.
ZUMTHOR, Paul (1975). *Langue, texte, énigme*. Paris: Seuil.

INDEX

access, 36, 118,-21, 251-2, 312-13, 399; and control of source, 128-30
accountability, 143-5
a-contextualization of technology, 50-6
actants, 89–90
Adorno, T.W., 35, 223, 235–6, 334
advertising, 47-9, 103, 108
Albion, R.G., 34, 318
Alfred, R.A., 295
alternative procedures of discourse, 239–308; discursive transformations, 249–54; feasibility, 239–41; freedom, 254–63; institutions, 246–7; interacationism, 263–72; interactivity, 272–83; legislation, 244–6; ombudsman, 305–7; policy, 241–3, 283–93; politics of communication, 294–7; public inquiries, 297–305: regulation, 248–9
analysis, 42–3; discourse, 2–5, 319–22, 335; of knowledge, 117–18
anonymity of power, 203–7
anthropomorphization of technology, 70–2
Antiope, 18, 111, 154
Apel, K.-O., 264, 334
Appelbaum-Hébert Commission Report, 289, 300
Arrow, K.J., 119
artificial intelligence (AI) research, 39–41, 57, 60, 70, 136–7
atomism, 117–18
autonomy: individual, 258–61; national, 38–9, 149, 155–6, 312–13, 338–9

Babe, R., 101
Bachelard, G., 324
Bacon, F., 72, 171
Baer, W.S., 29, 104
Bakcsy, A. de, 124
Bakhtin, M., 105, 264–5, 275–6, 303
Barthes, R., 71, 96, 101, 136, 159, 270, 335
Bateson, G., 58, 61, 67, 88, 165, 264
Bell, D., 6, 84, 135, 318, 321–2
Benveniste, E., 159
Bentham, J., 110
Berkley, E.C., 5
Berkman, D., 44, 81, 96–7, 100
Blumenthal, R., 101
Blumer, J., 316, 331
Boden, M., 71, 73–6, 95, 103
Boulding, K., 200
Bounine, J., 152–4
Braverman, H., 205, 207, 210
Broadcasting Act (1968), Canada, 252–3, 289
Brook, H., 54, 112
Brow, H., 95
Brownstein, C. N., 39
Burck, G., 152
bureaucracy, 183, 192–6, 212–14, 232, 310

Campus, R., 155
Canada: autonomy of communications, 149; Broadcasting Act, 252–3, 289; Computer/Communications Policy, 286; content and carrier, 38, 120; CRAB Report, 94, 101, 154, 287; Department of Communication (DOC), 66, 90, 98, 116, 180, 217, 302; free flow, 234–5; government intervention, 64, 154, 252–3; hardware industry, 161; information provision, 119; Institute for

Public Policy Research, 37, 146, 169–70; Kent Commission Report, 154, 234, 284, 293; Science Council study, 306; use of telecommunications, 92–3; *see also* Telidon
capitalism, 9, 36–7, 83, 203–6, 220
Cardoso, F. M., 65, 92
Carey, J., 29, 291; a-contextuality, 50, 54–6, 203, 326; communication, 21, 123–4, 178; futurology/idealization, 34, 49–50; mythos of electronics, 8, 27, 34, 55–6, 69
cartels, 115–17; *see also* institutions
Castoriadis, C., 17, 28, 88, 177, 260, 325
Cater, D., 303
centralization: of communications, 147–57, 285–7; of power, 198–201
Centre pour la Recherche sur l'Épistémologie et l'Autonomie (CREA), 258–60
charisma, 310
Chicago School, 264
Chisman, F. P., 303
class conflict, 180–1, 189, 234
Clausewitz, K. von, 185
'closed user groups', 131–4
Clyne commission, 293
Cohen, R. S., 52
Cohn-Bendit, D., 28, 88
Columbia School, 317
communication, 9; alternative approach to, 294–7, 308–39; competence, 277–9; as discourse, 11–14; rights, 250–3, 261–2, 266–7, 277–83, 291–3
communications technology, 3–4, 8; context of, 140–3, 323–8; as discourse (techne), 110–67 (access, 118–21; analysis/atomism, 117–18; cartels, 115–17; centralization, 147–57; 'closed user groups', 131–4; comp-unications, 114–15; context, 140–3; control of source, 128–30; efficiency, 145–6; knowledge reductionism, 134–6; networked privatization, 157–8; on-line-real-time mediation, 126–8; participation, 159–61; quantity *vs.* quality, 146–7; speed, efficiency, *etc.*, 145–6;static/dynamic language, 121–6; surveillance, 113–14; systems theory, 161–7; unaccountability, 143–5; unilateral control, 136–40); discursive critique (*q.v.*), 308–39; object of study, 321–3; ombudsman, 305–7; procedures of discourse on (episteme), 25–109 (a-contextualization, 50–6; anthropomorphization, 70–2; discursive approach, 26–8; double blind, 61–8; efficiency, 81–4; futurology, 45–50; individualization and atomization, 74–8; issues, 36–42; knowledge reductionism, 56–61; legitimation of authority, 102–5; man *vs.* machine, 72–4; mediation, 68–70; narrative senario, 43–5; order and analysis, 42–3; passification of the public, 89–95; public discredited, 96–102; referentiality, 29–36; reported discourse, 105–7; speed, efficiency, *etc.*, 81–4; surveillance, 79–81; trivilialization, 107–9; unilaterality of control, 84–9); and public inquiry, 297–305
competence, communicational, 14, 277–9, 281–3
competition, man *vs.* machine, 72–4, 188–9, 265
complexity of knowledge, 57–60; *see also* reductionism
comp–unications, 114–15
computers, 4, 57–9; *see also* artificial intelligence
Comte, A., 171
Condorcet, 171

conspiracy theory of power, 189–90, 204
content, 15–16, 38, 120
context: a-contextualization of technology, 50–6; contextualization of technology, 286–7, 326–8, 336–7; of dicourses of knowledge, 140–3; *see also* episteme
contradictions, 221–3
control: awareness of, 171–2; de-centred non-causalist theory of, 180–2; of discourse procedures, 169; and public interest, 252–3; of source of information, 128–30; unilaterality of, 84–9, 136–40; *see also* power; powermatics
'corporate individuals', 76, 91, 263
Cowlan, B., 250
CRAB Report, 94, 101, 154, 287
critical theory, 221–2, 255
critique, 169; discursive *see* discursive critique
cultural diversity, 141, 287–9, 312–13, 338–9
cybernetics, 164–5

Dagenais, A. 46, 48, 62, 79–80, 90, 108
d'Archy, J., 242, 266, 294
Darnton, R., 99, 175
data banks, 115–17
Davey commission, 293
decentralization: of communications, 147–57; of power, 198–203
de-contextualization of technology, 50–6
democracy, 89, 161, 239, 279–83; false, 267–70; and liberalism, 255–8; and rationality, 231–4; *see also* freedom
demystification, 8–10, 335
Denicoff, M., 41, 100, 129
Department of Communication, Canada, (DOC), 66, 90, 98, 116, 180, 217, 302
Dertouzos, M., 33, 47–8, 74–7, 82, 88, 103–4, 314
Descartes, R., 40, 72, 88
Dessauer, F., 49–51, 68–9, 71, 328
development, 6–7, 37, 50, 65, 83–4
deviance, 125
dialogue, 18–19, 272–6
Dick, P. K., 254
discipline, 182–3, 206
discourse, 11–15, 319–20; communication as, 11–14; defined, 15; and knowledge, 19–21, 174–7; procedures of, 14–16, 169, 185–6, 319–20 (*see also* alternative procedures); reported, 105–7
discourse analysis, 2–5, 322, 335; and object of study, 11–14, 319–22 discursive critique, 308–39; change, 332–4; contextualization, 326–8, 336–7; disintegration, 311–16; empiricism, 317; epistemic relativization, 335–9; historical context, 323–5, 336; idealization, 325–6; irregularities, 310–11; knowledge, 308–9, 331–2; myth, 334–5; object formulation, 319–23; power, 328–31; referentiality, 318–19
discursive hegemony, 199, 214–16, 219, 294–7
discursive procedures, 14–16, 169, 185–6, 319–20; alternative (*q.v.*), 239–307
'disemployment', 7
disintegration, 311–16
diversity, regional and cultural, 141, 287–91, 312–13, 338–9
Domenach, J. M., 259–60
double bind, 61–8
Dreyfus, H. L. 117, 124, 134, 136–7, 140, 142
Dumouchel, P., 259–60
Dupuy, J. -P., 249, 259–60
dynamic language, 121–6

economics, and power, 189–92, 196–7

Edelman, J. M., 270
education, 73–4, 141, 208, 212–13
efficiency, 81–4, 145–6, 217
electronic funds transfer (EFT), 145–6
'Eliza' program, 41, 87, 125, 129, 138–9, 271
Ellul, E., 30–1, 46, 50–2, 71, 84, 96–9, 331
empiricism, 317, 332, 336
encryption, 131–2, 143
engineering, scientific, 207–11
English, E., 269
episteme, 4, 21–3, 324, 328; dominant/classical, 75, 137, 206, 208, 219, 227, 231–2, 242, 249, 253, 310; and techne, 23–4; *see also* communication technology
epistemic relativization, 335–9
epistemology: alternative, 308–9, 317–18, 331–4; and autonomy, 258–61; and systems theory, 165–6; of technology, 239–307
exchange, 34–6, 193
exclusivity, 131–4

Faletto, E., 65, 92
Faulkner, H., 92
Fedoseyev, P. N., 52, 196
feedback, 107–9, 159–61, 268–9
Feuerbach, L. A., 73
Feyerabend, P., 20
Finlay-Pelinski, M., 218, 278
Fisher, D., 247, 250, 262–3, 267–8, 282–3
Foidart, D., 66, 98
Foote, J., 302
Forrester, T., 103, 296, 314
Foucault, M.:anthropocentric episteme, 72; change, 29, 196, 294, 296, 331; comentary, 25; discourse, 12–13, 203, 320–1; discursive procedures, 14–21, 131–2, 321, 324, 328, 331; dominant/classical episteme, 75, 137, 206, 208, 219, 227, 231–2, 242, 249, 253, 310; episteme, 22–3, 324, 328; exchange, 35, 193; field of truth, 102, 132, 171–2; knowledge, 19–20; panopticism, 110, 177, 179–80; participation, 296; power, 116, 169, 171–4, 178, 181–5, 190–1, 203, 237, 329–30; power and discourse, 169, 171–4, 181–2, 190–1, 203; sites of power, 198–9, 204–5, 215; subject of discourse, 75, 137; surveillance, 110, 115–6, 177, 179–80, 269;technologies, 16
Fox, F., 38, 116, 149, 180, 217
Frankfurt School, 221–2, 228, 255, 258, 317, 330
free flow, 37, 77, 82–3, 91, 234–5, 251–3
freedom, 77, 248–9, 254–63; autonomy and responsibility, 258–61; democracy and liberalism, 255–8; functional, 274; individualism, 262–3; and the right to communicate, 261–2; *see also* rights
Freud, S., 310
Freidrichs, G. V., 64, 93, 300
Fujitsu, 49, 107
Fulford, R., 46, 67, 91
funds, transfer of, 145–6
funnel effect, 150–1, 290
futureology, 45–50

'Gadget–philia', 30, 185, 294
Galbraith, G., 302
Galbraith, J. K., 205, 207, 214–16, 220
Galtung, J., 127
Garret, J., 81–2, 89
Gassman, H. P., 48, 85, 87, 121, 146, 148, 197
Geertz, C., 331
Gehlen, A., 40, 242
Geller, H., 197
Genette, G., 136
Gilpin, R., 66, 83, 198, 200–1, 205, 218–19
Godfrey D., 29, 36, 56–7, 104, 141, 239
Goldhaber, M., 275

Gotlieb, C. C. : double binds, 62–4; EFTS, 146; employment, 42, 84; industry, 39, 62–4, 82, 160, 211, 218–19, 285, 328; multinationals, 82, 211, 218–19; national autonomy, 156, 314
Gotshalk, D. W., 235
Gouldner, A.W., 68
government, 230; intervention, 55, 63–4, 91, 101–2, 155 (*see also* legislation); investment, 5, 111, 161, 285; monitoring and control, 143–4; use of technology, 114; *see also* power
Gramsci, A., 196
Grant, G., 168, 196
Greimas, A. J., 89–90, 316
growth, 82–3
Guité, M., 99, 101

Habermas, J.: capitalism, 83–4;communicational competence, 14, 277–9, 281–3; discourse, 13–14, 28, 306–7, 333; formal rationality, 194, 212–13, 228, 235–6, 256; freedom, 248, 255–7, 260; inter-actionism, 264; participation, 267–8, 292, 296, 303
Hall, E. T., 110
Hamelink, C.J., 311, 317, 330
Hammarskjold, D., 280, 283
hardware, 112, 285–6
Harms, L. S., 240–2, 247, 250, 261–2, 266, 277–82, 294, 302–3
Heath, S., 71
Hegel, G. W. F., 160, 308
hegemony, discursive, 199, 214–16, 219, 294–7
Heilmeier, G., 102
Heisenberg, W., 175, 318
hermeneutics, 309
Hi-Ovis, 18, 111
history: as context, 10 (*see also* episteme); of control, 171–2; of technology, 45–50, 323–5, 328, 336
Hitler, A., 67
Hobbes, T., 72, 223
Hocking, W. E., 77
Horkheimer, M., 221–2, 255–7
Hughes, P., 87
Husserl, E., 224–5
Hymer, S., 218

IBM, 69, 155, 198–9, 211, 218–19
'Ideal speech situation', 278–80
idealization of technology, 50–6, 325–6, 337
immediacy, 70, 126–8
individualism, 262–3
individualization, 74–8, 157–8
individuals, corporate, 76, 91, 263
Indonesia, 93
industrial revolution, 318–19
industry, 63, 160, 216–19, 285, 328; national, 39, 82, 156, 218–19; *see also* managerialism
Informart, 94, 119, 290
informatics, 164–5
information, 312–13; access to, 36–7; data banks, 115–17; theory, 161–7
Innis, H., 28; decentralization, 54, 201, 262, 287, 296; power, 169, 200–1, 203, 227, 262, 265, 287; 'space-binding', 123, 127, 147, 154, 200, 203
inquiries, public, 297–305
Instant World (Canadian Telecommission), 55, 78, 247, 262
Institute for Public Policy Research, Canada, 37, 146, 269–70
institutions, 208–11, 219–20, 246–7; *see also* government; sites of power
interactionism, 263–72; false participation, 267–72; man and machine, 265; right to communicate, 266–7; theory, 264–5
interactivity, 272–83; communicational competence, 277–9; dialogue, 272–5; 'ideal

speech situation', 278–80; procedures of dialogism, 275–6; right to communicate, 279–83
Intercom conference, Brazil, 292
investment, 5, 111, 161, 285
irregularities, 310–11
issues, 36–42, 184–5, 311–16

Jay M., 317
Jensen, J., 77, 235, 282
Johnson, B., 45
Juenger, F., 52

Katz, E., 316, 331
Kent Commission Report, 154, 234, 284, 293
Kidder, T., 125, 128–9, 141, 188, 211
Kierans, E., 74
Kierkegaard, S., 323
Knecht, H., 274
knowledge, 308–9; and discourse, 19–21, 174–7; reductionism, 56–61, 134–6; and value, 331–2, *see also* episteme
Kuhn, H.S., 118
Kumar, K., 46, 85, 171, 225, 233, 242, 310, 319
Kurchak, M., 118, 120, 130–1, 133–4, 161

Lacan, J., 32
laissez-faire, 37, 83; *see also* free flow
language, 174, 239; dynamic *vs.* static, 121–6; *see also* discourse
Laszlo, I., 162–7
Lazarsfield, P., 158
Le Bon, G., 98
legislation, 114, 244–6, 250–3, 284
legitimation, 102–5, 229–30, 236
Leiss, W., 74, 256
Lemelhstrich, N., 159
Lentricchia, F., 110, 204–6, 209–10, 215, 220
Lévi-Strauss, C., 101
Levitt, K., 217
Lewin, K., 119
liberalism, 234–5; and democracy, 255–8; neo-liberalist cynicism, 235–7
Licklider, J. C. R., 39, 47, 84–6, 100, 149, 219
Lippman, W., 127
Locke, J., 72, 223
Lowi, T., Jr: EFTS, 145–6; individualization, 77; institutions of power, 133, 186, 188, 199, 201–2, 205, 213, 218–20, 230, 236; localized communication, 296–7; managerialism, 199, 201–2, 218,, 230 26; manipulaton, 146, 176, 179; open *vs.* secret society, 102, 133; surveillance, 116
Luhman, N., 96–7
Lukasiewicz, J., 135
Lussato, B., 152–4

MacBride Commission Report, 37, 234, 248, 251–2, 267, 280–1
McCarthy, J., 278
MacClean, D., 184–5
McClean, J. M., 63–4, 318
machines, 4, 31; man *vs.*, 72–4, 188–9, 265; *see also* mediation
Machlup, F., 6, 218, 318
MacKinder, H., 83
Macoby, J., 216, 220, 324
Macpherson, C. B., 256
Madden, j.C. 29, 43, 81, 151, 217
madness, 20–1, 172, 177, 310
management, 218, 229–30
managerialism, scientific, 207–11
'manipulated consensus', 176
Marcuse, H., 173, 183, 196, 227–9, 248–9, 255
Marx, K., 53, 85, 180, 189, 197, 225
Marx, L., 50
Mattelart, A., 9, 11, 71, 121, 127, 158, 211, 216
Matthes, W. M., 5
Mead, G. H., 264
mediation, 68–70; on-line-real-time, 126–8

medicine, computerization of, 138–9
Melody, W. H. 154, 266, 296
Mendlesohn, H., 274
Menzies, H., 84, 154
Metz, C., 30
military-industrial complex, 86, 216–18, 312–13, 338–9
Mill, J.S., 36
Mills, C. W., 181–2, 201, 225–6, 232
Minc, A., 114, 141, 152–5, 176, 186, 201–2, 234, 314
Minsky, M., 134
Mitchell, W., 210
monitoring of data processing, 143–5
Mosco, V., 304
Moses, J., 71, 82, 103, 314
multinational enterprises, 69, 82, 155, 198–9, 211, 218–19
Mumford, L., 30–2, 40, 49–52, 69, 103, 170, 325–6
mythos, electronics, 8–10, 27, 34, 55–56, 69

narrative scenario, 43–5, 100
Nasselund, M., 247, 266
'naturalization', 335
Nehru, J., 87
neo-liberalist cynicism, 235–7
networks, 115–16, 148–58, 290
Newell, H., 41, 70
Neitzsche, F., 183, 191
Noble, D. F., 204–11, 219–20
nodes of power *see* sites
Nora, S., 114, 152, 154–5, 186, 314

occultation, 170–1
ombudsman, 305–7
on-line-real-time, 70, 126–8
opinion, public, 89–102, 268–70
Oppenheimer, D., 104
Organization for Economic Co-operation and Developmen (OECD), reports: centralization, 148, 155, 201–2; content and carrier, 120; control, 85, 87, 89, 143–4, 186; culture, 141; deregulation, 55, 91, 154–6; free flow, 37–8, 76, 332; futurology, 47–8; information collection, 145–6, 197; power, 170, 186; social conflict, 234
Ostry, B., 6, 103
Ouimet, A., 101, 154
Owen, K., 73

Palo Alto School, 61, 264
panopticism, 80, 110, 177–80
Papert, S., 134
paranoia, 203–7
Parker, D., 101
Parkhill, D. F., on: access, 36; authority derivation, 29, 104; democracy, 89, 239; deregulaton, 55, 77; development, 84; education,57, 141; networking, 150
Parsons, T., 225, 228
participation, 18, 39, 159–61; false, 267–72; public inquiries, 297–305; *see also* democracy
Peirce, C. S., 137, 143, 253, 264
Pergler, P., 159, 170, 202, 269–70, 289
perspective, 286; control of, 136–40 *see also* episteme
philosphy, 174, 264, 309
Piera, F., 48
'pig' principle, 36, 81–2, 146–7
Plato, 99
poetry, 5, 17
policy, 240–3; and automony, 261; contestualization, 286–7; legislation, 284; networking, 290; perspective, 286; regional and cultural diversity, 287–9; sites of power, 291; software, 285–6; unplugging, 291–3
politics of communication, 294–7
Pool, I. de S., on: corporations, 76, 90–1, 97; free flow, 37–8, 55, 76, 143–4, 154, 156, 332; participation, 267, 274, 282, 295–6; uses of technology, 240, 244

Popper, K., 161
Porat, M., 6, 84, 318, 322
power, 173–4; and alternative politics of communication, 294–7; centralization of, 198–203; and discourse procedures, 169; holders of, 186–8; and reason, 182–3; relations of, 196–7; repression and coercion, 183–4; sites of, 198–9, 203–9, 215, 291; study of, 9–10, 328–31
powermatics, 168–238; anonymity, paranoia, sites of power, 203–7; awareness of control, 171–2; bureaucracy, 192–6, 212–14, 232; centralizaton, 198–203; characteristics of power, 173–4; conspiracy theory, 189–93; critique of power, 1172–3, 221–4; de-centred non-causalist theory, 180–2; discourse and knowledge, 174–7; economics, 189–92, 196–7; external procedures of discourse, 185–6; extra-discursive relations, 218–19; formalist rationality, 211–14; holders of power, 186–8; issues, 184–5; liberalism, 234–7; man *vs.* machine, 188–9; military-industrial complex, 216–18; neo-liberalist cynicism, 235–7; occultation, 170–1; panopticism, 177–80; politics and science, 230–1; power and reason, 182–3; rationality and democracy, 231–4; rationality, critique of, 224–30; repression and coercion, 183–4; scientific engineering, 207–11; 'technostructure', 214–16, 219–20
Prague Symposium (1976), UNESCO, 52, 64, 68, 93, 300
Prestel, 18, 94, 111, 154
privacy, 37–8,. 48, 75–8, 253–4, 312–13, 339; *see also* surveillane
privatique networkg, 152–7
privatization of communication, 157–8
procedures, 2–4; of dialogism, 275–6; of discourse, 14–16, 169, 185–6, 319–20 (*see also* alternative procedures); and technologies, 16–19
professionalism, 103–4
programming, 128–30; *see also* software
progress, 45–50, 81–4, 145–6
prohibition, 132
Proulx, S., 33
public, the: discrediting of, 96–102; interests of, 252–3, 285–6; passification of, 89–95; and power, 230; public inquiry, 297–305
publishing, 120–1, 131

QUBE system, 33, 147, 154, 191
Quirk, J., 8, 27, 34, 49–50, 54–5, 69, 203, 326

rarefaction, 132
rationality, 168, 310; critique of, 224–5; and democracy, 231–4; formal, 193–6, 211–14; irrationality of, 226–9, 310; self-legitimation of, 229–30; substantive, 195–6
Reagan, R., 104–5
reality; control of, 194–5; mediation of, 126–8
reason, and power, 182–3
reductionism of knowledge, 56–61, 134–6
referentiality, 29–36, 318–20
regional diversity, 141, 287–9, 312–13, 338–9
Reid, A. L., 64, 94
responsibility, 258–61
revolution, 46; communications, 8–10, 33–4, 46, 184–5, 318; industrial, 318–19
Richstad, J., 266, 277, 302
Riesman, D., 71, 76
rights, human and civil, 7, 37–8, 48, 75–8, 248–9; communication, 250–3, 261–2, 266–7, 277–83; not

to communicate, 291–3
Robbe-Grillet, A., 71
Robinson, G. O., 29, 104, 314
Rosenblith, W., 125
Russel, R.A., 37, 84, 94, 214, 309

Saint-Simon, C. de, 171
Salter, L., 154, 265–6, 296–301, 305
Sarraute, N., 71
Sasson, R., 87
satellites, 66, 70, 197, 313
scenario, narrative, 43–5, 100
Schiller, H.I., 93, 216, 250, 262
Schumacher, F., 53, 328
science, 171, 230–4; *see also* empiricism; knowledge
Science Council of Canada, study, 306
scientific engineering, 207–11
'scopophilia', 30
Seeley, D., 139–40, 261, 274–6, 286, 335
Servan-Schreiber, J.-J., 11
Shannon, C., 164–5
SHRDLU program, 117, 134
Simon, H.A.: artificial intelligence, 40–1, 60, 70, 117; centralization, 54, 147–9; EFTS, 145; government, 55, 179; need for technology, 65, 85
sites of power, 198–9, 203–9, 215, 291
Slack, J. D., 75, 242, 247, 284
Slaco, D., 297–301, 305
Smith, Adam, 37, 223
Smith, Anthony, 50, 65, 93, 235
society, 8–10; *see also* power; public; *etc.*
software, 112, 129–30, 285–6, 290–1
Solomon, R.J., 37–8, 55, 76, 90–1, 97, 143–4, 154–6, 332
Southam, 119, 154, 272
'space-binding', 123, 127, 147, 154, 200, 203
speech-recognition, 116
static language, 121–6
Steele, D., 118
Steiner, G., 23
Stewart, W., 43, 67, 74, 96, 103, 105
surveillance, 78–81, 110, 113–17, 177, 179–80, 269, 213–13, 339
Sweden, 39, 154
systems theory, 161–7, 309

Tamec Report, 99, 154
Taylor, J., 198
'Taylorism' 209·10
techne, 17, 23–4; *see also* communication technology
technics, 31, 51–3, 68
technocracy, 97, 229, 232, 308
technologies, 16–19
technology: contextualization of, 326–8; control of, 207; epistemologies of, 239–307; history of, 45–50, 323–5; idealization of, 50–6, 325–6; legislation of, 244–6, 250–3; ombudsman, 305–7; *see also* communication technology
'technostructure', 214–17
telematics, 111–14, 137–9
telematique networking, 152–7
television, 46, 101, 108; cable, 146–7, 312–13, 338–9
Telidon, 18, 46, 90–1, 111, 119, 129, 147–9, 154–5, 268–9, 285
Thompson, G., 36–7, 74, 103
Toffler, A.: futurology, 46–7, 50, 69, 91, 108; on power, 39; reductionism, 57–8; referentiality, 31–3, 322; 'wave', 31–2, 91
tools, 31, 112
Torstar, 119, 154, 272
Tracy, D. de, 177
transformations, discursive, 249–54; communication rights, 250–2; control, 252–3; privacy, 253–4
trivialization: of technology, 107–9;of public opinion, 96–102, 268–9 truth, 20, 192–4, 132, 171–3, 242

unilaterality of control, 84–9, 136–40
United Nations: Charter of Human rights, 248, 262; Hammarskjold Report, 279–80, 283
UN Educational, Scientific and Cultural organization (UNESCO), 240, 247, 261, 281–2; MacBride Report, 37, 234, 248, 251–2, 267, 280–1; Prague Symposium, 52, 64, 68, 93, 300
unplugging, 291–3

Valaskakis, K., 6, 318
value, 331–2
Vandenberghe, H., 76, 170
Veyne, P., 324
videotex systems, 113, 121, 133, 145, 285, 313, 339
Vispac, 94, 119–20, 272, 290

Walker, E., 72
Watson, T., 103
Watzlawick, P., 61, 67
'wave' metaphor, 31–2, 91
Weaver, W., 164–5
Weber, M., on: bureaucracy, 196, 200–1, 203, 206, 209, 212, 225, 230–3, 237, 306, 310; instrumental knowledge, 17, 169; power, 182–3, 189–96; rationality, 168, 180–3, 194–6, 208, 212, 226–33, 253, 310
Weizenbaum, J., 28; on choice, 68; on 'computer revolution', 33–4, 318; on context, 55, 112; 'Eliza' program, 41, 87, 125, 129, 138–9, 271; on interaction, 265; on language, 174, 239; on mediation, 126; on power institutions, 64–5, 68, 219–20; on programming, 128; on reductionism, 56, 59–60, 228–9; on surveillance, 116
Werner, H., 164
Weston, T. J., 81
White, D. M., 119
Whitehead, N., 309
Whittaker, P. N., 47–8, 85, 89
Wilde, O., 107
Williams, R., 122–3, 271, 318–19
Winner, L., 88
Winograd, T., 41, 58, 103, 134, 136, 140–3, 274
winsbury, R., 119, 133
Wittgenstein, L., 174, 264, 309
Wohlstetter, A., 200
Woodbury, D. O., 5
work, 53, 93, 157, 212–14
Wright, G., 81–2, 89

Zeeman, Z. P., 39, 62–4, 82, 84, 156, 211, 218–19, 285, 328
Zumthor, P., 14